Venoms to Drugs

Venom as a Source for the Development of Human Therapeutics

RSC Drug Discovery Series

Titles in the Series:
1: Metabolism, Pharmacokinetics and Toxicity of Functional Groups
2: Emerging Drugs and Targets for Alzheimer's Disease; Volume 1
3: Emerging Drugs and Targets for Alzheimer's Disease; Volume 2
4: Accounts in Drug Discovery
5: New Frontiers in Chemical Biology
6: Animal Models for Neurodegenerative Disease
7: Neurodegeneration
8: G Protein-Coupled Receptors
9: Pharmaceutical Process Development
10: Extracellular and Intracellular Signaling
11: New Synthetic Technologies in Medicinal Chemistry
12: New Horizons in Predictive Toxicology
13: Drug Design Strategies: Quantitative Approaches
14: Neglected Diseases and Drug Discovery
15: Biomedical Imaging
16: Pharmaceutical Salts and Cocrystals
17: Polyamine Drug Discovery
18: Proteinases as Drug Targets
19: Kinase Drug Discovery
20: Drug Design Strategies: Computational Techniques and Applications
21: Designing Multi-Target Drugs
22: Nanostructured Biomaterials for Overcoming Biological Barriers
23: Physico-Chemical and Computational Approaches to Drug Discovery
24: Biomarkers for Traumatic Brain Injury
25: Drug Discovery from Natural Products
26: Anti-Inflammatory Drug Discovery
27: New Therapeutic Strategies for Type 2 Diabetes: Small Molecules
28: Drug Discovery for Psychiatric Disorders
29: Organic Chemistry of Drug Degradation
30: Computational Approaches to Nuclear Receptors
31: Traditional Chinese Medicine

32: Successful Strategies for the Discovery of Antiviral Drugs
33: Comprehensive Biomarker Discovery and Validation for Clinical Application
34: Emerging Drugs and Targets for Parkinson's Disease
35: Pain Therapeutics; Current and Future Treatment Paradigms
36: Biotherapeutics: Recent Developments using Chemical and Molecular Biology
37: Inhibitors of Molecular Chaperones as Therapeutic Agents
38: Orphan Drugs and Rare Diseases
39: Ion Channel Drug Discovery
40: Macrocycles in Drug Discovery
41: Human-based Systems for Translational Research
42: Venoms to Drugs: Venom as a Source for the Development of Human Therapeutics

How to obtain future titles on publication:
A standing order plan is available for this series. A standing order will bring delivery of each new volume immediately on publication.

For further information please contact:
Book Sales Department, Royal Society of Chemistry, Thomas Graham House, Science Park, Milton Road, Cambridge, CB4 0WF, UK
Telephone: +44 (0)1223 420066, Fax: +44 (0)1223 420247
Email: booksales@rsc.org
Visit our website at www.rsc.org/books

Venoms to Drugs

Venom as a Source for the Development of Human Therapeutics

Edited by

Glenn F. King
Institute for Molecular Bioscience,
The University of Queensland, Australia
Email: glenn.king@imb.uq.edu.au

ROYAL SOCIETY OF CHEMISTRY

RSC Drug Discovery Series No. 42

Print ISBN: 978-1-84973-663-3
PDF eISBN: 978-1-84973-787-6
ISSN: 2041-3203

A catalogue record for this book is available from the British Library

Published by The Royal Society of Chemistry,
Thomas Graham House, Science Park, Milton Road,
Cambridge CB4 0WF, UK

Registered Charity Number 207890

For further information see our web site at www.rsc.org

Printed and bound by CPI Group (UK) Ltd, Croydon, CR0 4Y

Preface

"You would do well to mark the various forms of the Viper.... And from the wound she makes there oozes a discharge like oil or, it may be, bloody or colourless, while the skin around starts up into a painful lump, often greenish, now crimson, or again livid of aspect. At other times it engenders a mass of fluid, and about the wound small pimples like blisters rise flabbily from the skin, which looks scorched. And all around spread ulcers, some at a distance, others by the wound, emitting a dark blue poison; and over the whole body the piercing bane eats its way with acute inflammation; and in the throat and about the uvula retchings following fast upon one another convulse the victim. The body is oppressed with failures of sense in every part, and forthwith in the limbs and loins is seated a burdening, dangerous weakness, and heavy darkness settles in the head".

From *Theriaca* by Nicander of Colophon, 2nd century BCE
(Translation from the Greek poem by
A.S.F. Gow and A.F. Scholfield)

Humans have been simultaneously enthralled and terrified by venomous animals for thousands of years. For some, the merest glimpse of a spider or snake provokes an irrational and debilitating fear. But for many the threat is all too real—venomous animals continue to have devastating effects on human populations, with an estimated 50 000 people dying each year from snake bite in India alone.

Yet, strangely, and perhaps perversely, venomous animals have proved to be a source of life-saving drugs. Captopril, the first blockbuster antihypertensive drug, was derived from a peptidic inhibitor of angiotensin-converting enzyme found in the venom of a Brazilian viper. Captopril has saved many more lives than this snake and its close relatives have killed in their millions of years on the planet. There are currently six FDA-approved drugs derived

RSC Drug Discovery Series No. 42
Venoms to Drugs: Venom as a Source for the Development of Human Therapeutics
Edited by Glenn F. King

Published by the Royal Society of Chemistry, www.rsc.org

from animal venoms, and they are used to treat a variety of ailments, including hypertension, diabetes, and chronic pain.

Research into venoms as a potential source of therapeutics greatly intensified over the past decade, with much activity in both academia and the pharmaceutical industry. There are several reasons for this. The dwindling output of new molecular entities combined with the recent success of biologics led to renewed interest in natural sources, such as venoms, as a source of drug leads. Increasing patient acceptance of injectable therapies created a more conducive environment for the development of peptide and protein drugs. And finally, our improved understanding of the chemical and pharmacological complexity of animal venoms facilitated the development of more focussed screens to discover venom molecules with desired biological activity. As a result of this increased research activity, the current pipeline of venom-derived drugs is larger than ever, with several venom peptides currently undergoing clinical trials and many more in various stages of preclinical development.

This book attempts to capture the current excitement and most important developments in the field of venoms-based drug discovery. The book begins with an overview of the extant suite of venomous animals and their toxin repertoires before drilling down to detailed case studies and issues surrounding optimisation and manufacturing of venom-derived peptide drugs.

Each venom has a specific ecological role, and therefore understanding the context in which a venom has evolved is important for focussing drug discovery efforts. It makes no sense, for example, to focus on neurotoxic spider venoms if one is seeking a drug that modulates blood pressure; haemotoxic snake venoms are likely to be a better bet. Thus, the book begins with a comprehensive overview of venom evolution and how this relates to biodiscovery. Chapter 2 provides an overview of the diverse range of disulfide-rich scaffolds found in animal venoms. These scaffolds have provided a robust structural framework for evolving peptides that are simultaneously potent, selective, and stable. Despite their potential for providing high target potency and selectivity, these disulfide-rich scaffolds have an inherently high degree of chemical and pharmacological plasticity, thereby facilitating the design of analogues with greater potential for therapeutic application.

Chapters 3 and 4 review the wide range of approaches that can be used for venoms-based drug discovery, including recent developments in proteomics, transcriptomics, high-throughput target-based screens, and automated electrophysiology. In addition, the case is made for the power of more traditional bioassays using whole animals or isolated tissues.

The complex venoms of reptiles, aquatic cone snails, scorpions, and spiders have evolved independently. Each of these venoms has a specific ecological purpose, with a chemical and pharmacological composition tuned for this role. In Chapters 5–8, experts on these venoms examine their potential as a source of drug leads.

Chapter 9 details the development of analgesic drugs derived from disulfide-rich peptides in the venom of cone snails. Chapter 10 provides a detailed case study of the ongoing and highly promising development of a sea anemone peptide for the treatment of a range of autoimmune and inflammatory diseases, including multiple sclerosis and psoriatic arthritis.

Most venom components being considered for drug development are disulfide-rich peptides, which creates significant challenges for large-scale production and drug delivery. Chapter 11 discusses chemical approaches that can be used to modulate the stability and bioavailability of disulfide-rich venom peptides, while the final chapter provides real-life examples of how synthetic challenges have been overcome in production of the venom-derived drugs eptifibatide, ziconitide, and exenatide.

This primary goal of this book is to provide a comprehensive resource for chemists and biologists who are interested in venoms as a source of new drugs, as well as members of the lay public who are fascinated by the notion that venomous creatures can also have positive impacts on human health. However, I also hope that this book will inspire the toxinology and medicinal chemistry communities to work together, with cross-fertilisation of expertise and knowledge, on creating a new generation of venom-derived therapeutics.

I would like to convey my sincere gratitude to the contributing authors who willingly gave their time and energy towards this project. My thanks and appreciation also go to the team at the Royal Society of Chemistry, including Gwen Jones, Rosalind Searle, Helen Prasad, Cara Sutton, Sarah Salter, and Rowan Frame, for their support and encouragement. Finally, I would like to thank my family and all the members of my laboratory for their patience and forbearance during the production of this book.

Glenn King

Contents

RSC Drug Discovery Series No. 42
Venoms to Drugs: Venom as a Source for the Development of Human Therapeutics
Edited by Glenn F. King

Published by the Royal Society of Chemistry, www.rsc.org

CHAPTER 1

Seeing the Woods for the Trees: Understanding Venom Evolution as a Guide for Biodiscovery

BRYAN G. FRY*[a,b], IVAN KOLUDAROV[a], TIMOTHY N. W. JACKSON[a,b], MANDË HOLFORD[c], YVES TERRAT[d], NICHOLAS R. CASEWELL[e], EIVIND A. B. UNDHEIM[a,b], IRINA VETTER[b], SYED A. ALI[a,b,f], DOLYCE H. W. LOW[a], AND KARTIK SUNAGAR[g,h]

[a]Venom Evolution Lab, School of Biological Sciences, The University of Queensland, St. Lucia, Queensland 4072, Australia; [b]Institute for Molecular Bioscience, The University of Queensland, St. Lucia, Queensland 4072, Australia; [c]The City University of New York-Hunter College & CUNY Graduate Center and The American Museum of Natural History, New York, New York, USA; [d]Montréal University, Research Institute in Plant Biology, Montreal Botanical Garden, Montreal, Québec, Canada; [e]Alistair Reid Venom Research Unit, Liverpool School of Tropical Medicine, Liverpool L3 5QA, UK; [f]HEJ Research Institute of Chemistry, International Center for Chemical and Biological Sciences (ICCBS), University of Karachi, Karachi 75270, Pakistan; [g]CIMAR/CIIMAR, Centro Interdisciplinar de Investigação Marinha e Ambiental, Universidade do Porto, Rua dos Bragas, 177, 4050-123 Porto, Portugal; [h]Departamento de Biologia, Faculdade de Ciências, Universidade do Porto, Rua do Campo Alegre, 4169-007 Porto, Portugal
*E-mail: bgfry@uq.edu.au

RSC Drug Discovery Series No. 42
Venoms to Drugs: Venom as a Source for the Development of Human Therapeutics
Edited by Glenn F. King

Published by the Royal Society of Chemistry, www.rsc.org

1.1 The Fundamental Problems

The majority of commercial drugs being used today in both developed and developing countries are based on natural products.[1] Most of these products are based upon plants, but research into animal venoms holds great potential for the discovery of novel medicinally useful natural products.[2,3] Knowledge of the evolutionary origins of venom proteins/peptides and the forces shaping the biodiversity seen today is crucial for efficient biodiscovery. In addition, efficient utilisation of venom toxins in drug design and development cannot be achieved without recognition of the true biochemical, ecological, morphological, and pharmacological diversity of venoms and associated venom systems. A major limitation of the use of venom proteins thus far has been the very narrow taxonomical range studied. Entire groups of venomous animals remain virtually ignored. Those that have been examined have apparently been selected due to their medical significance or ease of collection, rather than as a result of their ecological or evolutionary uniqueness.

Venom is defined as "a secretion, produced in a specialised gland in one animal and delivered to a target animal through the infliction of a wound (regardless of how tiny it may be), which contains molecules that disrupt normal physiological or biochemical processes in the victim so as to facilitate feeding or defence by the producing animal".[4] This definition encompasses creatures normally considered venomous (*e.g.*, scorpions, snakes, and spiders) as well as animals that have not been traditionally recognised as such (*e.g.*, leeches, ticks, and vampire bats). Acknowledgement of the evolutionary analogy of the recruitment and use of toxins in all these animals increases the number of known independent occasions in which venom has evolved independently. In addition, this acknowledgement improves our understanding of the factors underlying the evolution of venoms and their associated proteins while also drawing attention to the vast pool of unstudied toxins. Venom has been a key innovation in the evolutionary history of an incredibly diverse range of animals. Even using the traditional definition of venom, venom systems are believed to have evolved independently on at least 20 occasions in extant lineages (Figure 1.1). Intriguing fossil evidence has also led to speculation about the possibility of extinct venomous lineages represented by the theropod dinosaur *Sinornithosaurus*[5] and the extinct pantolestid mammal *Bisonalveus browni*.[5] If lineages such as ticks, leeches, vampire bats, *etc.* are rightfully recognised as venomous, the number of independent evolutionary events in which venom has arisen increases to over 30.

The evolutionary selection pressure upon defensive venoms (*e.g.*, those of fish and bees) is largely directed at the development of streamlined venom that has the primary action of immediate, intense localised pain.[6–8] In contrast, predatory venoms are shaped by a classic co-evolutionary arms race, where evolving venom resistance in prey and the evolution of novel venom composition exerts reciprocal selective pressures on one another in a situation that conforms to the Red Queen hypothesis of Van Valen.[9] Powerful

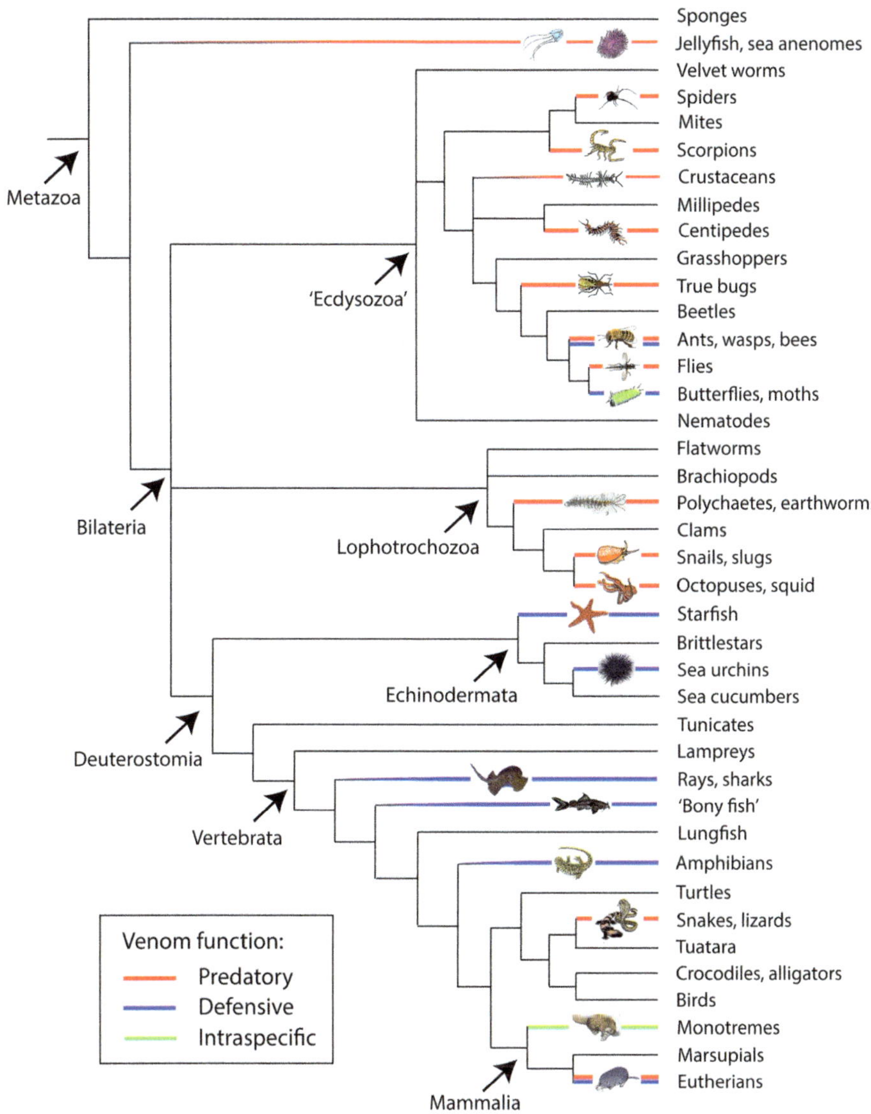

Figure 1.1 Schematic tree of venomous life in the animal kingdom. Coloured branches indicate lineages that include members with venom systems. Phylogeny based on the tree of life presented in Pennisi.[145] Note that a number of animal lineages have been pruned from the tree. Adapted from ref. 34.

purifying selection pressures acting on predatory venoms for millions of years have resulted in highly complex modern venom arsenals that consist of potent compounds with exquisite target specificity. Variation in venom composition is not only observed between different lineages, but also between the closely related species within a clade.[10] Intraclade differences in venom composition often arise as a result of the evolution of prey-specific toxins in species with specialised diets.[11–13] Significant variation in venom profile has even been demonstrated within individual species with widespread geographical distributions.[14,15] Venom can also vary intraspecifically as the result of numerous other factors, including sibling differences[16] and ontogenic changes in prey preference[17] or behaviour. In Sydney funnel-web spiders (*Atrax robustus*), juvenile male spiders and female spiders of all ages have similar insecticidal predatory venoms, whilst sexually mature males (who stop feeding and leave the burrows in search of females) have a vertebrate-specific defensive venom.[18] It is this adaptive complexity and innovation that makes predatory venoms ideal candidates for the discovery of therapeutic lead compounds.

The majority of venom components have evolved to target physiological systems reachable by the bloodstream. In particular, the neurological and haemostatic systems have been convergently targeted *via* a myriad of innovative pathways (Figure 1.2).[4] A consistent feature of venom proteins is a stable molecular scaffold of cross-linked cysteines[19] (see Chapter 2 for further details of disulfide-rich toxin scaffolds); this characteristic appears to facilitate modification of non-structural residues, which in turn facilitates protein neo-/sub-functionalisation. A remarkable degree of convergence exists not only in terms of toxin molecular scaffolding, but also in target specificity and bioactivity.[4] The superimposition of sequences from functionally convergent toxins reveals tremendously useful information regarding structure–function relationships. An example of this is the platelet-aggregation inhibiting RGD tripeptide motif. This motif has been independently derived on numerous occasions within a myriad of distinct protein scaffolds, ranging from snakes (two different occasions: disintegrins and three-finger toxins) to a wide variety of invertebrate species, including ticks (*e.g.*, *Ixodes* spp., *Argas* spp., *Rhipicephalus* spp., *Amblyomma* spp.), tabanid flies (*e.g.*, *Tabanus* spp.), true bugs (*e.g.*, *Triatoma* spp., *Rhodnius prolixus*), mosquitoes (*e.g.*, *Anopheles* spp., *Aedes* spp., *Culex* spp.), sand flies (*e.g.*, *Lutzomyia* spp., *Phlebotomus* spp.), leeches (*e.g.*, *Macrobdella* spp., *Placobdella* spp.), and worms (*e.g.*, *Ancylostoma* spp.).[4,20] This reinforces the fact that biological targets within prey animals are the primary drivers of the evolution of toxin structures.

Snakes, spiders, scorpions, marine cone snails, and sea anemones represent the majority of venomous organisms that have been studied, with other venomous lineages remaining neglected. Moreover, even within these well-studied lineages, there has been a significant taxonomical bias. Partly as a result of the limited taxonomic range studied, the majority of known venom components remain poorly understood, and it is likely that many more venom components await discovery. The complex nature of venom makes it energetically expensive to produce. Hence, most venomous organisms have evolved a highly sophisticated cocktail that can efficiently aid in predation and/or defence, even when secreted in very small quantities. The small amount of venom produced by many venomous organisms was a major obstacle that impeded venom exploration in the past. Even in snakes, which

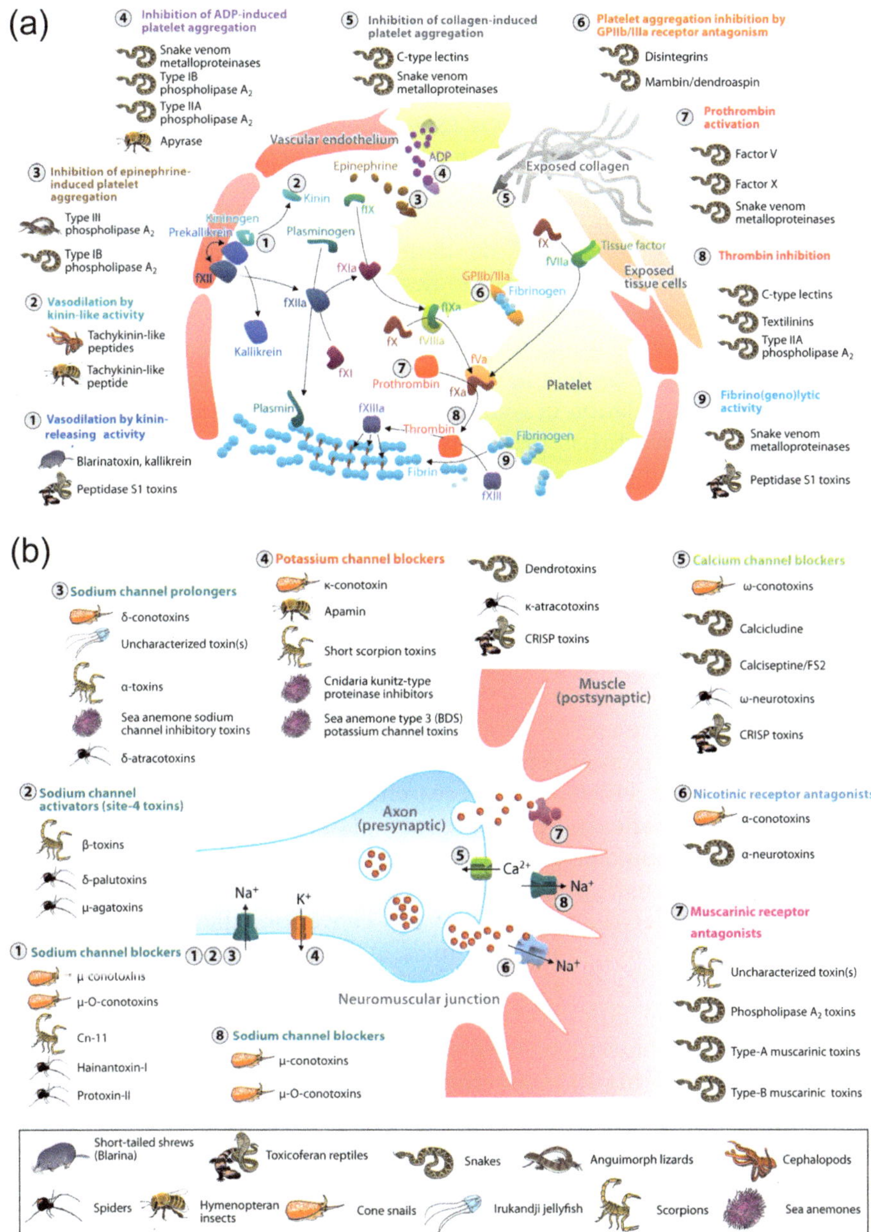

Figure 1.2 Convergence of toxin action in the animal kingdom. Sites of haemotoxic (A) and neurotoxic (B) convergence in animal venoms. Adapted from ref. 4.

may produce copious amounts of venom, particularly large amounts (multi-gram) were necessary for the discovery novel venom components that are secreted in miniscule amounts. For smaller animals such as spiders, the tiny amounts secreted made many species impossible to investigate using protein-based approaches.

Another impediment to venom exploration has been the difficulty of extracting venom from species that do not store secreted venom in readiness for delivery, or that have venom delivery systems that are difficult to access or stimulate. For instance, the venom delivery apparatus of non-front-fanged snakes is located at the back of the mouth and, unlike those of many front-fanged snakes, the venom glands do not contain an appreciable lumen for venom storage, instead only secreting venom as required. Hence, it was very difficult and time-consuming to obtain sufficient quantities of venom from such snakes for the "proteome-only" oriented venom research of the past. Even chemical stimulation of venom secretion (*e.g.*, injection of pilocarpine into the venom gland) has been unsuccessful in overcoming some of the aforementioned complications, impeding venom research in a large group of other organisms (*e.g.*, coleoids, centipedes, non-front-fanged snakes, spiders, and vampire bats). In some cases, these difficulties have been overcome through the application of considerable amounts of time and effort. For example, it took over 200 pilocarpine-stimulated milkings (venom extraction) of *Coelognathus radiatus* to obtain 110 mg of crude venom, which yielded 10 mg of pure α-colubritoxin in the first study of a three-finger toxin (3FTx) from a non-elapid snake.[21] In other cases, however, the difficulties described above have proven impossible to surmount.

With the advent of next-generation RNA sequencing, venom exploration has become more efficient, as researchers can now rapidly construct transcriptome libraries of entire venom glands, without depending on proteinaceous venomous secretions directly. While transcriptomics will rapidly yield full-length precursor sequences, the prediction of propeptide cleavage sites, other than those that are conventionally dibasic, may be impossible.[22] Moreover, transcriptomics alone cannot unravel post-translational modifications (PTMs), which are often crucial for the biological activities of venom components. For example, a sulfo-tyrosine PTM is required for the bioactivity of lizard venom cholecystotoxin; synthetic analogues lacking this PTM are completely inactive.[23] For these reasons, a combined proteomic–transcriptomic approach is essential for the most effective venom exploration (see Chapter 3).

Genome information remains scarce for species other than those routinely used as model organisms in genetics research. To date, genomes are available for only a few venomous animals. Knowledge of the location and organisation of venom-encoding genes can greatly increase our understanding of their molecular evolutionary history. Availability of genomic information will facilitate easier amplification of specific venom-encoding genes. Moreover, this will enable researchers to use small amounts of tissue or other non-destructive samples for sequencing of venom-encoding genes, not only making venom exploration in rare venomous organisms easier and more sustainable, but also overcoming the difficulties of obtaining permits for destructive sampling for research.

Another major problem that has affected venom research is the difficulty of obtaining the venomous animals themselves. Most well-studied venomous organisms have been those that are locally common in the regions in which the research was performed. Researchers often restrict their venom collection to species represented in local serpentariums or that are available from other venom suppliers. Samples acquired from third parties in this manner are often associated with uncertainties regarding geographical origin and sometimes even basic taxonomy. For example, the Sigma pharmaceutical catalogue entry for *Oxyuranus scutellatus* venom (see http://www.sigmaaldrich.com/catalog/product/SIGMA/V3129) states "This venom may be from subspecies *O. s canni* (Papuan taipan) or *O. s scutellatus* (Australian taipan) or a mixture from both"; with a note that "Physical characteristics are almost identical". This is despite the existence of abundant research showing that venoms may vary appreciably across a relatively short continuous geographical range, let alone the sort of variance that may occur between completely disjunct localities. Considerable differences in toxicity and antivenom coverage have recently been demonstrated for *O. s. canni* and *O. s. scutellatus*,[24] which highlights the fact that disregarding the geographical origin of samples is unacceptable in venom research.

The taxonomical bias in toxinology is starkly evident when sequenced toxins are mapped against organismal diversity. For example, in elapid snake venom research, two genera (*Bungarus* and *Naja*) account for almost 40% of all published sequences (Table 1.1). Moreover, almost 40% of all 3FTxs have been sequenced from *Naja* alone. Despite the diversity of toxin forms present, some toxin types are known from transcriptomic studies only. Similarly, of the 3FTxs known from the non-front fanged snake lineages, the majority are known from a single transcriptomic study.[25] Only three studies have characterised the bioactivity of fully-sequenced 3FTx from non-front-fanged snake venoms.[26–28] This bias is not unique to snakes, as the other venomous lineages that have received toxinological attention have suffered similar levels of taxonomical bias. For example, although scorpion venoms have received more research attention than the venoms of any other lineage, only 50 or so of the approximately 1700 species of scorpion recognised today have been examined. The major focus has been on basal families such as Buthidae, which account for more than 50% of all known scorpion toxin sequences. These basal families are known to have separated from all other scorpion families about 350 million years ago,[29,30] suggesting that there is likely to be a plethora of novel venom components that remain undiscovered in the other families. Similarly, despite spiders being the most speciose group of venomous animals, represented by ~45 000 recognised species, venom exploration remains primarily restricted to large mygalomorph species. For example, tarantulas account for more than one quarter of all spider toxins isolated to date, although they represent only ~2% of the

Table 1.1 Toxins sequenced from elapid venoms[a]

Genus	*Total*	*3FTx*	*AVIT*	*Acn*	*CRiSP*	*fV*	*fX*	*Kun*	*Lec*	*LAO*	*NP*	*NGF*	*PLA*$_2$	*SP*	*SVMP*	*Ves*	*Wap*
Acanthophis	**13**	4									1		8				
Acalyptophis	**2**	2															
Aipysurus	**13**												13				
Aspidelaps	**5**	3			1								1				
Astrotia	**4**	4															
Austrelaps	**63**	19			3			6					32				3
Boulengerina	**5**	3													2		
Bungarus	**218**	80		1	1			39	5	2	1	1	87	1			
Calliophis	**3**	1											2				
Cryptophis	**20**				1		1	6	3		2	1	4				2
Demansia	**40**	3			1		2	6	3	1		3	17				4
Dendroaspis	**59**	47	1					9			2						
Drysdalia	**38**	25			1			9					3				
Enhydrina	**3**	2											1				
Hemachatus	**11**	8						2					1				
Hoplocephalus	**16**	1			1		1	3			1		6		1	1	1
Hydrophis	**10**	9											1				
Laticauda	**60**	24			1								35				
Lapemis	**9**	6			2									1			
Micropechis	**2**												1		1		
Micrurus	**74**	44						2	5		3		20				
Naja	**310**	226			11			8	4	5	2	2	37	2	12	1	
Notechis	**44**	2			1		3	4	5	1	2	3	21				2
Ophiophagus	**57**	44			1			2	2	1	1		3	1	1	1	
Oxyuranus	**126**	15			5	2	2	12	9	3	12	3	60				3
Pelamis	**1**	1															
Pseudechis	**85**	3			2		1	13	12	1	6	4	37				6
Pseudonaja	**47**	12			1	2	1	7	5			2	15		1		1
Tropidechis	**20**	3			2		1	2	3			0	7				2
Walterinnesia	**9**	3						4					2				
Total	**1367**	**594**	**1**	**1**	**35**	**4**	**12**	**134**	**56**	**14**	**33**	**19**	**414**	**5**	**18**	**3**	**24**

[a]3FTx = three-finger toxin, Acn = acetylcholinesterase, fV = factor V, fX = factor X, Kun = kunitz, Lec = lectin, LAO = L-amino oxidase, NP = natriuretic peptide, NGF = nerve growth factor, PLA$_2$ = phospholipase A$_2$, SP = serine protease, SVMP = snake venom metalloprotease, Ves = vespryn, Wap = waprin

taxonomic diversity of spiders.[31] Furthermore, it is suggested that the currently recognised species constitute only ~25% of existing species.[32] Spiders evolved from the stem arachnid ancestor about 300 million years ago during the Carboniferous period. Spider venoms contain a range of low molecular weight peptides and proteins that are neurotoxic, haemotoxic or cytotoxic in activity. It is likely that the highly complex nature of the venom is responsible for the tremendous success and diversification of the spiders as a group. Continuing this theme, the venom has not been thoroughly characterised from a single species of centipede amongst the 3300 species of centipedes known today. The forcipules, or poison claws, which are modified front legs used for delivering venom into the prey, have been identified in centipede fossils dating back to the early Devonian period, 400 million years ago. This suggests that centipedes, along with scorpions, possess one of the most ancient venom delivery apparatuses.[33] Despite this, centipede venom research is very much in its infancy.

1.2 The Solutions

The most efficient venom exploration approach is multidisciplinary and encompasses various fields and techniques, including:

- Organismal selection based upon phylogenetic position and ecological niche occupied
- Transcriptomics and *in silico* studies
- Molecular evolution and phylogenetics of toxins
- Proteomics
- Bioactivity testing

1.2.1 Taxon Selection

In order to maximise the efficiency of biodiscovery efforts, researchers should endeavour to examine the most diverse range of taxa possible. As highlighted elsewhere in this chapter, toxinological research in the past has focussed on species that were easy to acquire venom from or that were the most "medically significant", where medical significance is defined as "the danger posed to a human by a bite/sting". This reliance on common species, or the tiny minority of venomous species that are dangerous to humans, has resulted in the fact that the vast majority of venomous animals remain under-researched or completely ignored.

There are two primary drivers of diversity in venom evolution, both of which should be taken into account. The first is phylogenetic distance—species that are distantly related are likely to have more divergent venom components than species that are closely related. For this reason, researchers can widen their net in the search for novel compounds by examining members of as many genera as possible within their chosen clade, rather than concentrating solely on one genus.

The second primary driver of venom diversity is ecological distance, specifically differences in prey type, foraging mode and prey-handling behaviour. The majority of venomous animals use their toxic secretions to aid in prey subjugation, and hence selection pressures driving the evolution of venom components originate, in large part, from the prey species.[34] It is likely that different venom components are more suited to subduing different types of potential prey animals, therefore venomous species, even those that are closely related, that feed on divergent prey types are likely to have divergent venom compositions. The foraging mode largely determines the type of prey available to a venomous animal, but it also determines prey condition. Prey conditions, such as body temperature, activity level, *etc.* may affect the physiological action of toxins; therefore venomous predators that feed on the same prey types, but feed when the prey is in a different condition, may have divergent venom compositions. For example, snakes that forage at night for sleeping diurnal lizards will encounter inactive prey with a low body temperature and may therefore have less need to quickly disable prey that are too cold to resist effectively but may also rely on toxins that have a temperature-independent activity. Snakes that forage for the same lizard species during the day will encounter active prey with high body temperatures and may therefore rely on toxins that quickly disable prey to prevent it escaping or injuring the snake, and may be able to utilise toxin types that have a temperature-dependent activity. Similarly, prey-handling behaviour may influence venom composition. Venomous predators that strike and release their prey may rely on toxins that rapidly disable motor functions in order to ensure that prey does not travel far after envenomation and is therefore easy to track down. Venomous predators that strike and hold onto prey may have less need to disable motor functions and may also have the option of physically subduing prey; that is, they may use a combined physical and chemical attack (*e.g.*, use of powerful pincers and venom in scorpions, the use of constriction and venom in snakes, or harpoon-shaped radulars filled with venom in marine snails). Good examples of the fact that investigation of atypical subjects of toxinological research yields "low-hanging fruit" for biodiscovery include recent studies on organisms as diverse as Antarctic octopods[35] and lizards previously considered "non-venomous".[23,36]

Although venom is ubiquitous within the Octopoda, the majority of toxinological work on this order has been focussed on the medically significant *Hapalochlaena* sp. (blue-lined octopods) that harbour tetrodotoxin (TTX) produced by endosymbiotic bacteria.[37] Prior to 2010, nothing was known about the composition of the venom of octopods from the waters of the Southern Ocean in the vicinity of Antarctica. Inhabiting waters with sub-zero temperatures has placed extreme selection pressures on the venoms of these octopods and not only were two new toxin classes not previously known from octopods identified in their venoms, but the activity of enzymatic toxins contained therein displayed extreme cold-adaption, with enzymes generally more active at 0 °C than 37 °C.[35]

Although it was previously believed that venom was restricted within lizards to the two species in the family Helodermatidae, the discovery that toxin-secreting oral glands are an ancestral trait of toxicoferan squamate reptiles[38,39] vastly increased the number of lizard species of potential interest to toxinological researchers. Subsequent investigation of the venom systems of these lizards has not only increased our understanding of the evolution of venom in squamate reptiles, but has also resulted in the discovery of a number of new peptide types, including three that affect the cardiovascular system and are thus of potential interest in drug design and development.[40]

As investigators continue to widen the scope of their toxinological research, it is inevitable that many new toxins of interest from a biodiscovery perspective will be uncovered. Understanding the evolutionary pathways that venom travels and the evolution and ecology of the whole organism in which venom evolves serves as an effective guide for biodiscoverers. Through this understanding they may gain valuable knowledge of the most efficient directions in which to cast their nets. Indeed, viewing the study of venom evolution and the search for novel toxins as separate disciplines has become an outmoded and unconstructive viewpoint. These areas of research are two sides of the same coin and complement one another to the extent that a discovery in the field of venom evolution will almost invariably uncover a valuable new resource for biodiscovery.

1.2.2 Transcriptomics

With the advent of high-throughput sequencing technologies, scientists can generate datasets of considerable size. These advances have had considerable benefit for the field of toxinology. Recent studies have demonstrated the effectiveness of applying these technologies to the tasks of identifying novel toxin components and exploring diverse venom-encoding gene families that have previously been characterised.[37,39–46] Several methods of toxin annotation have been employed for proteomic and transcriptomic datasets, mainly based on "BLAST gold standards".[47–51] Complementary phylogenetic analyses of toxin gene families have also revolutionised our understanding of the origins and evolution of toxins. However, phylogenetic analyses of toxin gene datasets have been hampered by significant problems: (i) several toxin types have very short sequences and consequently few informative phylogenetic positions are identified on alignments (*e.g.*, conopeptides from cone snails, crotamines in rattlesnakes); (ii) it has been shown that signal, pro and mature peptides may adopt different evolutionary pathways, leading to low statistical support among phylogenies; (iii) some toxin gene families experience tremendous diversifying selection pressures,[52] resulting in a lack of phylogenetic resolution as each amino acid site experiences multiple changes, which poses a major problem for multiple sequence alignments; (iv) performing phylogenetic analyses on large datasets can be problematic if strong statistical support is needed, especially for maximum likelihood and Bayesian inferences.

One way of addressing these issues is to employ genetic-network-based approaches to unravel the evolutionary histories of toxins. Genetic networks have been extensively used to study genes that follow reticulate evolutionary pathways, such as those that experience regular recombination and lateral gene transfers.[53] A genetic network (or a graph) is a mathematical model of pairwise relations among entities named nodes. The nodes in the network are linked by edges representing the connections or interactions between these entities. Building a network allows nodes to be grouped into connected components. Nodes and interactions can be of diverse nature (directed or undirected interactions, metabolic pathways, transcriptional regulation processes, *etc.*). In the case of toxin gene networks, the nodes represent amino acid or nucleotide sequences of toxins and the connections are represented as "% similarity" shared between two sequences. A new software called Evolutionary Gene & Genome Network (EGN) has recently been used for metagenomic analysis[54] and is now available to the public.[146] This software, which is written in PERL, has been used for building toxin gene networks by making use of the manually curated ToxProtDb database.[55] It begins with the comparison of all sequences from a given dataset with one another using sequence similarity tools such as BLAST or BLAT, which are implemented in the software itself. The tabular output generated by the software contains parameters such as percent ID, percent similarity, e-value and score, which are then processed by EGN to generate graphs. The graph can be further visualised using software such as Cytoscape[56] and Gephi.[57] The primary benefit of these analyses is their ability to process datasets generated with high-throughput techniques.

In order to discuss some of the main advantages of network-based analyses, we will describe a few examples.

The ToxProtDb dataset includes 5361 manually reviewed toxin sequences. We first excluded sequences that were less than 40 residues in length. Out of the remaining 4339 sequences, we were able to cluster 3071 sequences into 93 toxin groups or connected components (Figure 1.3A) using a BLAST similarity search (minimal e-value: 1E-05; minimal hit identity threshold: 20%). Using the aforementioned parameters, we were unable to find any matching sequences for the remaining 1268 sequences, which were therefore considered to be "singletons". Nodes were coloured according to the taxonomic classification of sequences. Graphical representation of the connected components not only reveals the diversity that exists among these examined groups (Figure 1.3B), but also the major convergent recruitment events of different toxins (Figure 1.3C and D). In our analyses, Kunitz proteases came out as one of the convergently recruited toxins in spiders, cone snails, insects, and snakes. The discovery of the similarity between U_{13}-theraphotoxin-Cj1a isolated from the spider *Chilobrachys jingzhao* and M-conopeptide isolated from *Conus vexillum,* however, was more surprising. As illustrated in Figure 1.3A, this similarity is not restricted to cysteine pattern alone. To our knowledge, this level of convergence has not been documented before, and its discovery reinforces the value of network-based analyses.

Network-based analyses can also be focussed on particular taxa. Figure 1.3D represents the evolutionary network analysis of snake venom proteins. Of 1701 sequences available on ToxProtDb, 1351 sequences form 24 different connected components (e-value threshold: 1E-05). This network perfectly illustrates the predominance of phospholipases and 3FTxs in snake venoms. The shape of each connected component also shows that some families such as the metalloproteinases are highly diverse, while others such as phospholipases are highly conserved.

Network-based analyses can help in processing large datasets quickly. Further, the network analyses offer numerous statistical frameworks for comparative purposes. Genetic diversity among each connected component could be evaluated using simple statistics such as the clustering coefficient, the sum of edges divided by total number of possible edges, or network diameter. Highly sophisticated parameters concerning whole network shape or individual nodes could also be estimated (degree distributions, neighbourhood connectivity, between-ness centrality, *etc.*) using the Cytoscape and Gephi plugins. A basic representation of comparative diversity is illustrated in Figure 1.3C. The most compact connected component includes highly similar sequences as shown by the Weblogo consensus.[58] A more diverse connected component is represented as well, with the sequences exhibiting a low level of similarity outside cysteine residues.

To summarise, network-based approaches offer several advantages when applied to toxin datasets. They: (i) allow efficient annotation of toxins in large datasets; (ii) efficiently process comparative analyses using previously identified toxins; (iii) help in identification of closest relatives for further phylogenetic analysis; (iv) significantly improve multiple sequence alignments and clustering; and (v) facilitate visualisation of reticulate evolutionary pathways that cannot be represented using bifurcating phylogenetic trees (nodes could have numerous connections instead of the usual tree nodes with only one parent).

1.2.3 Sequence Retrieval and Alignment

Newly sequenced toxins may be identified by comparison with previously characterised sequences using search programs such as BLAST[59] implemented in various public sequence databases like the UniProtKB protein database (http://web.expasy.org/blast) and NCBI (http:/ncbi.nlm.nih.gov/). This is often employed to retrieve additional homologous sequences, already deposited in the aforementioned databases, so as to increase the number of sequences in the dataset. Resultant sequence sets are aligned using programs such as CLC Main Workbench, CLUSTAL,[60] MUSCLE,[61] *etc.*

1.2.4 Phylogenetic Analyses

Phylogenetic analyses are performed to reconstruct the molecular evolutionary history of each toxin type. Datasets are analysed with methods such as Bayesian inference implemented in MrBayes[62] or maximum-likelihood

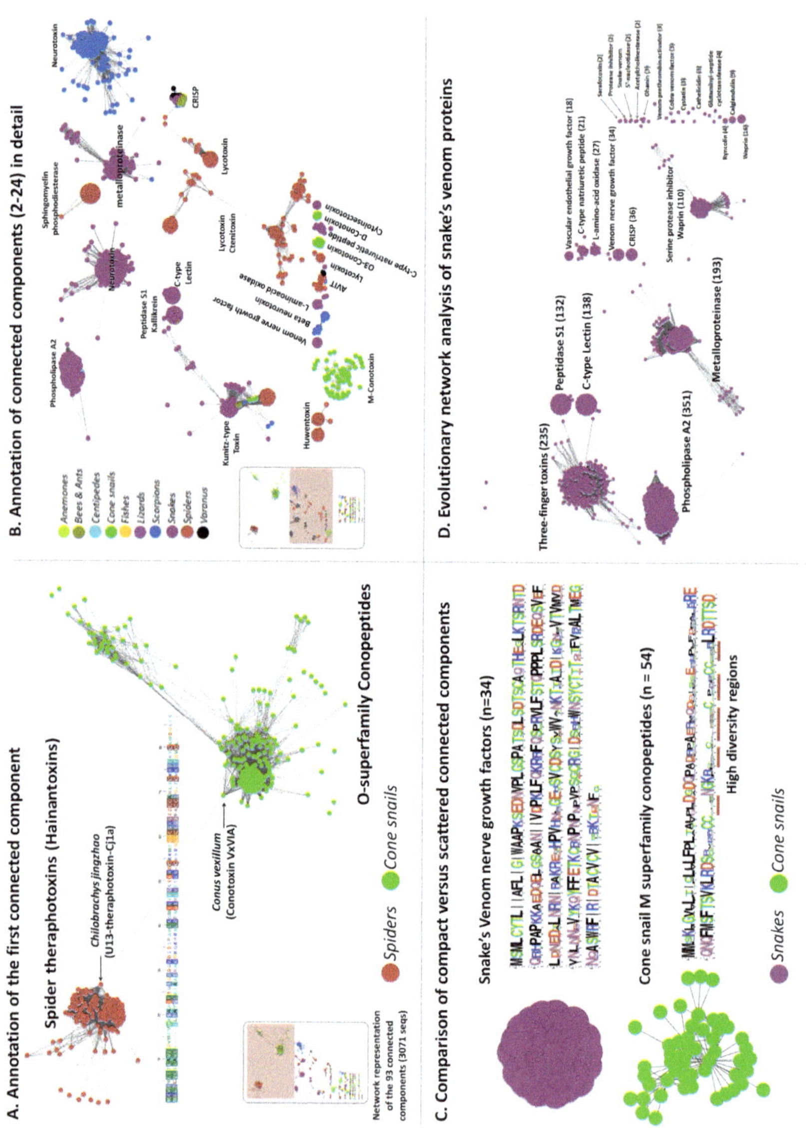
A. Annotation of the first connected component
Spider theraphotoxins (Hainantoxins)
Chilobrachys jingzhao (U13-theraphotoxin-Cj1a)
Conus vexillum (Conotoxin VxVIA)
O-superfamily Conopeptides
Network representation of the 93 connected components (3071 seqs)
Spiders
Cone snails
B. Annotation of connected components (2-24) in detail
Anemones
Bees & Ants
Centipedes
Cone snails
Fishes
Lizards
Scorpions
Snakes
Spiders
Varanus
Phospholipase A2
Kunitz-type Toxin
Peptidase S1 Kallikrein
C-type Lectin
Neurotoxin
Sphingomyelin phosphodiesterase
metalloproteinase
Neurotoxin
CRISP
Lycotoxin
Lycotoxin Ctenitoxin
Huwentoxin
M-Conotoxin
Venom nerve growth factor
Beta neurotoxin
L-aminoacid oxidase
AVIT
Lycotoxin
O3-Conotoxin
D-Conotoxin
Cytoinsectotoxin
C-type natriuretic peptide
C. Comparison of compact versus scattered connected components
Snake's Venom nerve growth factors (n=34)
Cone snail M superfamily conopeptides (n = 54)
High diversity regions
Snakes
Cone snails
D. Evolutionary network analysis of snake's venom proteins
Three-finger toxins (235)
Phospholipase A2 (351)
Peptidase S1 (132)
C-type Lectin (138)
Metalloproteinase (193)
Serine protease inhibitor Waprin (110)
CRISP (36)
Venom nerve growth factor (34)
L-amino-acid oxidase (27)
C-type natriuretic peptide (21)
Vascular endothelial growth factor (18)

implemented in various applications such as PhyML,[63] Randomized Axelerated Maximum Likelihood (RAxML),[64] *etc.* Although both the aforementioned methods can generate accurate phylogenetic trees, the Bayesian methods are widely preferred.

1.2.5 Test for Recombination

Recombination is a process in which two different molecules of DNA, usually homologous, exchange genetic material with one another. Recombination can mislead phylogenetic interpretations.[65] Hence it becomes essential to identify and eliminate recombinant sequences from datasets before conducting phylogenetic analyses. All of the contemporary algorithms that are designed to detect selection pressures assume the absence of recombination in the dataset. Moreover, the variations that result from recombination events may resemble those that result from adaptive evolution,[66] and the process also leads to apparent substitution rate heterogeneity.[67] Thus, recombination can have an impact on evolutionary selection analyses as well. One could either remove identified recombinant sequences from datasets altogether or allow the recombinant sequences to have independent phylogenetic histories while estimating selection pressures. The former can be done by employing various programs such as SIMPLOT,[68] RDP,[69] TOPALi,[70] *etc.* while the latter is achieved through methods such as Single Breakpoint algorithm (SBP) and Genetic Algorithm for Recombination Detection (GARD), implemented in the HyPhy package[71–73] Potential breakpoints can be

Figure 1.3 Network analysis of toxin genes. (A) The convergence between O-superfamily *Conus* venom peptide and typical spider-venom peptides. These two large groups of toxins (based on ToxProtDb data) are connected through the toxin U_{13}-theraphotoxin-Cj1a from the tarantula *Chilobrachys jingzhao* and conopeptide VxVIA from the venom of *Conus vexillum.* They share an identical cysteine framework (C–CC–C–C) and similar residues in the mature toxin sequence. (B) The convergent recruitment of various toxin groups. Some groups are exclusively found in a given taxonomic group (*e.g.*, huwentoxins in spiders) while highly similar cysteine-rich secretory proteins (CriSPs) are found in various taxa such as cone snails, bees, scorpions, spiders, and lizards. (C) Demonstration that gene network analysis also facilitates the characterisation of genetic diversity for a given family. The shape of the network shows the intrinsic diversity within each group. The M-conotoxin group from cone snails includes sequences with a low degree of conservation outside the cysteine framework, giving rise to a dispersed network shape, whereas snake-venom nerve growth factors are highly similar, giving rise to a compact shape. (D) The composition and the diversity of snake venom protein families. PLA_2s and three-finger toxins are the largest groups of snake venom proteins studied so far. This may reflect the composition of snake venom and/or a biased interest of scientists for certain types of toxins.

detected using the small sample Akaike information Criterion (cAIC) and sequences can be compartmentalised before conducting phylogenetic or evolutionary analyses.

1.2.6 Identifying Evolutionary Selection Pressures

It is essential to recognise the evolutionary selection pressures that shape venom components, including those that act on different regions of the same venom component. This not only reveals the functional importance of different toxin domains but also provides a wealth of information for drug design and development. Many essential genes that exist in single copy within the genome evolve through the regime of negative selection, whereby a single non-synonymous mutation in the functional domain can lead to the death of the organism (lethal mutations), resulting in the elimination of such mutations from the population. By contrast, the genes that encode immunoglobulins are required to incorporate variation to combat the diversity of pathogens the body encounters. Similarly, predatory venomous organisms benefit from producing a variety of toxin forms to stay ahead in the chemical arms race with their prey. Thus, negative selection eliminates mutations that have harmful effects on the fitness of the organism, while positive selection generates mutations/variations that increase the fitness of the organism. Parallel to these two mechanisms, neutral evolution causes non-functional synonymous and non-synonymous mutations to accumulate in equal proportions in neutral alleles (those that do not affect the fitness of the organism). It is worth noting that the neutral theory of molecular evolution considers neutral mutations to be far more common, and thus have more impact on sequence composition, than positive selection.[74]

Venomous predators stay ahead in the chemical arms race with their prey as a result of the rapidity with which toxins accumulate variations. Formation of toxin multigene families is one of the primary ways in which venom-encoding genes diversify. Recurrent duplication events result in the formation of new copies of venom-encoding genes, which evade pre-existing negative selection pressures. Following duplication, they are reinforced through positive selection and may be neofunctionalised to produce a myriad of different peptides with novel biochemical properties. Identifying regions that accumulate variation is not only important for understanding the evolutionary history of the toxin, but is also useful for novel methods of antivenom production. It is logical to target regions that are under the constraints of negative selection, and hence are least likely change over short periods of time, rather than targeting regions that accumulate tremendous variation and are thus less likely to cross-react with antibodies raised against other isoforms, limiting the usefulness of antivenoms based on these epitopes.

Historically, positive selection has been detected as a ratio (ω) of non-synonymous (Ka or dN or α) to synonymous (Ks or dS or β) substitutions. Synonymous substitutions are considered neutral in terms of selection, as they do not change the amino acid sequence, while non-synonymous substitutions are considered a function of selective pressure on the protein since they change the primary structure and may affect function as well. Negative selection pressure will reduce the accumulation of deleterious non-synonymous mutations, effectively reducing the ω value (dN/dS) to less than 1. By contrast, positive selection will increase non-synonymous mutations relative to synonymous mutations, increasing the ω value to more than 1. If the protein evolves neutrally, it will accumulate synonymous and non-synonymous mutations in equal proportions, resulting in an ω value of 1. Sophisticated likelihood models of coding-sequence evolution,[75,76] implemented in CODEML of the PAML[77] package, have been popularly utilised to evaluate selection pressures. Lineage-specific models can be utilised to identify selection pressures acting across lineages in a phylogenetic tree. The two-ratio model is often employed for identifying selection pressures across different lineages. However, this model requires the branches evolving under positive selection to be defined *a priori*, which is often not possible. Moreover, the lineage-specific models fail to identify regions in proteins that might be affected by episodic selection pressures, and hence they can underestimate the strength of selection. Hence, site-specific models are employed that estimate positive selection statistically as the non-synonymous-to-synonymous nucleotide-substitution rate ratio (ω), where a ratio of significantly greater than 1 indicates positive selection. While lineage-specific models estimate sites under positive selection across lineages and site-specific models along sites, the branch-site models identify codon sites under selection across lineages and along sites, making them valuable tools for estimation of selection pressures. However, similar to the lineage-specific branch models (two-ratio model), the branch-site model also requires the *a priori* definition of lineages under selection.

Unlike the lineage-specific branch and branch-site models, the GA-Branch Test implemented in the HyPhy[78] package does not require the foreground and background branches to be defined *a priori*. The algorithm works on the principle that there could be many models that fit the data more closely than a single *a priori* hypothesis. Separating lineages into foreground (positively selected) and background (negatively or neutrally evolving) groups *a priori* leads to high rates of false positives and false negatives, especially when the mode of evolution of the background branches is different from the modelling assumptions. To overcome these limitations, a branch-site random effects likelihood (REL) model has been proposed, which estimates variations over sites and branches within the REL framework.[79]

1.2.7 Structural Analyses

Domains under selection can be depicted by mapping mutations over the structure of the protein as determined using X-ray crystallography and nuclear magnetic resonance (NMR) spectroscopy. Because of the lack of structural information for most toxins, one may construct homology models using various servers such as the Swiss-model,[80–82] Phyre 2,[83] *etc.* Several webservers are also available for automatically measuring selection pressures, employing various models of estimating selection, and furnishing homology models with mapped evolutionary conservations and variations.[84–86]

1.2.8 Proteomics

The search for novel toxins and other bioactive compounds in organisms other than those routinely used as model organisms in genetic research has traditionally been an onerous task involving activity-guided purification followed by partial sequencing from the N-terminus by Edman degradation. Full toxin sequences were either obtained by proteolytic digestion of pure proteins and sequencing of purified fragments or rapid amplification of cDNA ends (RACE) from venom gland RNA using primers designed from N-terminal partial protein sequences, followed by sequencing of the amplified product using Sanger sequencing. Although these methods of obtaining toxin sequences are time-consuming and in many ways "out-dated", the coupling of transcriptomic and proteomic techniques nevertheless remains one of the most powerful approaches for investigation of venoms and their components.

High-throughput proteomics is heavily reliant upon the existence of sequence templates, with the sequencing of entirely novel peptides and proteins still requiring substantial manual interpretation. Similarly, most bioinformatic pipelines for the processing and annotation of the ever growing next-generation sequencing (NGS) datasets still rely on sequence homology to tease out toxin sequences. While there are algorithms available for identification of putative toxin sequences without the use of homology searches, the physical presence of the predicted mature toxin in the venom cannot be confirmed without proteomic evidence. Furthermore the presence of PTMs, which can have a significant effect on activity and specificity of toxins,[25] is generally not reliably deduced from sequence information alone and must be confirmed by examination of the pure native toxin.

The ability to rapidly attain information on toxin sequences and PTMs is also useful beyond high-throughput description of venoms. For example, it is useful in identifying hits from activity screens, a process that could otherwise quickly become expensive and time consuming. By using transcriptomic data as a sequence template, the interpretation of top-down proteomic data can be largely automated and toxin identification, along with post-translational cleavage sites and any modifications, can generally be obtained within short time.

Thus, in isolation, both approaches are prone to becoming laborious when dealing with novel samples, but in disparate ways; one requiring a sequence template and the other requiring sequence validation. With the advent of NGS and the continuing development of faster, more sensitive mass spectrometers, a complementary approach allows for high-throughput verification of mature venom component sequences from transcriptomic data, including those of entirely novel toxins and any PTMs (see Chapter 3 for further details). The number of software suites (*e.g.*, Tandem, PEAKS, ProteinPilot) available means that integration of the two datasets is largely automated, maintaining the high-throughput rate all the way through to data interpretation. Further investigation of sequences from the resulting venomic library can then be pursued on the basis of novelty, biochemical and structural properties, or homology to previously identified bioactive compounds.

1.2.9 Bioactivity Testing

Venoms have evolved over millions of years to target vital physiological processes within prey and predators with exquisite specificity and potency. The vast pharmacological cornucopia contained in venoms has: (i) contributed substantially to our understanding of human physiology and the pathophysiology of disease; (ii) provided important pharmacological modulators and tool compounds; and (iii) most importantly, led to the discovery of new drugs that have had significant positive impact on human health.

As we begin to appreciate the full complexity of venoms, with several hundreds to thousands of bioactive molecules contained in the venom of a single spider or cone snail,[87,88] understanding and delineating the biological activity of venoms and venom components becomes increasingly important. However, determining the molecular targets of venom components or isolating specific molecules with particular activity from venoms has proven challenging, as discussed in further detail in Chapter 4.

Early studies focussed on delineating the physiological and pharmacological effects of venoms resulting from *in vivo* administration to animals. Such studies were able to provide immediate insight into the physiological consequences of envenomation and led to the discovery of some of the most notable examples of drugs derived from venoms. For example, conotoxin GVIA was isolated following the observation that intracerebroventricular administration of venom to mice caused a shaking phenotype.[89–91] This was determined to occur as a result of inhibition of neuronal calcium channels, in particular $Ca_v2.2$,[92] and a related $Ca_v2.2$-selective conopeptide, MVIIA (ziconotide or Prialt®), was subsequently developed as a novel treatment for intractable pain[93,94] (see Chapter 9 for further details). Similarly, the observation that injection of venom from a Brazilian viper (*Bothrops jararaca*) caused a marked, sudden drop in blood pressure mediated through inhibition of angiotensin-converting enzyme (ACE), led to the development of modern antihypertensive ACE-inhibiting drugs (see Chapter 5 for further details).[95–97]

However, depending on the site of administration, such *in vivo* assays can produce false negative results. This was the case initially for several peptides from *Conus geographus* venom, which produced no effects after intraperitoneal injection in mice, but elicited profound behavioural changes after intracerebroventricular administration.[89] Similarly, *in vivo* studies are unlikely to detect biological activity that is not associated with overt behavioural or physiological changes. In addition, such studies may be no longer ethically acceptable, are time-consuming, low-throughput and clearly ill suited to large-scale screening of venoms for specific biological activity.

As an extension to *in vivo* studies, *ex vivo* (organ bath or isolated tissue preparation) experiments have been used for many years to characterise the physiological and pharmacological effects of venoms and venom components (see Chapter 4 for further details). *Ex vivo* preparations facilitate detailed pharmacological interrogation of the mechanisms of action underlying the biological activity of venom components. Exemplifying this strategy is the discovery of the χ and ρ classes of conotoxins from the venom of *Conus tulipa* and *Conus marmoreus*, which displayed inhibition and prolongation of electrically excited contractions, respectively, for ρ-TIA and χ-MrIA in a rat vas deferens tissue preparation[98] (see Chapter 9 for further details).

However, like *in vivo* studies, bioactivity testing in *ex vivo* tissue preparations is time-consuming and not compatible with high-throughput identification of venom components with specific action at selected therapeutic targets. Thus, in order to accelerate drug discovery efforts, high-throughput approaches are increasingly being applied to the isolation of bioactive components from venoms.

Taking advantage of the rich chemical diversity present in venom, high-throughput activity-guided fractionation has been used as an alternative strategy for the discovery and isolation of novel venom components. The aim of high-throughput screening (HTS) is to systematically isolate and characterise bioactive molecules from the complex mixture of peptides, proteins and small molecules present in venoms. This approach relies on the establishment of specific assays to examine the effect of novel compounds on a particular pharmacological target of interest, usually in cell lines heterologously expressing this target (see Chapter 4 for further details). As a minimum, it requires high assay sensitivity, accuracy, robustness, and reproducibility, often defined by a high Z' score.[99] Assays developed for and amenable to HTS are as varied as the potential therapeutic targets themselves, and are able to assess activity of G-protein-coupled receptors, ligand- and voltage-gated ion channels, enzymes, or transporters, to name a few. Accordingly, the repertoire of HTS assays available now expands on more traditional approaches such as electrophysiological recordings and radioligand binding studies to include assays based on the detection of absorbance, bioluminescence, fluorescence, fluorescence polarisation, fluorescence-resonance energy transfer (FRET), and bioluminescence-resonance energy transfer (BRET) as well as label-free technologies and image-based high content screening (reviewed in ref. 100). The ability to miniaturise such assays, allowing the assessment of activity in 96-, 384- or 1536-well format, enables screening of activity from venom samples that are only available in very limited quantities, including venoms from cone snails, spiders, centipedes, and scorpions. In combination with transcriptomic approaches, thus circumventing the need for time-consuming and resource-intensive sequence analysis, HTS has the potential to vastly accelerate the discovery of novel venom components with defined biological activities. Indeed, high-throughput Ca^{2+} assays have recently been reported to have been instrumental in the discovery, isolation and characterisation of bioactive peptides from snakes and cone snails.[101–103]

In addition to technical limitations inherent to HTS assays,[100,103] the most notable disadvantage of this strategy for discovery of novel bioactive venom components lies in the high specificity of these assays for their intended molecular target. A lack of activity in such assays does not necessarily imply lack of biological activity, but rather incomplete knowledge of the evolutionary role and thus putative molecular target of such compounds. Accordingly, a multidisciplinary approach, incorporating transcriptomic, evolutionary, and pharmacological analysis will likely be needed to guide the discovery and characterisation of novel venom components with biological activity.

1.3 Case Studies

The multidisciplinary approach described above can yield a wealth of information regarding the evolution of various venom encoding genes. Here, we illustrate this fact with a few examples (Figure 1.4).

1.3.1 Differential Evolution of *Psammophis mossambicus* SVMP Domains[101]

Most viper venoms are haemotoxic in nature, disrupting homeostasis through destruction of red blood cells, coagulation of blood, necrosis, oedema, haemorrhage, *etc.* This results in severe localised and systemic tissue damage and/or organ failure. A subfamily of zinc-dependent matrix-metalloproteinases called the snake venom metalloproteinases (SVMPs) is largely responsible for these pathological effects of viperid snake envenomation. Since the present-day antivenom fails to neutralise these rapidly developing symptoms effectively, a focus of venom research in the recent past has been on studying the effects of SVMP-induced inflammatory reactions.

Snake venom metalloproteinases belong to the reprolysin subfamily of the M12 family of proteinases. They are grouped into three major types based on their domain organisation:[104] (i) P-I SVMPs have only the metalloproteinase domain; (ii) P-II SVMPs contain a disintegrin-like domain in addition to the former; and (iii) P-III SVMPs have an additional cysteine-rich domain. The P-III class of SVMPs have been demonstrated to be more potent in causing haemorrhage than the P-I and P-II classes.[104,105] Hence it is hypothesised that the additional cysteine-rich domain absent from both P-I and P-II SVMPs but present in the P-III SVMP is responsible for the increased potency of the latter. SVMPs have been demonstrated to destroy blood vessels by the degradation of basement membrane proteins and to impair coagulation through fibrinogenolysis. However, several SVMPs devoid of haemorrhagic activity have also been isolated.

The multidomain, multiproduct SVMP gene exhibits some fascinating characteristics such as the selective expression of domains. The selective expression of the disintegrin-like domain from P-II SVMP is well known in viper venoms. Recently, a similar phenomenon was also described in

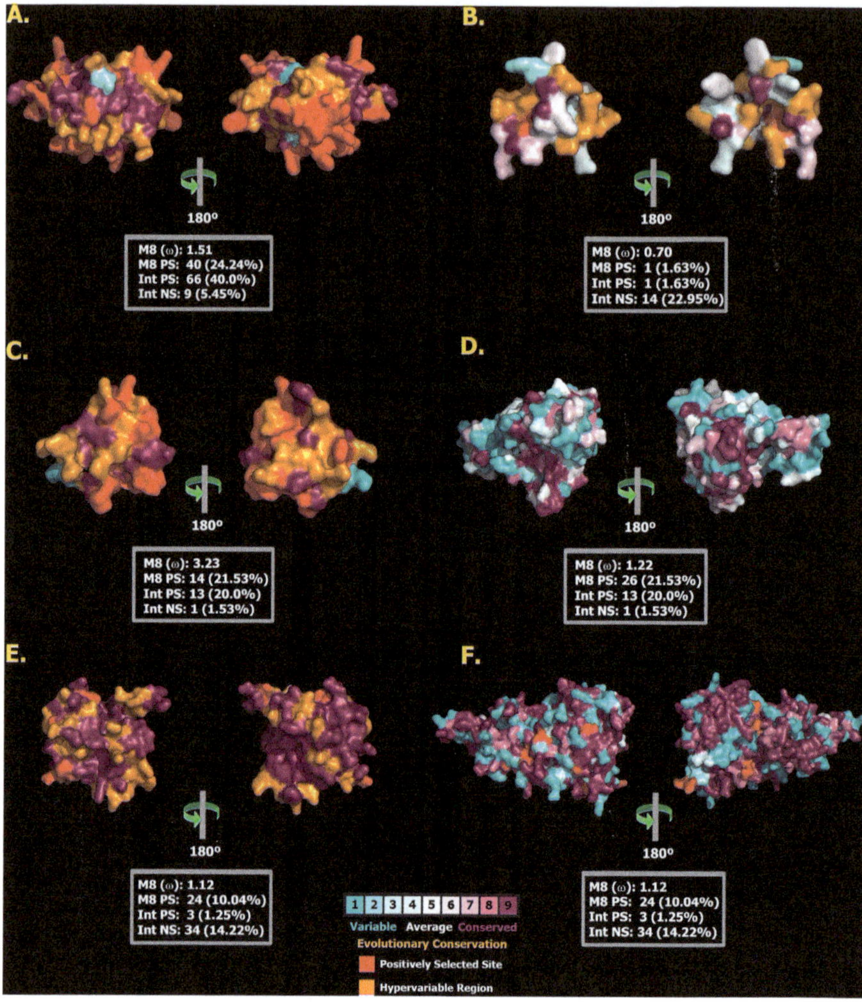

Figure 1.4 Molecular evolution of diverse venom components. (A) Basal snake lectins; (B) Iguania crotamines; (C) Crotalinae crotamines; (D) Elapidae CRiSPs; (E) Homelopsidae veficolins and (F) Vampire bat (*Desmodus rotundus*) plasminogen activator. The homology models show the locations of positively selected sites in red (Model 8, PP $\geq$ 0.95, Bayes-Empirical Bayes approach), along with the site-model 8 computed ω value and the number of positively and negatively selected sites detected using the HyPhy integrative approach [Single Likelihood Ancestor Counting (SLAC), Fixed-Effects Likelihood (FEL), Mixed Effects Model Evolution (MEME): 0.05 significance; Random Effects Likelihood (REL): 50 Bayes factor; Fast, Unconstrained Bayesian AppRoximation (FUBAR): 0.9 posterior probability].

Psammophis mossambicus, in which the P-III SVMP propeptide was shown to be selectively expressed.[41] We have shown that this has put the pre-protein domain under significant selection pressure in these lamprophiid snakes. As a result, the prepro domain of *Psammophis* SVMP has accumulated a pool of mutations at a rapid frequency. Positive selection pressure was found to have influenced evolution of the *Psammophis* monodomain SVMP prepro region more than that of the highly lethal *Echis coloratus* multidomain SVMPs. However, a few species of *Echis* that express similar pre-pro only domains also express the regular multidomain SVMPs. Hence, they do not require similar variations like the *Psammophis*. As a result, they do not accumulate similar variations in the pre-pro region, which remains non-functional. Evolutionary selection analyses conducted on different domains of *Echis* multidomain SVMPs also indicated that the peptidase domain, hypothesised to play a major role in haemotoxicity, accumulates more variations than the disintegrin and cysteine-rich domains. Bioassays revealed that some of the *Psammophis* pre-pro SVMPs inhibit the post-synaptic α7 nicotinic acetylcholine receptor in a similar manner to the α-neurotoxins of Elapidae. Thus, our multidisciplinary approach involving transcriptomics, proteomics, bioassays and bioinformatics allowed us to identify various interesting evolutionary aspects of the *Psammophis* venom arsenal.

1.3.2 Evolution of Vampire Bat Venom[106]

With a level of infamy that extends far beyond the boundaries of science, blood-sucking vampire bats (Chiroptera, Desmodotinae) have been the subject of folk tales, superstitions and stories associated with the legendary Count Dracula for centuries.[107,108] All three species of vampire bat are confined to Central and South America and typically live in caves, tree hollows, and abandoned mines.[109] The relatively rare hairy-legged vampire bat (*Diphylla ecaudata*) feeds exclusively on avian hosts, while the white-winged vampire bat (*Diaemus youngi*) thrives on both mammalian and avian blood, but most likely favours the latter.[107,110] By contrast, the common vampire bat (*Desmodus rotundus*) feeds exclusively on mammals and has established itself in large colonies over an extensive, and apparently widening, distribution.[107,111,112] The expanding population of these bats is attributed to the increasing human population and the associated large number of domesticated animals and livestock, which provide a constant, high-density food supply.[109,112]

All three species of vampire bat are highly specialised for a haematophagous lifestyle. In order to facilitate blood feeding, the vampire bats must be capable of interfering with their prey's natural haemostatic response during both feeding and digestion.[113,114] A typical haemostatic response produces a fibrin clot within minutes of the infliction of a wound, preventing further blood loss. The response commences with the constriction of blood vessels, restricting blood flow to the wound, and is followed by the adhesion of activated platelets to the site of injury and the conversion of fibrinogen to insoluble fibrin, forming a blood clot.[114] In contrast to this normal response to injury, bleeding from a wound induced by vampire bats may be prolonged from minutes to hours, ensuring a constant flow of blood for the bat to feed upon.[114] The submaxillary gland of vampire bats has been shown to secrete venom that has strong anticoagulant and proteolytic activities.[115] The venom delays the onset of blood clotting by interfering with fibrin formation or acting upon fibrin as it is converted from fibrinogen. In addition it has a strong proteolytic action that breaks up any blood clots that may form. This proteolytic action is accomplished through activation of the host's fibrinolytic system, which converts plasminogen to plasmin, solubilising and removing fibrin clots to prevent excessive fibrin build-up at the site of the wound.

We recently elucidated the evolution of some of the common vampire bat venom encoding genes.[106] By utilising a combined proteomic, transcriptomic, bioinformatics, and phylogenetic approach, we not only discovered a much richer suite of secreted proteins than had been previously recognised, but also unravelled interesting aspects of venom evolution in these majestic flying mammals. For example, it was previously unclear whether the anterior and posterior lobes of the *Desmodus rotundus* submaxillary glands were evolving on different evolutionary trajectories, or if they remained under shared genetic control. Recovery of identical transcripts from the cDNA libraries in this study provides the first evidence that they remain a single expression system. Multiple transcripts of the majority of each protein type were recovered from the cDNA libraries, a pattern consistent with accelerated diversification in toxin multigene families as observed in other venoms.[19,116,117]

1.3.3 Evolution of the Venom Apparatus and Peptide Toxin Characterisation in Terebrid and Turrid Marine Snails

While the progress made in characterising cone snail venom is significant, it is only the tip of the iceberg when considering the extensive biodiversity of venomous marine molluscs. The 700 known species of cone snail are a relatively small component of the biodiversity of venomous molluscs in the superfamily Conoidea, which includes sister groups of the family Terebridae, ~400 species, and Turridae, >10 000 species.[118,119] However, unlike cone snails and their venom, the Terebridae and Turridae have not been as extensively characterised. Recent investigations to characterise terebrid[120–122,147] and turrid[123–125] venom as well as their phylogeny[126–130] suggest that the peptide toxins found in terebrids and turrids, termed teretoxins and turritoxins, respectively, are similar to conopeptides in molecular structure. Namely, teretoxins and turritoxins are expressed as a gene product comprising a signal sequence and an intervening pro-region followed by the mature disulfide-rich peptide toxin. Despite their gene organisation similarity there are several notable differences in size and PTMs between conopeptides, teretoxins, and turritoxins. Teretoxins and turritoxins are generally larger in size compared with conopeptides. Conopeptides are typically 10–40 amino acid residues long, whereas terebrids and turrids can range up to 70–80 amino acid residues in size. In this respect, teretoxins and turritoxins are similar to peptides toxins from snakes and scorpions. Analysis of terebrid and turrid venom and cDNA transcriptome data further enhance the diversity from conopeptides (Table 1.2). With regard to PTMs, unlike conopeptides and turritoxins, teretoxins do not appear to be extensively post-translationally modified apart from the formation of disulfide bonds. The recent biochemical characterisation of turritoxin cce9a from the turrid species *Crassispira cerithina* revealed a distinctive age-dependent behavioural phenotype when it was injected intracranially into mice.[125] Similar to conantokin G from *Conus geographus*, turritoxin cce9a caused lethargy in mice 12–14 days old and hyperactivity in mice older than 16 days, suggesting that cce9a may target the *N*-methyl-D-aspartate (NMDA) subtype of glutamate receptors. However, bioassays of cce9a on NMDA receptors were negative, indicating that while the phenotypic behaviour induced by cce9a is similar to that of conantokin G, the molecular mechanism of action is different. This result is indicative of what has been observed generally in the characterisation of terebrid and turrid venom. While cysteine patterns similar to conopeptides are present in teretoxins and turritoxins, it is not a general indicator for comparative function. Even the disulfide connectivity can vary between homologous toxins from these different Conoidea families. For example, Tv1 from the venom of *Terebra variegata* has an M-like conopeptide arrangement of CC–C–C–CC, but it has a disulfide connectivity unlike any M conopeptide previously described.[147]

Although there are striking similarities to conopeptides, teretoxins and turritoxins have a divergent story to tell in terms of venom composition and functional targets in the nervous system, as is evidenced by the fact that unlike cone snails, not all species of terebrids and turrids hunt prey using a venom apparatus.[120,130] Molecular analyses based on cytochrome c oxidase subunit I (COI) and ribosomal 12S, 16S and 28S genes, combined with anatomical morphometric data, revealed a correlation between the appearance of a venom apparatus and clade delineation in the Terebridae. The Terebridae appear to have lost the venom gland at least eight times during their evolution. The foregut anatomy in the family Terebridae is as varied as the range of variability within the entire superfamily Conoidea. Assuming the diversity of foregut structures in the Terebridae is linked to the diversity of feeding types and prey, it follows that the species diversity of the Terebridae could be linked to the prey diversity, suggesting that terebrids have adapted to dietary changes that may have resulted from predator–prey relationships. As terebrids are largely worm hunters, while cone snails hunt fish, worms, and other molluscs, variations in predator–prey relationships may account for the differences in conopeptides and teretoxins.

The complexity of terebrid and turrid venom is a compelling tale to enhance what has been discovered from cone snails. The possibilities for discovering novel bioactive compounds with unique molecular targets from

Table 1.2 Examples of peptide toxins identified in Turridae and Terebridae venoms

Genus	*Species*	*Peptide toxin*	*Peptide sequence*	*References*
			Turridae	
Crassispira	*cerithina*	Cce9a	GSCGLPCHENRRCGWACYCDDGICKPLRV	125
Gemmula	*diomedea*	Gdm9.1	GDOORFCVHKICYγDSγCNQWCTGGCNOTQGKCDTA	148
Gemmula	*kieneri*	Gkn9.1	GVOORFCRDKNCNγDSγCNOWCTGGCSSVKGNCγS	148
Gemmula	*sogodensis*	Gsg9.1	LDOOGYCTHKICYγDGγCNQWCTAGCNODTGKCDTT	148
Gemmula	*speciosa*	Gsp9a	IDOORYCNHIICYγDSγCSQWCTAGCNSITSKCDT	148
		Gsp9b	GDOORFCRDKLCSGDGDCSVWCTAGCNHDMGKCDTL	148
Lophiotoma	*olangoensis*	OL105	FNTCPGQYDECGNGPEEGECCGTHNYCCKNACTTYHHCHGGRDAGKLLRSLKKLTTH	124
		OL108	QHDCTCVEEGQPCQGQACAAECEYNQCLFDGR	124
		OL11	CLSVCSAEYEPVCGSDGKTYANKCHLMTEACWSPTSITLVHEGKC	124
		OL127	YYICESCWTCESCAGSTESSCVSACNACDLCPNK	124
		OL135	VECNETCEEFCTYCDDNNAEETENCRTDQTDHSRCVDFYTANNLPT	124
		OL139	QEGNVCHRPCFRCHVCGETIAACAACSICIGCEEVVEDACAGNPCYWCDNCG VNDGSHRTTRDTADKTHGGQRDRFFQSIA	124
		OL142	QYCPSNTCNERDECPGNCNDCEYVPGTGPNKRCVKK	124
		OL172	QEECDDPDGLMCCTISEMPTC	124
		OL184	AESCDPYQACVLLSAEGRRVPLCSCAGRDCPNTDSHKIQSMYFCEDVSVVY	124
		OL22	TSCETHQICGRKIIYRDGTTNTQ/EEIDYCRCSGDTDCPKDDVNEISFVD/ EWSYWEVSYYTCL	124
		OL25	SLVCDLECSAEVTTCCETGTCHGITTYNCVGGTEPET	124
		OL38	SEVLECYFECGNWEGTCCDTGICVGIHNCKIPEN	124
		OL47	TSCNAATGRSPGCFCNNDNNCRDTCCPRSDTEKKCTGGPDPCPPRQWPD	124
		OL49	LLCISCCVSITECCQLMSGCAVEIKS	124
		OL55	QACSETSDCLEGLECSGNQCLIPYDGDDSCVIGVGCVYDNGNPGCIRDHRC KGDKKDICTNPATECDEDKVCGYKEGETCYGPCRKGLTCRNTRCQK	124
		OL57	LTCLTKMVECIQLPLDVEDSSDTLCC	124

		OL67	SNTCPGYFGECGDGPEEGECCGMYNYCCKGRCLMLASCQKRRDAGRLL RSLKKLKLTTH	124
		OL71	ASCAARACALSGSNITVTYCSCTDGTTHVCPDGSHETNDMYFCENISGVA	124
		OL78	TQAGNYHYCNEGAAVQPGAGPIIYRQNNVEHPYKMNCRCNPWINKARCCTPGT	124
			Terebridae	
Terebra	*subulata*	Agx-s11a	DCEQHTDCSAASGGPVYCCQDSDCCGGVDYICTNYGQCVRHF	121
		Agx-s6a	SLDEELKSNDCPEYCPHGNECCEHHECRYDPWSRELKCLDS	121
		Agx-s7a	ATNRHQCDTNDDCEEDECCVLVGGNVNNPGVQTRICLACS	121
Hastula	*hectica*	hhe1a	GECCTDCAQTAAANYC	122
		Hhe53	GLSQSGCQAFTGRWCVGCERLRSRVVWECSPKRVVNSI	122
		Hhe6.1	GMGIGINLPPCIKNGEYCNPWTGSIILGGACCGTCTDYECH	122
		Hhe6.2	ALPCPYGCPLRCCHMTDGVCLRNKQGC	122
		Hhe6.3	VLFTPPELLGCGNRCSDDCCKWGRCQPGCTD	122
		Hhe6.4	SSLHCGDDPWCPTGCCENEDCDIGCKRDWEKSRSQP	122
		hhe7a	ARCEQCPSYCCQSDSPPECDGCE	122
		Hhe9	ALSVLLQUSCTMCLFCCYL	122
		Hhe9.2	DEEVGCFPNVCKNDGNCSIETSTGMTRCQCLEGYTGHVCENPL	122
		hhe9a	YEENCGTEYCTSKIGCPGRCVCKEYNYNGEITRRCRA	122
		hheTx1	DCLPCGHDVCC	122
		hheTx2	SCSSGCSDCNSDDSCQCTLNQFTNSDSCCC	122
		hheTx3	KQCTSNMCSADCSPGCCIIDKLEWCTCDC	122
		hheTx4	NEVCPPGECQQYCCDLRKCKCINLSFYGLTCNCDS	122
		hheTx5	NEVCPPGRCEPYCCDPRKCKCLSIDFYGLVCNCDS	122
Terebra	*variegata*	Tv1	TR(I/L)CCGCYWNGSKDVCSQSCC	147

terebrid and turrid snails is a promising scientific exploration that requires an integrated approach. A strategy combining genomic, proteomic, and transcriptomic data is not only a promising way to identify the factors that led to diversification of the Terebridae and Turridae, but it also advances the characterisation of terebrid and turrid peptide toxins with novel function and potentially new therapeutic applications.

1.4 Significance of the Combined Approach

1.4.1 Mutation of the Surface Chemistry

Bioinformatics and evolutionary analyses have shown that most of the mutations that accumulate within venom components affect changes on the molecular surface (Figure 1.5).[131–133] Mutation of the surface chemistry ensures that the enzymatic activity, when present, is conserved while variation is still generated in residues responsible of interaction with different target cells and receptors in the prey. This results in the generation of exquisite variation,

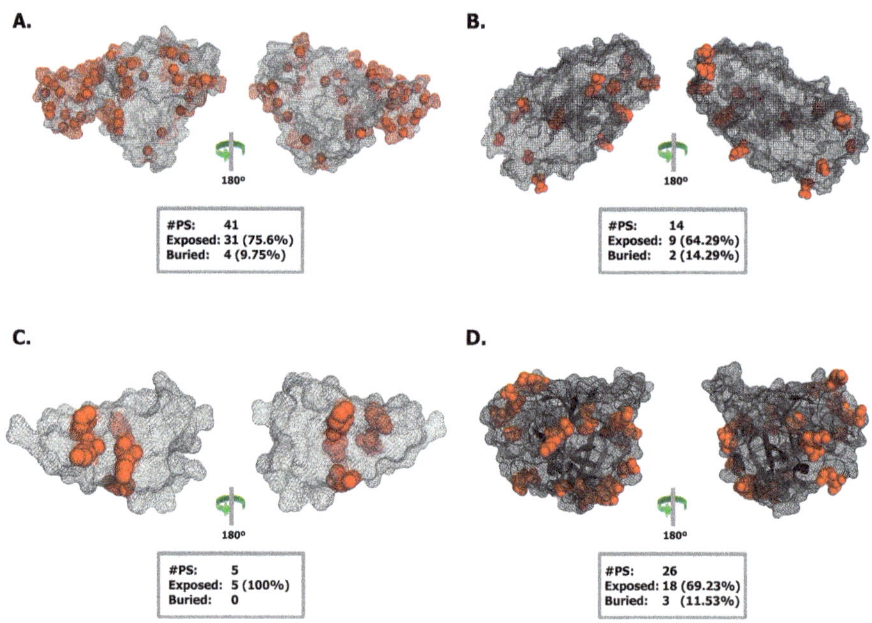

Figure 1.5 Surface accessibility of venom components. A depiction of locations of positively selected sites in different toxins and their surface accessibility: (A) Caenophidian snake CRiSPs; (B) *Crotalus oreganus helleri* L-amino-acid oxidase; (C) Vampire Bat (*Desmodus rotundus*) kunitz protease inhibitor domain I and (D) Coleoid serine protease. Total number of positively selected sites (#PS) detected by PAML (Model 8, Bayes Empirical Bayes approach, Posterior probability ≥ 0.95), total number of exposed and buried positively selected sites are also indicated.

utilising the existing venom arsenal. Identification of such phenomena associated with toxin evolution is extremely important in drug design and antivenom production. Antivenom can be targeted at the conserved structural residues, thus efficiently neutralising the severe effects of envenomation.

1.4.2 Alternative (Differential) Splicing

Alternative splicing is a post-transcriptional process that involves the modification and rearrangement of exons, resulting in multiple protein products (isoforms) encoded by the same genetic sequence. It is the most common phenomenon through which eukaryotes generate proteomic complexity and diversity. Alternative splicing can reduce the maintenance cost associated with multiple-gene families and at the same time increase the genetic potential for encoding complex proteomes. For instance, the *Drosophila melanogaster* DSCAM gene alone can code for a hypothetical maximum of approximately 38 106 peptides utilising the same stretch of nucleotides.[134] Molecular evolution research has shown that several venom-encoding genes adopt alternative splicing to generate an array of toxin types, without increasing the number of copies of the gene itself.[135,136] These isoforms can target a wide range of cells and receptors[137] and could potentially acquire completely different biochemical functions from one another.

1.4.3 Post-Translational Modification

To further increase the variation of the venom arsenal, many venom components undergo PTMs such as glycosylation, phosphorylation, disulfide bridge formation, proteolysis, *etc.* PTMs are also important for achieving a functional state, stabilisation, proper folding of the mature protein, protection from lytic enzymes of the target animal that may disrupt the venom component's function, and exposure of hidden residues that could enable the recognition of new biological targets. PTMs are essential for the modulation of biochemical activities. For instance, glycosylation of the *Naja kouthia* 3FTx decreases its cytotoxicity and expression by two-fold, relative to the unglycosylated form.[138] PTMs can only be detected through proteomic analyses.

1.5 Concluding Remarks

The last two decades has seen a surge in projects exploiting the extraordinary biological potency of venom components to develop novel drugs and diagnostics for human diseases, or as probes to study cells and their receptors.[3,139] Encouraged by the substantial medicinal and fiscal success of the Bristol-Myers Squibb angiotensin-converting enzyme inhibitor, captopril,[140] many pharmaceutical companies have invested in venom-based drug discovery programs.[3] The majority of the currently approved products were developed from snake venom proteins with distinct cardiovascular specificities, particularly those that target thrombin, fibrinogen, and integrin receptors.[3,141] Rapid advances in proteomics, genomics, and transcriptomics have since resulted in affordable technology platforms[100,142–144] that enable mining of venom proteins/peptides for drug discovery from species which, unlike snakes, produce venom in very small quantities. For example, the toxin repertoires of spiders are estimated to contain more than 10 million compounds available for bioprospecting.[142] These reports illustrate that venom peptides, particularly those enriched with the molecular stability imparted by extensive disulfide bonds, are driving the development of, amongst other things, new analgesics, anti-tumour agents and even insecticides. Drug bioprospecting activity is likely to continue to rise as largely unstudied venomous animal lineages are exploited for discovery of novel lead compounds. Venoms are now beginning to receive a great deal of attention as natural sources of novel diagnostic and therapeutic compounds.[140] The venom pool studied to date, often with particular focus on certain toxin types through selective assaying, represents an infinitesimally small representation of the true diversity available. An understanding of the evolutionary and ecological biology of different venomous animal lineages is therefore fundamental to the optimal selection of biological targets for future drug discovery programs.

Acknowledgements

BGF was funded by an Australian Research Council Future Fellowship and by The University of Queensland. This research was supported in part by Fundação para a Ciência e a Tecnologia (FCT) through the PhD grant conferred to KS (SFRH/BD/61959/2009). EABU would like to acknowledge funding from the University of Queensland (International Postgraduate Research Scholarship, UQ Centennial Scholarship, and UQ Advantage Top-Up Scholarship) and the Norwegian State Education Loans Fund. SAA was the recipient of a postdoctoral fellowship (PDRF Phase II Batch-V) from the Higher Education Commission (HEC Islamabad) of Pakistan. NRC was funded by the UK Natural Environment Research Council.

References

1. D. J. Newman, G. M. Cragg, S. Holbeck and E. A. Sausville, *Curr. Cancer Drug Targets,* 2002, **2**, 279–308.
2. R. J. Lewis and M. L. Garcia, *Nat. Rev. Drug Discovery,* 2003, **2**, 790–802.
3. G. F. King, *Expert Opin. Biol. Ther.,* 2011, **11**, 1469–1484.
4. B. G. Fry, K. Roelants, D. E. Champagne, H. Scheib, J. D. A. Tyndall, G. F. King, T. J. Nevalainen, J. A. Norman, R. J. Lewis, R. S. Norton, C. Renjifo and R. C. Rodríguez de la Vega, *Annu. Rev. Genomics Hum. Genet.,* 2009, **10**, 483–511.
5. E. Gong, L. D. Martin, D. A. Burnham and A. R. Falk, *Proc. Natl. Acad. Sci. U.S.A.,* 2010, **107**, 766–768.

6. J. E. Church and W. C. Hodgson, *Toxicon,* 2002, **40**, 1083–1093.
7. D. C. de Graaf, M. Aerts, E. Danneels and B. Devreese, *J. Proteomics,* 2009, **72**, 145–154.
8. N. Peiren, F. Vanrobaeys, D. C. de Graaf, B. Devreese, J. Van Beeumen and F. J. Jacobs, *Biochim. Biophys. Acta,* 2005, **1752**, 1–5.
9. L. Van Valen, *Evol. Theor.,* 1973, **1**, 1–30.
10. S. P. Mackessy, *Toxicon,* 2010, **55**, 1463–1474.
11. N. Jorge da Silva, Jr. and S. D. Aird, *Comp. Biochem. Physiol., Part C: Toxicol. Pharmacol.,* 2001, **128**, 425–456.
12. A. Barlow, C. E. Pook, R. A. Harrison and W. Wüster, *Proc. Biol. Sci.,* 2009, **276**, 2443–2449.
13. H. L. Gibbs and S. P. Mackessy, *Toxicon,* 2009, **53**, 672–679.
14. J. L. Glenn and R. Straight, *Toxicon,* 1978, **16**, 81–84.
15. J. J. Calvete, L. Sanz, P. Cid, P. de la Torre, M. Flores-Diaz, M. C. Dos Santos, A. Borges, A. Bremo, Y. Angulo, B. Lomonte, A. Alape-Giron and J. M. Gutierrez, *J. Proteome Res.,* 2010, **9**, 528–544.
16. M. C. Menezes, M. F. Furtado, S. R. Travaglia-Cardoso, A. C. Camargo and S. M. Serrano, *Toxicon,* 2006, **47**, 304–312.
17. D. V. Andrade and A. S. Abe, *Herpetologica,* 1999, **55**, 200–204.
18. S. K. Sutherland and J. Tibballs, *Australian animal toxins: the creatures, their toxins and care of the poisoned patient*, Oxford University Press, 2001.
19. B. G. Fry, *Genome Res.,* 2005, **15**, 403–420.
20. T. C. Assumpcao, J. M. Ribeiro and I. M. Francischetti, *Toxins,* 2012, **4**, 296–322.
21. B. G. Fry, W. Wüster, R. M. Kini, V. Brusic, A. Khan, D. Venkataraman and A. P. Rooney, *J. Mol. Evol.,* 2003, **57**, 110–129.
22. J. A. Veenstra, *Arch. Insect Biochem. Physiol.,* 2000, **43**, 49–63.
23. B. G. Fry, K. Winter, J. A. Norman, K. Roelants, R. J. Nabuurs, M. J. van Osch, W. M. Teeuwisse, L. van der Weerd, J. E. McNaughtan, H. F. Kwok, H. Scheib, L. Greisman, E. Kochva, L. J. Miller, F. Gao, J. Karas, D. Scanlon, F. Lin, S. Kuruppu, C. Shaw, L. Wong and W. C. Hodgson, *Mol. Cell. Proteomics,* 2010, **9**, 2369–2390.
24. M. Herrera, J. Fernandez, M. Vargas, M. Villalta, A. Segura, G. Leon, Y. Angulo, O. Paiva, T. Matainaho, S. D. Jensen, K. D. Winkel, J. J. Calvete, D. J. Williams and J. M. Gutierrez, *J. Proteomics,* 2012, **75**, 2128–2140.
25. B. G. Fry, H. Scheib, L. van der Weerd, B. Young, J. McNaughtan, S. F. Ramjan, N. Vidal, R. E. Poelmann and J. A. Norman, *Mol. Cell. Proteomics,* 2008, **7**, 215–246.
26. B. G. Fry, N. G. Lumsden, W. Wüster, J. C. Wickramaratna, W. C. Hodgson and R. M. Kini, *J. Mol. Evol.,* 2003, **57**, 446–452.
27. J. Pawlak, S. P. Mackessy, N. M. Sixberry, E. A. Stura, M. H. Le Du, R. Ménez, C. S. Foo, A. Ménez, S. Nirthanan and R. M. Kini, *FASEB J.,* 2009, **23**, 534–545.
28. J. Pawlak, S. P. Mackessy, B. G. Fry, M. Bhatia, G. Mourier, C. Fruchart-Gaillard, D. Servent, R. Ménez, E. Stura, A. Ménez and R. M. Kini, *J. Biol. Chem.,* 2006, **281**, 29030–29041.

29. M. Goyffon and C. Landon, *C. R. Seances Soc. Biol. Ses Fil.*, 1998, **192**, 445–462.
30. H. Rochat, P. Bernard and F. Couraud, *Adv. Cytopharmacol.*, 1979, **3**, 325–334.
31. V. Herzig and G. F. King, in *Spider Ecophysiology*, ed. W. Nentwig, Springer Berlin Heidelberg, 2013, pp. 203–215.
32. J. A. Coddington and H. W. Levi, *Annu. Rev. Ecol. Systemat.*, 1991, **22**, 565–592.
33. L. I. Anderson and N. H. Trewin, *Palaeontology*, 2003, **46**, 467–509.
34. N. R. Casewell, W. Wüster, F. J. Vonk, R. A. Harrison and B. G. Fry, *Trends Ecol. Evol.*, 2013, **28**, 219–229.
35. E. A. Undheim, D. N. Georgieva, H. H. Thoen, J. A. Norman, J. Mork, C. Betzel and B. G. Fry, *Toxicon*, 2010, **56**, 897–913.
36. I. Koludarov, K. Sunagar, E. A. Undheim, T. N. Jackson, T. Ruder, D. Whitehead, A. C. Saucedo, G. R. Mora, A. C. Alagon, G. F. King, A. Antunes and B. G. Fry, *J. Mol. Evol.*, 2012, **75**, 168–183.
37. B. G. Fry, K. Roelants and J. A. Norman, *J. Mol. Evol.*, 2009, **68**, 311–321.
38. N. Vidal and S. B. Hedges, *C. R. Biol.*, 2005, **328**, 1000–1008.
39. B. G. Fry, N. Vidal, J. A. Norman, F. J. Vonk, H. Scheib, S. F. R. Ramjan, S. Kuruppu, K. Fung, S. B. Hedges, M. K. Richardson, W. C. Hodgson, V. Ignjatovic, R. Summerhayes and E. Kochva, *Nature*, 2006, **439**, 584–588.
40. B. G. Fry, K. Roelants, K. Winter, W. C. Hodgson, L. Griesman, H. F. Kwok, D. Scanlon, J. Karas, C. Shaw, L. Wong and J. A. Norman, *Mol. Biol. Evol.*, 2010, **27**, 395–407.
41. B. G. Fry, H. Scheib, L. van der Weerd, B. Young, J. McNaughtan, S. F. Ryan Ramjan, N. Vidal, R. E. Poelmann and J. A. Norman, *Mol. Cell. Proteomics*, 2008, **7**, 215–246.
42. L. Junqueira-de-Azevedo Ide and P. L. Ho, *Gene*, 2002, **299**, 279–291.
43. S. C. Wagstaff and R. A. Harrison, *Gene*, 2006, **377**, 21–32.
44. R. S. Rodrigues, J. Boldrini-Franca, F. P. Fonseca, P. de la Torre, F. Henrique-Silva, L. Sanz, J. J. Calvete and V. M. Rodrigues, *J. Proteomics*, 2012, **75**, 2707–2720.
45. D. R. Rokyta, A. R. Lemmon, M. J. Margres and K. Aronow, *BMC Genomics*, 2012, **13**, 312.
46. Y. Terrat, D. Biass, S. Dutertre, P. Favreau, M. Remm, R. Stocklin, D. Piquemal and F. Ducancel, *Toxicon*, 2012, **59**, 34–46.
47. S. Gotz, J. M. Garcia-Gomez, J. Terol, T. D. Williams, S. H. Nagaraj, M. J. Nueda, M. Robles, M. Talon, J. Dopazo and A. Conesa, *Nucleic Acids Res.*, 2008, **36**, 3420–3435.
48. S. Gotz, R. Arnold, P. Sebastian-Leon, S. Martin-Rodriguez, P. Tischler, M. A. Jehl, J. Dopazo, T. Rattei and A. Conesa, *Bioinformatics*, 2011, **27**, 919–924.
49. A. Conesa, S. Gotz, J. M. Garcia-Gomez, J. Terol, M. Talon and M. Robles, *Bioinformatics*, 2005, **21**, 3674–3676.
50. A. Conesa and S. Gotz, *Int. J. Plant Genomics*, 2008, **2008**, 619832.

51. N. R. Casewell, R. A. Harrison, W. Wüster and S. C. Wagstaff, *BMC Genomics,* 2009, **10**, 564.
52. D. R. Rokyta, K. P. Wray, A. R. Lemmon, E. M. Lemmon and S. B. Caudle, *Toxicon,* 2011, **57**, 657–671.
53. T. Dagan, *Trends Microbiol.,* 2011, **19**, 483–491.
54. S. Halary, J. W. Leigh, B. Cheaib, P. Lopez and E. Bapteste, *Proc. Natl. Acad. Sci. U.S.A.,* 2010, **107**, 127–132.
55. F. Jungo and A. Bairoch, *Toxicon,* 2005, **45**, 293–301.
56. M. E. Smoot, K. Ono, J. Ruscheinski, P. L. Wang and T. Ideker, *Bioinformatics,* 2011, **27**, 431–432.
57. M. Bastian, S. Heymann and M. Jacomy, *International AAAI Conference on Weblogs and Social Media*, 2009.
58. G. E. Crooks, G. Hon, J. M. Chandonia and S. E. Brenner, *Genome Res.,* 2004, **14**, 1188–1190.
59. S. F. Altschul, T. L. Madden, A. A. Schaffer, J. Zhang, Z. Zhang, W. Miller and D. J. Lipman, *Nucleic Acids Res.,* 1997, **25**, 3389–3402.
60. D. G. Higgins and P. M. Sharp, *Gene,* 1988, **73**, 237–244.
61. R. C. Edgar, *Nucleic Acids Res.,* 2004, **32**, 1792–1797.
62. F. Ronquist and J. P. Huelsenbeck, *Bioinformatics,* 2003, **19**, 1572–1574.
63. S. Guindon, J. F. Dufayard, V. Lefort, M. Anisimova, W. Hordijk and O. Gascuel, *Syst. Biol.,* 2010, **59**, 307–321.
64. A. Stamatakis, P. Hoover and J. Rougemont, *Syst. Biol.,* 2008, **57**, 758–771.
65. D. Posada and K. A. Crandall, *J. Mol. Evol.,* 2002, **54**, 396–402.
66. G. A. McVean, *Heredity (Edinb.),* 2001, **87**, 613–620.
67. M. Worobey, *Mol. Biol. Evol.,* 2001, **18**, 1425–1434.
68. K. S. Lole, R. C. Bollinger, R. S. Paranjape, D. Gadkari, S. S. Kulkarni, N. G. Novak, R. Ingersoll, H. W. Sheppard and S. C. Ray, *J. Virol.,* 1999, **73**, 152–160.
69. D. Martin and E. Rybicki, *Bioinformatics,* 2000, **16**, 562–563.
70. I. Milne, F. Wright, G. Rowe, D. F. Marshall, D. Husmeier and G. McGuire, *Bioinformatics,* 2004, **20**, 1806–1807.
71. S. D. Frost, Y. Liu, S. L. Pond, C. Chappey, T. Wrin, C. J. Petropoulos, S. J. Little and D. D. Richman, *J. Virol.,* 2005, **79**, 6523–6527.
72. W. Delport, A. F. Poon, S. D. Frost and S. L. Kosakovsky Pond, *Bioinformatics,* 2010, **26**, 2455–2457.
73. S. L. Kosakovsky Pond, D. Posada, M. B. Gravenor, C. H. Woelk and S. D. Frost, *Mol. Biol. Evol.,* 2006, **23**, 1891–1901.
74. M. Kimura, Cambridge University Press, 1983, Cambridge.
75. N. Goldman and Z. Yang, *Mol. Biol. Evol.,* 1994, **11**, 725–736.
76. Z. Yang, *Mol. Biol. Evol.,* 1998, **15**, 568–573.
77. Z. Yang, *Mol. Biol. Evol.,* 2007, **24**, 1586–1591.
78. S. L. Pond and S. D. Frost, *Mol. Biol. Evol.,* 2005, **22**, 478–485.
79. S. L. Kosakovsky Pond, B. Murrell, M. Fourment, S. D. Frost, W. Delport and K. Scheffler, *Mol. Biol. Evol.,* 2011, **28**, 3033–3043.

80. K. Arnold, L. Bordoli, J. Kopp and T. Schwede, *Bioinformatics,* 2006, **22**, 195–201.
81. F. Kiefer, K. Arnold, M. Kunzli, L. Bordoli and T. Schwede, *Nucleic Acids Res.,* 2009, **37**, D387–D392.
82. M. C. Peitsch, *Biochem. Soc. Trans.,* 1996, **24**, 274–279.
83. L. A. Kelley and M. J. Sternberg, *Nat. Protoc.,* 2009, **4**, 363–371.
84. A. Armon, D. Graur and N. Ben-Tal, *J. Mol. Biol.,* 2001, **307**, 447–463.
85. A. Doron-Faigenboim, A. Stern, I. Mayrose, E. Bacharach and T. Pupko, *Bioinformatics,* 2005, **21**, 2101–2103.
86. A. Stern, A. Doron-Faigenboim, E. Erez, E. Martz, E. Bacharach and T. Pupko, *Nucleic Acids Res.,* 2007, **35**, W506–W511.
87. J. Davis, A. Jones and R. J. Lewis, *Peptides,* 2009, **30**, 1222–1227.
88. P. Escoubas, B. L. Sollod and G. F. King, *Toxicon,* 2006, **47**, 650–663.
89. B. M. Olivera and L. J. Cruz, *Toxicon,* 2001, **39**, 7–14.
90. B. M. Olivera, J. M. McIntosh, L. J. Cruz, F. A. Luque and W. R. Gray, *Biochemistry,* 1984, **23**, 5087–5090.
91. C. Clark, B. M. Olivera and L. J. Cruz, *Toxicon,* 1981, **19**, 691–699.
92. E. W. McCleskey, A. P. Fox, D. H. Feldman, L. J. Cruz, B. M. Olivera, R. W. Tsien and D. Yoshikami, *Proc. Natl. Acad. Sci. U.S.A.,* 1987, **84**, 4327–4331.
93. G. P. Miljanich, *Curr. Med. Chem.,* 2004, **11**, 3029–3040.
94. R. J. Lewis, S. Dutertre, I. Vetter and M. J. Christie, *Pharmacol. Rev.,* 2012, **64**, 259–298.
95. L. J. Greene, A. C. Camargo, E. M. Krieger, J. M. Stewart and S. H. Ferreira, *Circ. Res.,* 1972, **31**(Suppl. 2), 62–71.
96. S. H. Ferreira, *Br. J. Pharmacol. Chemother.,* 1965, **24**, 163–169.
97. M. A. Ondetti, B. Rubin and D. W. Cushman, *Science,* 1977, **196**, 441–444.
98. I. A. Sharpe, J. Gehrmann, M. L. Loughnan, L. Thomas, D. A. Adams, A. Atkins, E. Palant, D. J. Craik, D. J. Adams, P. F. Alewood and R. J. Lewis, *Nat. Neurosci.,* 2001, **4**, 902–907.
99. J. H. Zhang, T. D. Chung and K. R. Oldenburg, *J. Biomol. Screen.,* 1999, **4**, 67–73.
100. I. Vetter, J. L. Davis, L. D. Rash, R. Anangi, M. Mobli, P. F. Alewood, R. J. Lewis and G. F. King, *Amino Acids,* 2011, **40**, 15–28.
101. A. Brust, K. Sunagar, E. A. Undheim, I. Vetter, D. Yang, N. R. Casewell, T. Ruder, T. N. Jackson, I. Koludarov, P. F. Alewood, W. C. Hodgson, R. J. Lewis, G. F. King, A. Antunes, I. Hendrikx and B. G. Fry, *Mol. Cell. Proteomics,* 2012, **12**, 651–663.
102. F. Marcon, M. Leblanc, I. Vetter, R. J. Lewis, P. Escoubas and G. M. Nicholson, *Biochem. Pharmacol.,* 2012, **84**, 851–863.
103. I. Vetter, *Adv. Exp. Med. Biol.,* 2012, **740**, 45–82.
104. L. A. Hite, L. G. Jia, J. B. Bjarnason and J. W. Fox, *Arch. Biochem. Biophys.,* 1994, **308**, 182–191.
105. J. B. Bjarnason and J. W. Fox, *Pharmacol. Ther.,* 1994, **62**, 325–372.

106. D. H. W. Low, K. Sunagar, E. A. B. Undheim, S. A. Ali, A. C. Alagon, T. Ruder, T. N. W. Jackson, S. P. Gonzalez, G. F. King, A. Jones, A. Antunes and B. G. Fry, *J. Proteomics*, 2013, **89**, 95–111.
107. M. D. Tuttle in *Natural history of vampire bats*, ed. A. M. Greenhall and U. Schmidt, CRC Press, USA, 1988, pp. 1–6.
108. C. B. Villa and R. M. Canela in *Natural history of vampire bats*, ed. A. M. Greenhall and U. Schmidt, CRC Press, USA, 1988, pp. 233–240.
109. C. E. Ray, O. J. Linares and G. S. Morgan in *Natural history of vampire bats*, ed. A. M. Greenhall and U. Schmidt, CRC Press, USA, 1988, pp. 19–30.
110. K. F. Koopman in *Natural history of vampire bats*, ed. A. M. Greenhall and U. Schmidt, CRC Press, USA, 1988, pp. 7–18.
111. K. P. Bhatnagar in *Natural history of vampire bats*, ed. A. M. Greenhall and U. Schmidt, CRC Press, USA, 1988, pp. 4–70.
112. A. M. Greenhall in *Natural history of vampire bats*, ed. A. M. Greenhall and U. Schmidt, CRC Press, USA, 1988, pp. 111–132.
113. A. V. Basanova, I. P. Baskova and L. L. Zavalova, *Biochemistry (Mosc.)*, 2002, **67**, 143–150.
114. C. M. Hawkey in *Natural history of vampire bats*, ed. A. M. Greenhall and U. Schmidt, CRC Press, USA, 1988, pp. 133–142.
115. P. E. Disanto, *J. Morphol.*, 1960, **106**, 301–335.
116. T. F. Duda, Jr. and S. R. Palumbi, *Proc. Natl. Acad. Sci. U.S.A.*, 1999, **96**, 6820–6823.
117. B. G. Fry, W. Wüster, R. M. Kini, V. Brusic, A. Khan, D. Venkataraman and A. P. Rooney, *J. Mol. Evol.*, 2003, **57**, 110–129.
118. P. Bouchet, P. Lozouet, P. Maestrati and V. Heros, *Biol. J. Linnean Soc.*, 2002, **75**, 421–436.
119. P. Bouchet, Y. I. Kantor, A. Sysoev and N. Puillandre, *J. Molluscan Stud.*, 2011, **77**, 273–308.
120. N. Puillandre and M. Holford, *BMC Chem. Biol.*, 2010, **10**, 7.
121. J. S. Imperial, M. Watkins, P. Chen, D. R. Hillyard, L. J. Cruz and B. M. Olivera, *Toxicon*, 2003, **42**, 391–398.
122. J. S. Imperial, Y. Kantor, M. Watkins, F. M. Heralde, 3rd, B. Stevenson, P. Chen, K. Hansson, J. Stenflo, J. P. Ownby, P. Bouchet and B. M. Olivera, *J. Exp. Zool., Part B.*, 2007, **308**, 744–756.
123. E. Lopez-Vera, E. P. Heimer de la Cotera, M. Maillo, J. R. Riesgo-Escovar, B. M. Olivera and M. B. Aguilar, *Toxicon*, 2004, **43**, 365–374.
124. M. Watkins, D. R. Hillyard and B. M. Olivera, *J. Mol. Evol.*, 2006, **62**, 247–256.
125. A. B. Cabang, J. S. Imperial, J. Gajewiak, M. Watkins, P. S. Corneli, B. M. Olivera and G. P. Concepcion, *Toxicon*, 2011, **58**, 672–680.
126. N. Puillandre, S. Samadi, M. C. Boisselier, A. V. Sysoev, Y. I. Kantor, C. Cruaud, A. Couloux and P. Bouchet, *Mol. Phylogenet. Evol.*, 2008, **47**, 1122–1134.
127. M. Holford, N. Puillandre, Y. Terryn, C. Cruaud, B. Olivera and P. Bouchet, *Mol. Biol. Evol.*, 2009, **26**, 15–25.

128. M. Holford, N. Puillandre, M. V. Modica, M. Watkins, R. Collin, E. Bermingham and B. M. Olivera, *PLoS One,* 2009, **4**, e7667.
129. N. Puillandre, C. P. Meyer, P. Bouchet and B. M. Olivera, *Zool. Scr.,* 2011, **40**, 350–363.
130. M. Castelin, N. Puillandre, Y. I. Kantor, M. V. Modica, Y. Terryn, C. Cruaud, P. Bouchet and M. Holford, *Mol. Phylogenet. Evol.,* 2012, **64**, 21–44.
131. K. Sunagar, W. E. Johnson, S. J. O'Brien, V. Vasconcelos and A. Antunes, *Mol. Biol. Evol.,* 2012, **29**, 1807–1822.
132. R. M. Kini and Y. M. Chan, *J. Mol. Evol.,* 1999, **48**, 125–132.
133. N. R. Casewell, S. C. Wagstaff, R. A. Harrison, C. Renjifo and W. Wüster, *Mol. Biol. Evol.,* 2011, **28**, 2637–2649.
134. D. L. Black, *Cell,* 2000, **103**, 367–370.
135. T. J. Fujimi, T. Nakajyo, E. Nishimura, E. Ogura, T. Tsuchiya and T. Tamiya, *Gene,* 2003, **313**, 111–118.
136. T. Tamiya, S. Ohno, E. Nishimura, T. J. Fujimi and T. Tsuchiya, *Toxicon,* 1999, **37**, 181–185.
137. D. Servent, V. Winckler-Dietrich, H. Y. Hu, P. Kessler, P. Drevet, D. Bertrand and A. Ménez, *J. Biol. Chem.,* 1997, **272**, 24279–24286.
138. A. V. Osipov, M. V. Astapova, V. I. Tsetlin and Y. N. Utkin, *Eur. J. Biochem.,* 2004, **271**, 2018–2027.
139. C. J. Bohlen, A. T. Chesler, R. Sharif-Naeini, K. F. Medzihradszky, S. Zhou, D. King, E. E. Sanchez, A. L. Burlingame, A. I. Basbaum and D. Julius, *Nature,* 2011, **479**, 410–414.
140. D. W. Cushman and M. A. Ondetti, *Hypertension,* 1991, **17**, 589–592.
141. C. Y. Koh and R. M. Kini, *Toxicon,* 2012, **59**, 497–506.
142. P. Escoubas and G. F. King, *Expert Rev. Proteomics,* 2009, **6**, 221–224.
143. B. M. Olivera, D. R. Hillyard, M. Marsh and D. Yoshikami, *Trends Biotechnol.,* 1995, **13**, 422–426.
144. S. Liang, *Expert Rev. Proteomics,* 2008, **5**, 731–746.
145. E. Pennisi, *Science,* 2003, **300**, 1694.
146. S. Halary, J. O. McInerney, P. Lopez and E. Bapteste, *BMC Evol. Biol.,* 2013, **13**, 146.
147. P. Anand, A. Grigoryan, M. H. Bhuiyan, B. Ueberheide, V. Russell, J. Quinoñez, P. Moy, B. T. Chait, S. F. Poget and M. Holford, *PloS One,* 2014, **9**, e94122.
148. Heralde III and M. Francisco, *et al.*, *Toxicon,* 2008, **51**, 890–897.

CHAPTER 2

The Structural Universe of Disulfide-Rich Venom Peptides

VINCENT LAVERGNE*, PAUL F. ALEWOOD, MEHDI MOBLI, AND GLENN F. KING*

Institute for Molecular Bioscience, The University of Queensland, St Lucia, QLD 4072, Australia
*E-mail: glenn.king@imb.uq.edu.au; v.lavergne@imb.uq.edu.au

2.1 Introduction

2.1.1 Venom Peptides

Animal venoms are heterogenous composite secretions designed for predatory and defensive purposes. Their function is to disrupt homeostasis in target organisms following traumatic injection of the venom *via* a specialized envenomation apparatus. Venoms usually contain inorganic salts, small organic molecules, as well as a broad range of high molecular weight proteins including enzymes. However, cysteine-rich bioactive peptides (<50 amino acid residues) and polypeptides (50–150 residues) are the most abundant class of molecules found in these secretions.

In recent decades, peptide toxins have taken centre stage because of their modest size and compact structure, which is often stabilized by conserved disulfide-rich scaffolds that provide physicochemical resistance against enzymatic degradation and tissue clearance. Moreover, their remarkable structural and functional diversity make them an abundant source of templates that can be used as pharmacological probes or for the design of therapeutic agents.[1,2] Indeed, most venom peptides have the ability to

RSC Drug Discovery Series No. 42
Venoms to Drugs: Venom as a Source for the Development of Human Therapeutics
Edited by Glenn F. King

Published by the Royal Society of Chemistry, www.rsc.org

selectively modulate heterologous receptors or ion channels with high potency or inhibit enzymes that mediate key biochemical processes.

Small, disulfide-rich secreted proteins are found across the phylogenetic spectrum. Recently, we described the range of disulfide-rich peptide frameworks found in humans.[3] Here we describe the diversity of disulfide scaffolds present in venom peptides isolated from well-studied venomous taxa such as snakes, scorpions, spiders, and cone snails, as well as more exotic venomous creatures such as the duck-billed platypus.[4–7]

2.1.2 Disulfide Frameworks: Structural Skeletons Sustaining Bioactivity

Venom peptides can dramatically vary in size, ranging from the 7–9 residue contryphans isolated from marine cone snails[8] to the 14–18 kDa phospholipase A_2 (PLA_2) enzymes found in a variety of snake venoms.[9] The most conserved structural feature of these peptides is the presence of intra-chain disulfide bonds that stabilize their tertiary structure; even the small contryphans contain a single disulfide bond.[10]

We will show in this chapter that only a limited number of disulfide frameworks and three-dimensional (3D) folds are found among more than 2000 mature venom mini-proteins reported to date. The structural stability imparted by the disulfide bridges, which allows venom peptides to reach their site of action when delivered into prey or predators, is not the only reason for their remarkable efficiency. The disulfide framework directs the 3D fold of these peptides to such an extent that it allows extreme hypervariability of the non-cysteine residues within the inter-cystine loops, thereby allowing the most selective and potent venom peptides to be selected throughout the course of evolution. The combined structural and functional properties of venom peptides directed toward heterologous receptors such as G-protein coupled receptors (GPCRs),[11,12] voltage- and ligand-gated sodium, potassium, calcium, and chloride channels,[13–16] transporters,[17–19] or specific subtypes of enzymes[20–22] makes them a source of molecular ligands of unparalleled richness.[1,2,23,24]

2.2 Diversity of Disulfide Scaffolds and Tertiary Structures

In this chapter we define a disulfide-rich mini-protein as a peptide or polypeptide up to 150 residues in length displaying a cysteine-rich framework in which all the cysteine residues are involved in *intra*-chain disulfide bonds and where all Cys–Cys connectivities are known. Thus, for the purposes of this analysis, we discarded amino acid sequences containing an odd number of cysteine residues (where often one cysteine residue is involved in an *inter*-chain disulfide bond) as well as polymeric disulfide-rich mini-proteins. We describe each class of disulfide-rich venom peptides, with an emphasis on their secondary structure, 3D fold, and function.

2.2.1 Overview of Framework Richness

To present the mosaic of the diverse frameworks of disulfide-rich venom peptides we surveyed the UniProtKB/Swiss-Prot and TrEMBL databases and retained only proteins that were isolated at the protein level and located exclusively in extracellular compartments (Figure 2.1).[25] We then used a high-throughput algorithm developed in-house to select only the mature region of proteins and their associated meta-data (*i.e.* amino acid sequence, cysteine connectivity, domain(s), subcellular location, and 3D structure references). We then used PLA_2 from *Pseudonaja textilis*, one of the longest venom polypeptides, as a reference (with a 5% tolerance threshold) to retain venom peptide sequences that were less than 162 residues long and contained an even number of cysteine residues ranging from 2 to 16. By selecting proteins with at least 9% cysteine content in order to highlight the role of the cysteine framework and to avoid molecules with long inter-cysteine sequences, we obtained a set of 2022 secreted mature proteins. A final sorting of these peptides and polypeptides according to their taxonomic affiliation gave rise to 31 different groups. The great majority of these 31 venom-peptide classes belong to venomous eumetazoans from diverse taxa such as marine invertebrates (cone snails, sea anemones, stingrays, octopus), arthropods (centipedes, hymenopterans, spiders, scorpions), reptiles (snakes, lizards), and mammals (platypus).

Interestingly, despite the broad diversity of disulfide frameworks uncovered, as illustrated in Table 2.1, many belong to defined protein families sharing similar folds or structural motifs. This observation has led us to describe, in the first part of this chapter, protein groups from different taxa characterized by identical domains and clusters of secondary structures. In the second part, we focus on venom peptides with unique disulfide frameworks that tend to be taxon-specific.

2.2.2 Consensus Architectures of Venom Peptides

In this section we review some typical disulfide frameworks found in peptides isolated from animal venoms. We explain how the disulfide scaffolds are used to stabilize these molecules in a compact and well-defined form by describing the shapes and secondary structures usually encountered. We also briefly describe how the functionally critical residues (*i.e.* the peptide pharmacophore), where known, are displayed on the 3D framework.

2.2.2.1 Inhibitor Cystine Knot Toxins

The inhibitor cystine knot (ICK) motif,[26] also known as a "knottin" fold,[27] is particularly abundant in spider-venom peptides,[28,29] although it can also be found in venom peptides from predatory marine cone snails and scorpions.[16,30] The ICK motif is composed of six cysteine residues distributed in a C–C–CC–C–C pattern, in which the thiol groups of C^{I}–C^{IV}, C^{II}–C^{V}, and C^{III}–C^{VI} form three distinct disulfide bonds (Figure 2.2A–C).[31] The C^{I}–C^{IV} and C^{II}–

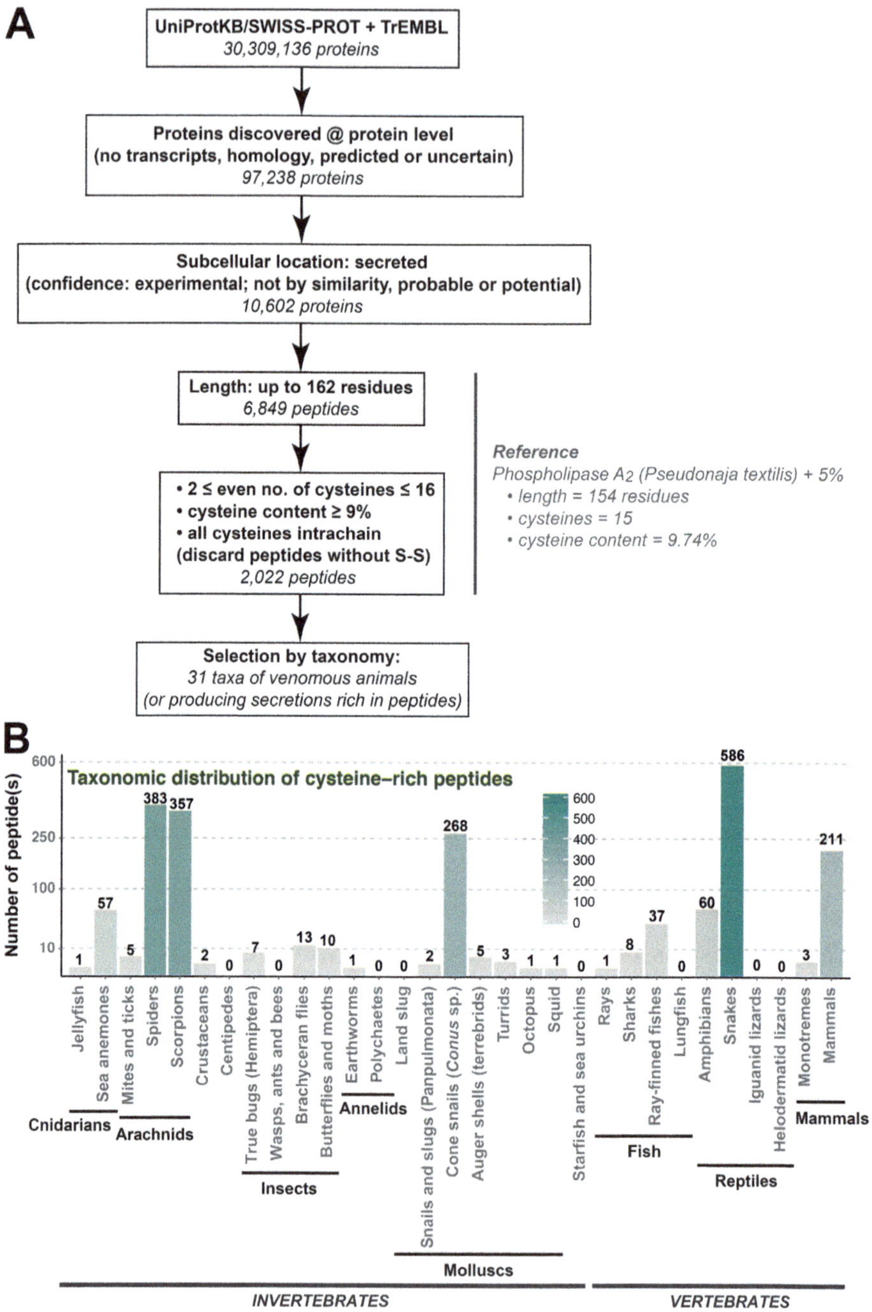

Figure 2.1 Pipeline for the selection of secreted cysteine-rich peptides. (A) The UniProtKB/SWISS-PROT and TrEMBL databases were searched for entries supported by proteomic evidence (based on Edman sequencing, mass spectrometry, X-ray or NMR structure, as well as

C^V disulfide bonds and the intervening sections of the peptide backbone form a ring that is bisected by the third disulfide bridge (Figure 2.2B), resulting in four hypervariable inter-cystine loops (Figure 2.2A). This "knotted" architecture confers on ICK peptides quite remarkable thermodynamic stability and resistance to proteases.[29] The four peptides described below illustrate the core ICK architecture and how it can be elaborated with additional disulfide bonds.

ω-Conotoxin MVIIA (Figure 2.2D) is a 25-residue peptide from the venom of the cone snail *Conus magus*. A synthetic version of this peptide is marketed under the trade name ziconitide as an analgesic drug for the treatment of intractable chronic pain (see Chapter 9 for more details).[32] MVIIA contains an ICK motif supported by a triple-stranded β-sheet involving residues Ala6–Cys8 (β-strand 1), Cys20–Arg21 (β-strand 2), and Lys24–Cys25 (β-strand 3).[33,34] MVIIA exerts its anti-nociceptive action by potently and selectively inhibiting the voltage-gated calcium (Ca_v) channel $Ca_v2.2$ *via* an interaction mediated largely by residues Arg10 and Arg21 in loops 2 and 4, respectively.[35,36]

Huwentoxin IV from the venom of the Chinese tarantula *Haplopelma schmidti* also contains an ICK motif (Figure 2.2E). Its 3D structure is very similar to that of ω-conotoxin MVIIA, with two antiparallel β-strands connected by a loop that contains the positively charged Arg26 residue that is essential for inhibition of the human voltage-gated sodium (Na_V) channel $Na_V1.7$.[37–39] Loss-of-function mutations in the gene encoding this channel lead to a congenital insensitivity pain,[40] without any other sensory deficits except anosmia,[41,42] and consequently this channel is considered an excellent analgesic target.[43–46] Thus, there is much interest in the potential of this class of spider-venom-derived ICK toxins as potential analgesic drugs.[46,47]

Psalmotoxin 1 (π-theraphotoxin-Pc1a or PcTx1; Figure 2.2F) is a 40-residue peptide isolated from the venom of the Trinidad chevron tarantula (*Psalmopoeus cambridgei*). It is a potent and selective inhibitor ($IC_{50} \sim 0.5$ nM) of acid sensing ion channel (ASIC) 1a.[48,49] The structure of PcTx1 is dominated by a β-hairpin (Leu21 to Lys35) made of two antiparallel β-strands (Leu21–Trp24 and Val32–Lys35), plus a single turn of 3_{10} helix in the second loop

detection by antibodies) and isolated from any extracellular compartment. A step-wise algorithm was designed for retaining only mature fragments from the parent precursor proteins, extracting their cysteine framework and their related meta-data (sequences, domains, 3D structure references, structural homologies, phylogenetic groups). The PLA_2 enzyme from *Pseudonaja textilis* was used as a reference to apply length and cysteine-content thresholds in our query (for more flexibility we added a 5% window in the selection criteria). Thus, we selected monomeric proteins up to 162 residues in length with an even number of 2 to 16 cysteine residues engaged in intra-chain disulfide bonds and with an overall cysteine content of at least 9%. (B) Histogram showing the taxonomic distribution of protein hits that were generated using this pipeline.

Table 2.1 Cysteine frameworks found in venom peptides. Eighty-six unique cysteine frameworks were identified in venom peptides and polypeptides (see Figure 2.1 for details about the selection process). Cysteine frameworks are displayed according to the number of cysteines, cysteine pattern, and their connectivities. Taxa containing each cysteine scaffold are listed along with the number of family members. The number of residues found between each pair of cysteine residues is shown in parentheses. The number of three-dimensional structures resolved per taxon is also listed in the table.

No. of cysteine residues	*No. of disulfide bond(s)*	*Cysteine pattern*	*Cysteine connectivity*	*Taxon*	*Inter-cysteine length*	*No. of toxins/taxon*	*No. of 3D structure(s)*
2	1	C–C	I–II	Actinopterygii	(0–5)C(4–15)C(0–3)	23	–
				Amphibia	(0–14)C(4–15)C(0–4)	48	–
				Batoida	C(4)C(3)	1	–
				Brachycera	(8)C(2)C(4)	3	–
				Colubridea	(3–8)C(3–10)C(4–9)	5	–
				Conus	(0–4)C(4–5)C(0–3)	15	4
				Decapoda	(2)C(5)C	1	–
				Dytrisia	(2)C(5)C	1	1
				Eutheria	(0–5)C(4–15)C(0–3)	19	4
				Hemiptera	(10)C(6)C(3)	1	1
				Lumbricina	C(4)C(3)	1	–
				Monotrema	C(4)C(3)	1	–
				Octopidadae	C(4)C(3)	1	–
				Panpulmonata	C(4)C(3)	1	–
				Selachii	(0–6)C(4–15)C(0–5)	8	–
4	2	C–C–C–C	I–III, II–IV	Actinaria	(6)C(2)C(8)C(5)C(3)	1	–
				Colubridea	(2)C(11)C(6)C(16)C(2)	1	–
				Conus	(2–6)C(3–10)C(1–10) C(1–3)C(0–1)	2	1
				Eutheria	(3–16)C(2–5)C(4)C(2–4) C(0–7)	6	4
				Scorpiones	(10)C(5)C(5)C(5)C(7)	2	2

4	2	C–C–C–C	I–IV, II–III	Araneae	(1)C(3)C(4)C(3)C(3)	1	1
				Colubridea	C(1)C(7)C(3)C(6–10)	4	1
				Conus	(5–6)C(3)C(11)C(3)C(1)	6	–
				Eutheria	(5)C(1)C(4)C(1)C(1-3)	4	3
				Scorpiones	(1–5)C(3–5)C(5–9)C(3)C(0–6)	16	6
4	2	C–C–CC	I–III, II–IV	Conus	(1)C(1)C(5)CC	1	1
4	2	C–C–CC	I–IV, II–III	Conus	(1–2)C(1–3)C(4–6)CC(0–2)	3	–
				Terebridae	(1)C(2)C(4)CC	1	–
4	2	CC–C–C	I–III, II–IV	Conus	(0–4)CC(3–4)C(3–7)C(0–2)	50	19
4	2	CC–C–C	I–IV, II–III	Conus	(0–3)CC(4)C(2–3)C	9	3
4	2	CC–CC	I–III, II–IV	Conus	(0–21)CC(4–6)CC(0–4)	25	1
6	3	C–C–C–C–C–C	I–III, II–IV, V–VI	Actinaria	(5)C(8)C(5)C(10)C(1)C(8)C(5)	1	–
				Eutheria	(7)C(7)C(4)C(10)C(1)C(8)C(7)	3	1
6	3	C–C–C–C–C–C	I–III, II–V, IV–VI	Araneae	(2–3)C(3)C(8–9)C(4–7)C(5) C(4)C(3)	34	1
6	3	C–C–C–C–C–C	I–IV, II–III, V–VI	Scorpiones	(2)C(4)C(3)C(9)C(4)C(1)C(3)	1	–
6	3	C–C–C–C–C–C	I–IV, II–V, III–VI	Acari	(3)C(6)C(3)C(9)C(7)C(1)C(3)	2	–
				Araneae	(2–3)C(3)C(8–13)C(7–11) C(4–5)C(4)C(3)	5	1
				Brachycera	(2–12)C(6–12)C(3)C(9) C(4–5)C(1)C(2)	7	3
				Conus	(1–5)C(3–4)C(5)C(2–3) C(1)C(4)C(4)	5	1
				Dytrisia	(6)C(10)C(3)C(9)C(7) C(1)C(1–2)	4	2
				Hemiptera	(2)C(16)C(3)C(9)C(4) C(1)C(2)	2	–
				Scorpiones	(1–28)C(2–9)C(2–3) C(5–11)C(3–8) C(1–2)C(0–11)	115	34
				Terebridae	(4)C(4)C(5)C(3)C(1)C(12)C(2)	1	–
				Turridae	(2–6)C(4)C(4)C(3)C(1–3) C(6–10)C(2–4)	3	–

(continued)

Table 2.1 (*continued*)

No. of cysteine residues	*No. of disulfide bond(s)*	*Cysteine pattern*	*Cysteine connectivity*	*Taxon*	*Inter-cysteine length*	*No. of toxins/taxon*	*No. of 3D structure(s)*
6	3	C–C–C–C–C–C	I–V, II–IV, III–VI	Actinopterygii	(11)C(8)C(7)C(10)C(2)C(17)C	1	–
				Amphibia	(5)C(7)C(7)C(10)C(2)C(21)C	2	–
				Araneae	(5)C(15)C(2)C(6)C(3)C(8)C(5)	3	1
				Brachycera	(2)C(3)C(7)C(17)C(6)C(14)C(1)	1	–
				Dytrisia	(13)C(3)C(2)C(12)C(3)C(10)C(13)	2	–
				Eutheria	(6–13)C(6–9)C(7)C(10)C(2)C(17)C(0–8)	14	5
				Scorpiones	(1)C(4)C(3)C(8)C(3)C(1)C(2–3)	2	1
6	3	C–C–C–C–C–C	I–VI, II–IV, III–V	Actinaria	(1–7)C(6–8)C(4–15)C(7–10)C(3–12)C(2–3)C(0–7)	17	3
				Araneae	(3)C(8)C(13)C(7)C(12)C(3)C(3)	1	1
				Colubridea	(1–6)C(8–9)C(15)C(7)C(12)C(3)C(2–10)	33	7
				Eutheria	(4–6)C(8)C(15)C(7)C(12)C(3)C(3–9)	5	1
				Panpulmonata	(6)C(8)C(15)C(7)C(12)C(3)C(1)	1	–
				Scyphozoa	(2)C(8)C(6)C(13)C(3)C(2)C	1	1

6	3	C–C–C–C–CC	I–V, II–IV, III–VI	Actinaria	(2–4)C(1)C(14–21) C(6–9)C(6–9) CC(1–5)	35	6
				Colubridea	(3)C(6)C(6–7)C(11)C(5) CC(4–8)	12	1
				Eutheria	(4–10)C(6)C(4)C(9)C(6) CC(1–4)	20	4
				Monotrema	(8)C(6)C(7)C(7)C(6) CC(2)	2	2
6	3	C–C–C–C–CC	I–VI, II–IV, III–V	Colubridea	(1)C(17)C(22)C(1) C(6)CC(7)	1	–
				Eutheria	(0–36)C(1)C(3–5)C(9) C(6–9)CC(0–3)	46	13
6	3	C–C–C–CC–C	I–II, III–IV, V–VI	Conus	(3)C(6)C(3)C(9) CC(14)C(1–2)	2	1
6	3	C–C–CC–C–C	I–III, II–V, IV–VI	Araneae	(1)C(6)C(6)CC(19)C(2)C(1)	1	–
				Conus	(3)C(6)C(6)CC(4)C(3)C(1)	1	–
				Dytrisia	(5–9)C(3)C(9)CC(10)C(5–6) C(1–3)	2	–
6	3	C–C–CC–C–C	I–IV, II–V, III–VI	Araneae	(0–6)C(4–6)C(4–9)CC(2–10) C(3–14)C(1–16)	227	28
				Brachycera	(10)C(6)C(5)CC(3)C(6)C(2)	1	–
				Conus	(0–6)C(2–7)C(2–9)CC(2–9) C(3–10)C(0–7)	78	14
				Hemiptera	(4–5)C(6)C(6)CC(4–5) C(6)C(1–3)	3	2
				Scorpiones	(1–4)C(6)C(5)CC(3–5)C(8–10) C(1–6)	4	2
				Terebridae	(2–10)C(2–5)C(3–5)CC(4–16) C(2–10)C(1–3)	3	–
6	3	CC–C–C–C–C	I–III, II–V, IV–VI	Conus	(0–1)CC(6)C(2)C(1)C(3) C(0–5)	3	–

(continued)

Table 2.1 (*continued*)

No. of cysteine residues	*No. of disulfide bond(s)*	*Cysteine pattern*	*Cysteine connectivity*	*Taxon*	*Inter-cysteine length*	*No. of toxins/ taxon*	*No. of 3D structure(s)*
6	3	CC–C–C–C–C	I–IV, II–V, III–VI	Eutheria	(3)CC(9)C(12)C(15) C(10)C(9)	1	1
6	3	CC–C–C–C–C	I–V, II–III, IV–VI	Conus	(1)CC(7)C(2)C(1)C(6–7) C(2–6)	2	2
6	3	CC–C–C–C–C	I–V, II–IV, III–VI	Actinaria	(2)CC(1)C(4)C(5)C(4)C(5)	1	1
6	3	CC–C–C–CC	I–IV, II–V, III–VI	Conus	(1–3)CC(3–6)C(4–9)C(1–5) CC(0–3)	17	6
6	3	CC–C–C–CC	I–V, II–IV, III–VI	Conus	(0–2)CC(4)C(1–5)C(1) CC(0–4)	12	1
6	3	CC–C–C–CC	I–VI, II–IV, III–V	Conus	(0–3)CC(2–4)C(2–4) C(2–3)CC(0–2)	20	–
6	3	CCC–C–C–C	I–VI, II–IV, III–V	Conus	(1–7)CCC(3)C(3–5) C(2–3)C(0–1)	2	–
8	4	C–C–C–C–C–C–C–C	I–II, III–VI, IV–VII, V–VIII	Scorpiones	(1)C(2)C(4)C(5)C(3) C(10)C(4)C(1)C(2)	4	1
8	4	C–C–C–C–C–C–C–C	I–III, II–VII, IV–V, VI–VIII	Araneae	(3)C(3)C(8)C(4)C(1)C(10)C(5) C(4)C(3)	2	–
8	4	C–C–C–C–C–C–C–C	I–IV, II–V, III–VI, VII––VIII	Scorpiones	(12)C(9)C(3)C(10)C(11) C(1)C(10)C(7)C(1)	1	–
8	4	C–C–C–C–C–C–C–C	I–V, II–IV, III–VIII, VI–VII	Dytrisia	(2)C(10)C(3)C(8)C(7)C(1) C(11)C(5)C	1	–
8	4	C–C–C–C–C–C–C–C	I–V, II–VI, III–IV, VII–VIII	Scorpiones	(2)C(5)C(3)C(5)C(4)C(4) C(1)C(2)C(0–1)	3	2
8	4	C–C–C–C–C–C–C–C	I–V, II–VI, III–VII, IV–VIII	Scorpiones	(2–6)C(5)C(3)C(4–5)C(4) C(4)C(1) C(2–3)C(0–2)	8	6

8	4	C–C–C–C–C–C–C–C	I–VIII, II–V, III–VI, IV–VII	Brachycera	(1)C(8)C(7)C(3)C(9)C(5) C(1)C(2)C	1	1
				Scorpiones	(9–12)C(3)C(3–9)C(3) C(8–11)C(4–9) C(1)C(13–17)C(0–4)	167	28
8	4	C–C–C–C–C–CC–C	I–III, II–IV, V–VI, VII–VIII	Amphibia	(2)C(13)C(6)C(12)C(5) C(14)CC(4)C(2)	3	–
				Colubridea	(2)C(8–16)C(4–6)C(13–20) C(1–3) C(5–11)CC(2–5)C(0–8)	165	34
8	4	C–C–C–C–CC–C–C	I–III, II–VI, IV–VII, V–VIII	Conus	(2)C(7)C(4)C(1)C(8)CC(4) C(5)C(6)	2	–
8	4	C–C–C–C–CC–C–C	I–VI, II–VII, III–V, IV–VIII	Colubridea	(6–9)C(7–12)C(3)C(5) C(5)CC(3)C(3–4)C(5–7)	2	2
				Eutheria	(15)C(6)C(8)C(5)C(5) CC(3)C(3)C(4)	1	1
8	4	C–C–C–CC–C–C–C	I–IV, II–VI, III–VII, V–VIII	Scorpiones	(0–15)C(2–10)C(3–10)C(2–10) CC(3–9)C(1–4)C(1–19) C(0–8)	34	6
8	4	C–C–C–CC–C–C–C	I–V, II–VI, III–VIII, IV–VII	Araneae	C(7)C(3)C(1)CC(5)C(12) C(26)C(1)	2	–
8	4	C–C–C–CC–C–C–C	I–V, II–VII, III–VI, IV–VIII	Araneae	(1)C(10)C(3)C(1)CC(5) C(8)C(28)C(8)	1	–
8	4	C–C–CC–C–C–C–C	I–IV, II–V, III–VI, VII–VIII	Araneae	C(6)C(5)CC(3)C(5—6) C(4–6)C(3)C(2–3)	4	–
8	4	C–C–CC–C—C–C–C	I–IV, II–V, III–VIII, VI–VII	Araneae	(1–11)C(4–7)C(4–10) CC(4–13)C(1) C(4–16)C(1)C(1–30)	49	8
8	4	C–C–CC–C–C–C–C	I–IV, II–VI, III–VIII, V–VII	Araneae	(2)C(6)C(7)CC(2)C(1) C(15)C(1)C(4)	2	–
8	4	C–C–CC–CC–C–C	I–IV, II–VI, III–VII, V–VIII	Conus	(0–4)C(6)C(5)CC(1–5) CC(3–7) C(3–10)C(3–8)	10	1

(continued)

Table 2.1 (*continued*)

No. of cysteine residues	*No. of disulfide bond(s)*	*Cysteine pattern*	*Cysteine connectivity*	*Taxon*	*Inter-cysteine length*	*No. of toxins/taxon*	*No. of 3D structure(s)*
8	4	C–C–CC–CC–C–C	I–VI, II–VII, III–IV, V–VIII	Araneae	(2–3)C(6)C(2)CC(1)CC(4)C(9–10)C(3–5)	4	1
				Conus	(2)C(6)C(5)CC(3)CC(2)C(3)C(4)	1	1
8	4	C–C–CCC–C–C–C	I–IV, II–VI, III–VII, V–VIII	Araneae	C(6)C(5)CCC(3)C(10)C(10–11)C(0–2)	7	3
8	4	C–CC–C–C–C–C–C	I–IV, II–VI, III–VII, V–VIII	Colubridea	(0–6)C(4)CC(2)C(8)C(9–11)C(4)C(1)C(5–11)	16	3
8	4	C–CC–CC–C–CC	I–VIII, II–IV, III–VI, V–VII	Actinopterygii	(1)C(2)CC(1)CC(5)C(2)CC(2)	1	1
				Eutheria	(1–6)C(2)CC(1)CC(4)C(2)CC(2)	3	2
10	5	C–C–C–C–C–C–C–C–C–C	I–V, II–VI, III–VII, IV–VIII, IX–X	Eutheria	(11)C(13)C(9)C(3)C(19)C(14)C(15)C(1)C(2)C(6)C(3)	1	–
10	5	C–C–C–C–C–C–C–C–C–C	I–VI, II–VII, III–VIII, IV–IX, V–X	Conus	(1)C(3)C(4)C(3)C(1)C(5)C(1)C(10)C(1)C(1)C(1)	1	–
10	5	C–C–C–C–C–C–C–C–CC	I–X, II–IX, III–VI, IV–VII, V–VIII	Eutheria	(31)C(11)C(8)C(1)C(3)C(10)C(1)C(1)C(9)CC(3)	1	–
10	5	C–C–C–C–C–C–C–CC–C	I–III, II–VI, IV–V, VII–VIII, IX–X	Colubridea	(2)C(9–12)C(5–6)C(4–5)C(3)C(10)C(3)C(10–11)CC(4)C(2–18)	46	8
10	5	C–C–C–C–C–C–C–CC–C	I–V, II–III, IV–VI, VII–VIII, IX–X	Colubridea	(2–9)C(2)C(4–7)C(5–7)C(6)C(14–20)C(3)C(9–11)CC(4)C(1–3)	27	8
				Eutheria	(2)C(2)C(6–8)C(5–6)C(6–7)C(12–22)C(2–5)C(15–17)CC(4–5)C(3–8)	5	1

10	5	C–C–C–C–C–CC–C–C––C	I–VII, II–V, III–IX, IV–VI, VIII–X	Colubridea	(1)C(15–17)C(19)C(2) C(1)C(6)CC(11) C(8)C(13)C(2)	2	–
				Eutheria	(1–2)C(7–15)C(17–18)C(2)C(1) C(6–7)CC(12–13)C(8)C(13)C(7–25)	4	2
				Loliginidae	(1)C(19)C(18)C(2)C(1)C(6)CC(13) C(6)C(12)C(3)	1	–
10	5	C–C–C–CC–C–C–C–C–C	I–V, II–VI, III–VII, IV–IX, VIII–X	Araneae	(3)C(6)C(3)C(1)CC(4)C(1)C(24)C(1) C(11)C(5)	2	–
10	5	C–C–C–CC–C–C–C–C–C	I–V, II–VI, III–X, IV–IX, VII–VIII	Araneae	(0–3)C(6)C(3)C(1)CC(4–5)C(1) C(3–11)C(1)C(8–16)C(0–6)	13	–
10	5	C–C–C–CC–C–C–C–C–C	I–XII, IX–VI, III–X, IV–IX, VII–VIII	Araneae	(1)C(6)C(6)C(1)CC(5)C(1)C(9)C(1) C(11)C(3)	1	–
10	5	C–C–C–CC–CC–C–C–C	I–IV, II–VII, III–VIII, V–IX, VI–X	Actinopterygii	(10)C(2)C(16–17)C(2)CC(26) CC(18–20)C(1)C(2)C(5)	4	–
				Amphibia	(10)C(2)C(18)C(2)CC(26)CC(21) C(1)C(2)C(5)	1	–
				Eutheria	(4–10)C(2)C(17)C(2)CC(26)CC(21) C(1)C(2)C(5)	11	1
10	5	C–C–CC–C–C–C–C–C–C	I–IV, II–V, III–VII, VI–IX, VIII–X	Amphibia	(6)C(5)C(4)CC(11)C(9)C(17)C(1)C(5) C(8)C(1)	4	1
				Araneae	(2–7)C(4–5)C(4)CC(7–8)C(9)C(12–15) C(1)C(5)C(5–6)C(3–5)	8	–
				Colubridea	(6)C(5)C(4)CC(11)C(9)C(17)C(1) C(5)C(9)C(4)	1	1
				Decapoda	(2)C(6)C(4)CC(11)C(9)C(15)C(1)C(5) C(6)C(13)	1	–
				Eutheria	(6–11)C(5)C(3–4)CC(10–11)C(9) C(11–38)C(1)C(5)C(9–17)C(4–9)	7	1
10	5	C–C–CC–C–C–C–C–C-C	I–IV, II–V, III–VIII, VI–VII, IX–X	Araneae	(1)C(6)C(5)CC(13)C(1)C(6)C(1) C(3)C(3)C	1	–
10	5	C–C–CC–C–C–C–C–C–C	I–IV, II–VI, III–IX, V–VIII, VII–X	Araneae	(2)C(6)C(6)CC(2)C(2)C(1)C(7)C(1) C(6)C(3)	1	–

(continued)

Table 2.1 (*continued*)

No. of cysteine residues	*No. of disulfide bond(s)*	*Cysteine pattern*	*Cysteine connectivity*	*Taxon*	*Inter-cysteine length*	*No. of toxins/ taxon*	*No. of 3D structure(s)*
10	5	C–C–CC–C–C–C–C–C–C	I–IV, II–VI, III–IX, V–X, VII–VIII	Araneae	(2–3)C(6)C(5)CC(2)C(2)C(1)C(6–7)C(1)C(6–8)C(1–2)	6	–
				Eutheria	(65–70)C(6)C(6)CC(2)C(2)C(1)C(6)C(1)C(6–9)C(0–3)	2	1
10v	5	C–C–CC–C–C–C–CC–C	I–V, II–III, IV–VI, VII–VIII, IX–X	Colubridea	(2)C(2)C(9)CC(6)C(26)C(3)C(13)CC(4)C(7)	2	–
12	6	C–C–C–C–C–C–C–C–C–C–C–C	I–V, II–IV, III–VI, VII–XI, VIII–X, IX–XII	Eutheria	(7–15)C(11)C(7)C(10)C(2)C(17)C(3)C(6)C(7)C(10)C(2)C(17)C(0–3)	8	–
				Hemiptera	(5)C(1)C(7)C(10)C(3)C(16)C(8)C(2)C(8)C(10)C(3)C(16)C(2)	1	1
12	6	C–C–C–C–C–C–C–C–C–C–C–C	I–VI, II–IV, III–V, VII–XII, VIII–X, IX–XI	Acari	(5–9)C(8)C(14–15)C(7)C(12)C(3)C(10–17)C(8)C(15)C(7)C(12)C(3)C(1–3)	3	1
				Eutheria	(4–6)C(8)C(15)C(7)C(12)C(3)C(5)C(8)C(15)C(7)C(12)C(3)C(12)	2	–
12	6	C–C–C–C–C–C–C–C–C–C–C–C	I–VI, II–VII, III–XII, IV–VIII, V–IX, X–XI	Actinopterygii	(1–5)C(13)C(2)C(7)C(3)C(18–20)C(14)C(15)C(1)C(2)C(6)C(9–11)C(6–21)	8	–
				Amphibia	(0–3)C(13)C(2)C(7)C(3)C(18)C(14)C(15)C(1)C(2)C(6)C(9)C(5–7)	2	–
				Eutheria	(0–10)C(13)C(2)C(7)C(3)C(18–20)C(14)C(15)C(1)C(2)C(6)C(9)C(7–11)	17	2
12	6	C–C–C–CC–C–C–C–C–C–C–C	I–V, II–VII, III–XI, IV–X, VI–XII, VIII–IX	Araneae	(1)C(6)C(3)C(1)CC(1)C(3)C(1)C(13)C(1)C(7)C(6)C(3)	1	–
12	6	C–C–C–CC–C–C–C–C–C–C–C	I–XII, IX–VI, III–X, IV–IX, VII–VIII, XI–XII	Araneae	(1)C(6)C(6)C(1)CC(5)C(1)C(10)C(1)C(11)C(6)C(7)C(9)	2	–
12	6	C–C–CC–C–C–C–C–C–C–C–C	I–VII, II–IV, III–XII, V–XI, VI–IX, VIII–X	Actinaria	(10)C(15)C(14)CC(5)C(6)C(17)C(3)C(6)C(4)C(1)C(6)C(12)	1	–

12	6	C–C–CC–C–C–C–C–C–C–C–C	I–XII, II–IV, III–XI, V–X, VI–VIII, VII–IX	Actinaria	(24)C(1)C(14)CC(5)C(11)C(16)C(8)C(4)C(1)C(6)C(17)C	1	1
				Eutheria	(25)C(1)C(14)CC(5)C(8)C(17)C(5)C(4)C(1)C(6)C(19)C(1)	1	–
12	6	C–C–CC–C–C–C–C–CC–C–C	I–IV, II–V, III–VI, VII–X, VIII–XI, IX–XII	Araneae	(1)C(6)C(5)CC(6)C(7)C(12)C(6)C(5)CC(4)C(7)C(8)	1	–
12	6	C–C–CC–C–C–CC–C–C–C–C	I–III, II–IV, V–VIII, VI–XI, VII–IX, X–XII	Colubridea	(0–8)C(1)C(6)CC(4)C(7)C(4)CC(2)C(8)C(10–11)C(6)C(5–7)	35	4
12	6	C–C–CC–C–C–CC–C–C–C–C	I–V, II–IV, III–IX, VI–VIII, VII–XI, X–XII	Colubridea	(3)C(1)C(6)CC(4)C(7)C(4)CC(2)C(8)C(11)C(6)C(6)	1	1
12	6	C–C–CC–CC–C–C–C–C–C–C	I–XI, II–IV, III–X, V–XII, VI–IX, VII–VIII	Colubridea	(25)C(1)C(14)CC(4)CC(24)C(10)C(1)C(6)C(19)C(6)C	3	–
14	7	C–C–C–C–CC–C–C–C–C–C–C–C	I–VI, II–VII, III–V, IV–VIII, IX–XIV, X–XII, XI–XIII	Eutheria	(11)C(6)C(7)C(5)C(5)CC(3)C(3)C(7)C(8)C(15)C(7)C(12)C(3)C(6–7)	2	–
14	7	C–C–C–C–CC–C–C–C–C–CC–C–C	I–XIV, II–VI, III–V, IV–VII, VIII–XII, IX–XI, X–XIII	Eutheria	(5)C(1)C(10)C(9)C(4)CC(10)C(11)C(9)C(9)C(4)CC(10)C(8)C(2)	2	1
14	7	C–C–C–C–CC–C–C–CC–C–C–C–C	I–IV, II–VII, III–VI, V–XI, VIII–X, IX–XIII, XII–XIV	Colubridea	(4)C(10)C(1)C(5)C(3)CC(4)C(7)C(4)CC(2)C(8)C(11)C(6)C(4)	1	–
14	7	C–C–C–CC–C–C–C–C–C–C–C–C	I–V, II–VI, III–X, IV–IX, VII–VIII, XI–XII, XIII–XIV	Araneae	(1–2)C(6)C(6)C(1)CC(5)C(1)C(9)C(1)C(11)C(5)C(5–12)C(3)C(4)C(0–4)	2	–
14	7	C–C–C–CC–C–C–C–C–C–C–C–C	I–V, II–VI, III–X, IV–IX, VII–VIII, XI–XIII, XII–XIV	Araneae	(5)C(6)C(6)C(1)CC(5)C(1)C(6)C(1)C(11)C(5)C(10)C(3)C(3)C(1)	2	–
14	7	C–C–C–CC–C–C–C–C–C–C–C–C	I–VIII, II–XIV, III–V, IV–XIII, VI–XII, VII–X, IX–XI	Colubridea	(10)C(13–15)C(1)C(13–14)CC(5)C(8–9)C(10–15)C(5–8)C(5–6)C(4)C(1)C(6)C(17–20)C(0–3)	74	20
				Eutheria	(8–10)C(15)C(1)C(14)CC(5)C(9)C(15)C(6)C(6)C(4)C(1)C(6)C(17–18)C(0–2)	8	3
14	7	C–C–C–CC–C–C–C–C–C–CC–C–C	I–VIII, II–IV, III–VII, V–VI, IX–XI, X–XIV, XII–XIII	Conus	(5)C(2)C(4)C(5)CC(4)C(9)C(9)C(6)C(11)C(5)CC(4)C(5)C(1)	1	–
				Eutheria	(6–9)C(2)C(7)C(5)CC(5)C(9)C(9)C(6)C(5)C(5)CC(4)C(5)C(7–10)	8	1

(continued)

Table 2.1 (*continued*)

No. of cysteine residues	*No. of disulfide bond(s)*	*Cysteine pattern*	*Cysteine connectivity*	*Taxon*	*Inter-cysteine length*	*No. of toxins/ taxon*	*No. of 3D structure(s)*
14	7	C–C–CC–CC–C–C–C–C–C–C–C–C	I–XIII, II–IV, III–XII, V–XIV, VI–XI, VII–IX, VIII–X	Colubridea	(24–25)C(1)C(13–15)CC(4)CC(6)C(16–19)C(4–6)C(3–4)C(1)C(6)C(17–20)C(5–7)C	153	42
				Eutheria	(24)C(1)C(14)CC(4)CC(8)C(17)C(5)C(4)C(1)C(6)C(19)C(6)C	2	–
16	8	C–C–C–C–C–CC–C–C–C–C–C–C–C–C–C	I–X, II–III, IV–XVI, V–VII, VI–XV, VIII–XIV, IX–XII, XI–XIII	Colubridea	(18)C(3)C(3)C(7)C(1)C(14)CC(5)C(9)C(15)C(6)C(6)C(4)C(1)C(6)C(18)C(1)	1	–
16	8	C–C–C–C–CC–C–C–C–C–C–C–C–C–C–C	I–X, II–XVI, III–VI, IV–VII, V–XV, VIII–XIV, IX–XII, XI–XIII	Colubridea	(10)C(14)C(1)C(1)C(11)CC(3)C(1)C(9)C(10)C(6)C(6)C(4)C(1)C(6)C(18)C(1)	1	1
16	8	C–C–C–C–CC–C–C–C–C–C–C–CC–C–C	I–VI, II–VII, III–V, IV–VIII, IX–XIV, X–XV, XI–XIII, XII–XVI	Eutheria	(5–14)C(7–10)C(2–7)C(5)C(5)CC(3–4)C(3)C(9–17)C(6–12)C(3–8)C(5)C(5)CC(3)C(3–4)C(5–14)	7	2
16	8	C–C–C–CC–CC–C–C–C–C–C–C–C–C–C	I–IX, II–XV, III–V, IV–XIV, VI–XVI, VII–XIII, VIII–XI, X–XII	Eutheria	(10)C(13)C(1)C(14)CC(4)CC(8)C(10)C(6)C(6)C(4)C(1)C(6)C(17)C(6)C(1)	1	–
			86 frameworks			**2022**	**404**

(His14–Asp16), and a β-turn in the third loop (Cys18–Leu21).[48,49] The overall structure of PcTx1 is stabilized by a typical ICK framework. PcTx1 is currently in preclinical studies for treatment of chronic pain[50] and stroke.[51]

Delta-hexatoxin-Hv1a (δ-HXTX-Hv1a; Figure 2.2G) from the Blue Mountains funnel-web spider (*Hadronyche versuta*) is a member of the lethal δ-hexatoxin family, which is responsible for deaths from envenomation by this family of spiders. δ-HXTX-Hv1a slows down inactivation of insect and vertebrate Na_V channels and causes a hyperpolarizing shift in the voltage-dependence of activation.[13] This induces spontaneous repetitive firing and prolongation of action potentials, which results in neurotransmitter release from somatic and autonomic nerve endings.[13] In humans, this leads to a severe envenomation syndrome that includes lachrymation, salivation, skeletal muscle fasciculation, sweating, nausea, vomiting, diarrhoea, pulmonary oedema, and disturbances in respiration, blood pressure and heart rate, followed by severe hypotension and respiratory or circulatory failure.[52]

δ-HXTX-Hv1a is a 42-residue peptide with a cysteine framework comprising eight cysteine residues that form a C–C–CCC–C–C–C pattern with a unique triplet of consecutive cysteine residues.[53] The eight cysteine residues are engaged in four disulfide bonds with C^{I}–C^{IV}, C^{II}–C^{VI}, C^{III}–C^{VII}, C^{V}–C^{VIII} connectivity.[52] Although, the cysteine pattern is unique, δ-HXTX-Hv1a is simply a variant ICK toxin in which an additional disulfide bond has been added to stabilize an extended C-terminal region. Three of the four disulfide bonds in δ-ACTX-Hv1a (Cys1–Cys15, Cys8–Cys20 and Cys14–Cys31) form a classical ICK motif in which the Cys14–Cys31 disulfide bond passes through a 14-residue ring formed by the peptide backbone and the Cys1–Cys15 and Cys8–Cys20 disulfide bonds.[52] The core ICK region contains a classical three-stranded β-sheet comprising Asn6–Trp7 (β-strand 1), Met18–Val21 (β-strand 2), and Ser30–Ser33 (β-strand 3) (Figure 2.2E). The C-terminal region beyond the last Cys residue of the ICK core is unusually long and forms several turns of 3_{10} helix (residue Ile35–Lys41) (Figure 2.2E). The fourth, atypical disulfide bond between the C-terminal Cys42 residue and Cys16 in the ICK core region effectively acts as a molecular staple that limits the flexibility of the C-terminal helical extension.[52]

δ-HXTX-Hv1a is an example of the way in which ICK toxins are commonly elaborated *via* the addition of non-core disulfide bonds. There are numerous examples of ICK toxins in which an extended β-hairpin loop (loop 4; see Figure 2.2A) is held in place by an additional disulfide bond positioned at the tip of the β-strands; examples include the spider-venom peptides Aps III,[54] ω-agatoxin IVA,[55] and δ-palutoxin IT2.[56] In some cases, more exotic disulfide bonds are present, such as the vicinal disulfide bridge found in loop 2 of κ-HXTX-Hv1c,[57] which is essential for its high-affinity block of insect calcium-activated potassium (K_{Ca}) channels.[58,59] Since the disulfide bonds strongly direct the 3D fold of ICK toxins, they are very permissive to sequence variations in the inter-cystine loops. As a result, ICK toxins have evolved to target a wide range of ion channels and receptors, including Ca_V, Kv, Na_V, and K_{Ca} channels, ASICs, transient receptor potential (TRP) channels, and P2X3

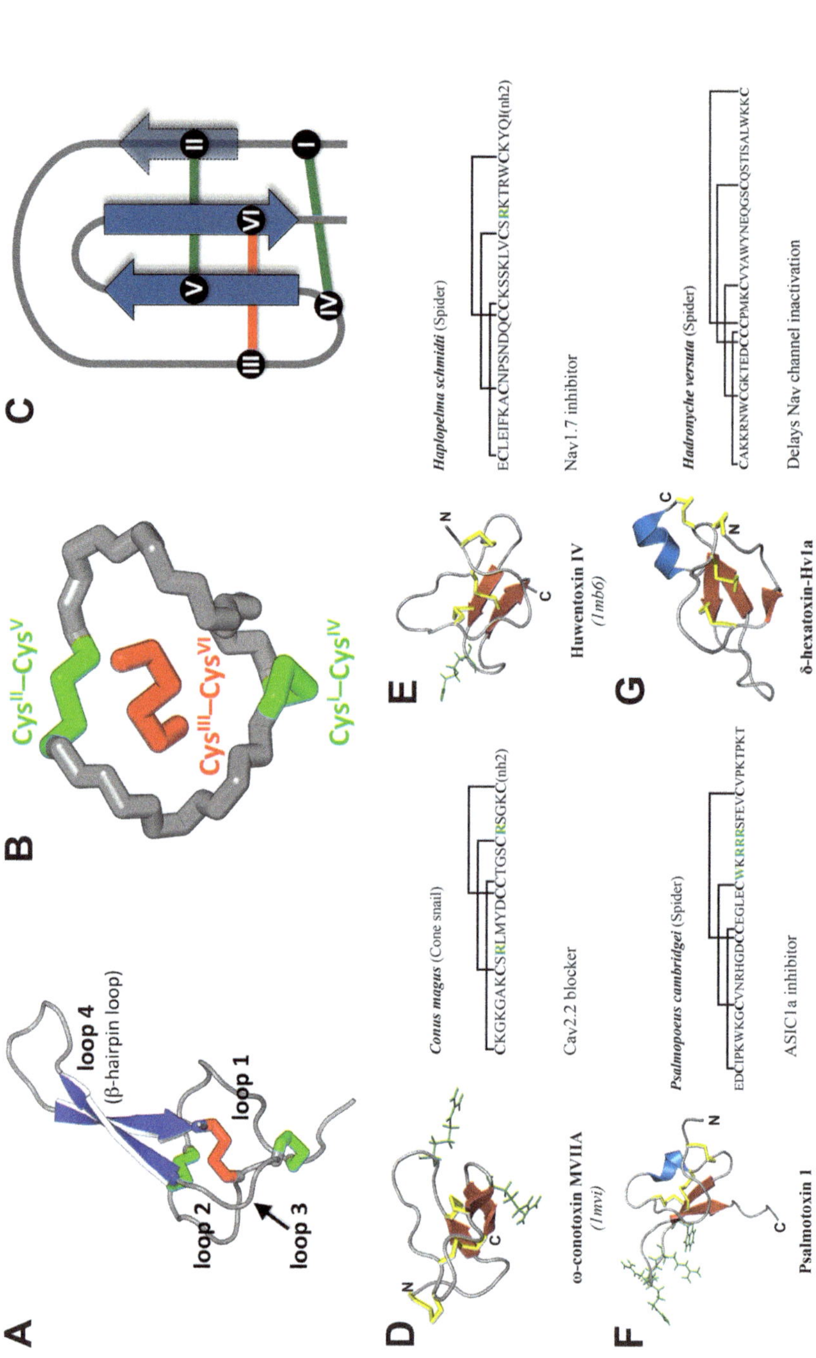
A
loop 4
(β-hairpin loop)
loop 1
loop 2
loop 3
B
Cys^II–Cys^V
Cys^III–Cys^VI
Cys^I–Cys^IV
C
I
II
III
IV
V
VI
D
N
C
ω-conotoxin MVIIA
(1mvi)
Conus magus (Cone snail)
CKGKGAKCSRLMYDCCTGSCRSGKC(nh2)
Cav2.2 blocker
E
N
C
Huwentoxin IV
(1mb6)
Haplopelma schmidti (Spider)
ECLEIFKACNPSNDQCCKSSKLVCSRKTRWCKYQI(nh2)
Nav1.7 inhibitor
F
N
C
Psalmotoxin 1
(2kni)
Psalmopoeus cambridgei (Spider)
EDCIPKWKGCVNRHGDCCEGLECWKRRRSFEVCVPKTPKT
ASIC1a inhibitor
G
C
N
δ-hexatoxin-Hv1a
(1vtx)
Hadronyche versuta (Spider)
CAKKRNWCGKTEDCCCPMKCVYAWYNEQGSCQSTISALWKKC
Delays Nav channel inactivation

receptors. Moreover, this sequence plasticity has enabled ICK toxins to serve as templates for the design of diagnostic agents and drugs in which residues in the inter-cystine loops have been substituted in order to optimize bioavailability or bioactivitiy.[60–62]

Two-disulfide toxins with a knottin-like fold have also been described in spiders and scorpions.[63] In this fold, denoted a disulfide-directed β-hairpin (DDH),[57] the N-terminal Cys^{I}–Cys^{IV} disulfide bond found in knottins is absent. At this stage, there is conflicting evidence about whether the ICK fold is derived from the DDH fold by addition of a disulfide bond (making the DDH fold the plesiotypic state) or *vice versa*.[57,63,64] So-called "double-knot" toxins have also been described in which two ICK domains are joined in a head-to-tail fashion; it appears that these toxins evolved either by duplication and fusion of a single ancestral ICK gene[65] or by transcription-mediated fusion of distinct ICK-encoding genes.[66]

2.2.2.2 BPTI/Kunitz Inhibitor Domain

The Kunitz/Bovine pancreatic trypsin inhibitor (BPTI) family are a ubiquitous group of cysteine-rich proteinase inhibitors.[67] Members of this family contain a conserved cysteine framework (C–C–C–C–C–C pattern with

Figure 2.2 Venom peptides containing an inhibitor cystine knot motif. (A) Structure of the spider-venom peptide ω-hexatoxin-Hv1a (pdb code 1axh)[174] with the three disulfide bonds that form the ICK motif highlighted in red and green. The disulfide bridges result in four inter-cystine loops, with the final loop emanating from the C-terminal β-sheet (β-strands shown in blue). (B) The Cys^{I}–Cys^{IV} and Cys^{II}–CysV disulfide bonds in ω-hexatoxin-Hv1a (shown in green) and the intervening sections of the peptide backbone (shown in grey) form a closed loop that is bisected by the Cys^{III}–Cys^{VI} disulfide bond. (C) Topology of the ICK motif, which comprises an antiparallel β-sheet stabilized by a cystine knot. β-strands are shown in blue and the six cysteine residues that form the cystine knot are labelled I–VI. The β-sheet comprises two mandatory C-terminal β-strands (shown in blue) that house Cys^{V} and Cys^{VI}. A third N-terminal β-strand (shown in translucent blue) that encompasses Cys^{II} is sometimes present. The two "outer" disulfide bonds are shown in green and the "inner" disulfide bridge that bisects the cystine-knot loop is shown in red. (D–G) Amino acid sequence and ribbon representations of the three-dimensional structure of the ICK toxins ω-conotoxin MVIIA (D), huwentoxin IV (E), psalmotoxin 1 (F), and δ-hexatoxin-Hv1a (G). In these images and all subsequent structures shown in this chapter, α-helices, β-strands, and disulfide bonds are shown in blue, red, and yellow, respectively. Residues that have been experimentally demonstrated to be critical for toxin function are highlighted in green in the amino acid sequence and their side chains are displayed in the 3D structures. The N- and C-termini are labelled, PDB accession numbers are indicated in parentheses, and the source organism and molecular target are indicated. Cysteine connectivities are shown above the primary structures. (nh2) indicates that the C-terminus is amidated.

connectivity C^{I}–C^{VI}, C^{II}–C^{IV}, C^{III}–C^{V}) and a consensus α/β fold, called the Kunitz domain, responsible for their protease inhibitory activity, mainly against serine proteases such as trypsin, chymotrypsin, cathepsin, and plasmin.[68,69] Kunitz domain peptides have been recruited into a number of animal venoms. For example, the Kunitz-type proteinase inhibitor SHPI-1 isolated from the sea anemone *Stoichactis helianthus* (Figure 2.3A),[70–72] κ-theraphotoxin-Hh1a (also called huwentoxin-11; Figure 2.3B) from the venom of the Chinese bird spider,[73] and calcicludin from the venom of the Eastern green mamba snake (Figure 2.3C)[74,75] all contain Kunitz domains. They share a similar elongated structural organization with an α-helix in their C-terminal region connected to one of the two (SHPI-1, calcicludin) or three (huwentoxin-11) central β-strands, as well as to the N-terminal extremity of the toxin through the C^{I}–C^{VI} and C^{III}–C^{V} disulfide bonds. The remaining disulfide bridge C^{II}–C^{IV} stabilizes the flexible portion of the toxin located at the opposite side of the linked N- and C-termini. In SHPI-1 and huwentoxin-11 the key pharmacophoric Arg and Lys residues are located in the first and second loop, respectively.[73,76] However, although these three peptide toxins have a similar 3D structure, they have different specificities of

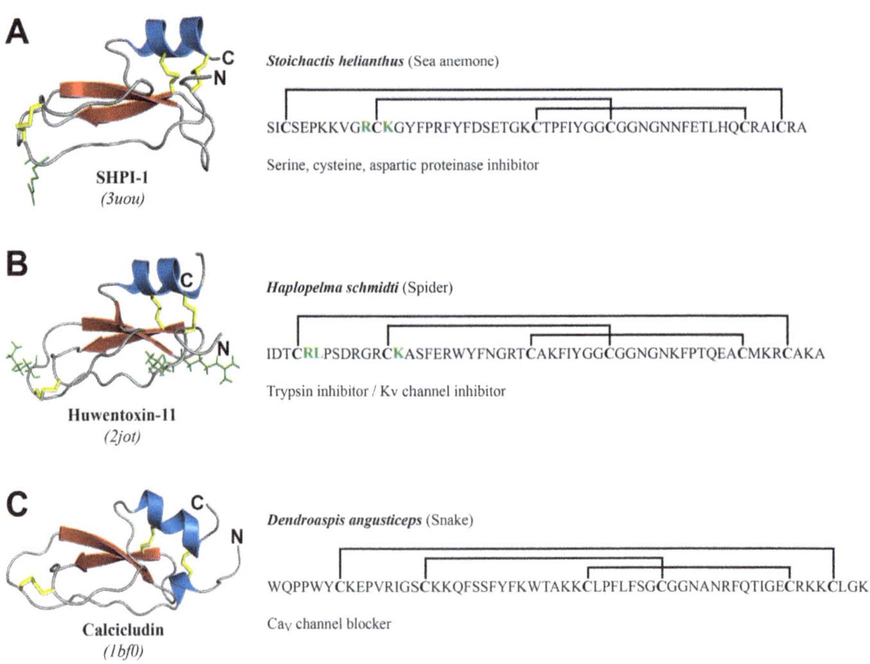

Figure 2.3 Venom peptides containing a BPTI/Kunitz inhibitor motif. Shown on the left are the 3D structures of (A) SHPI-1, (B) huwentoxin-11, and (C) calcicludin. The N- and C-termini are labelled and PDB accession numbers are indicated in parentheses. The primary structure of these peptides along with their disulfide framework, source organism, and molecular function are shown in the right panels of the figure.

action. While SHPI-1 appears to be primarily a protease inhibitor (acting on vertebrate trypsin and chymotrypsin),[76] huwentoxin-11 also strongly inhibits the Kv1.1 channel,[73,77] and calcicludin inhibits Ca_V1 (L-type) calcium channels with an IC_{50} in the nanomolar range.[74] This observation highlights the importance of hypervariablity in the protein primary structure and the difficulty of predicting the function of venom peptides based solely on structural homology.

2.2.2.3 Kazal-Like Domain

According to the MEROPS database,[78] members of this family of protease inhibitors contain one or multiple consensus Kazal domains in reference to the pancreatic secretory trypsin inhibitor SPINK1 originally discovered by Kazal *et al.* in 1948.[79] This domain contains a conserved cysteine framework (C–C–C–C–C–C pattern) with disulfide connectivity C^{I}–C^{V}, C^{II}–C^{IV}, C^{III}–C^{VI} compared to the C^{I}–C^{VI}, C^{II}–C^{IV}, C^{III}–C^{V} connectivity in the Kunitz domain protease inhibitors and C^{I}–C^{IV}, C^{II}–C^{V}, C^{III}–C^{VI} connectivity in ICK toxins. The consensus core of the Kazal domain usually consists of an α-helix and a small β-hairpin cross-linked by two disulfide bonds.[80] As Kazal domain-containing mini-proteins can be found in many divergent species, this characteristic structure is also present in venom peptides like rhodniin, an extremely potent inhibitor of thrombin (Figure 2.4). This 103-residue (11 kDa) polypeptide was isolated from the assassin bug *Rhodnius prolixus* and is presumed to be expressed in the saliva of this blood-feeding insect in order to prevent blood clotting and thus allow prolonged feeding.[81] Homologues are present in many other triatomine bugs. Rhodniin contains N- and C-terminal Kazal domains separated by an acidic linker. The 2.6 Å resolution crystal

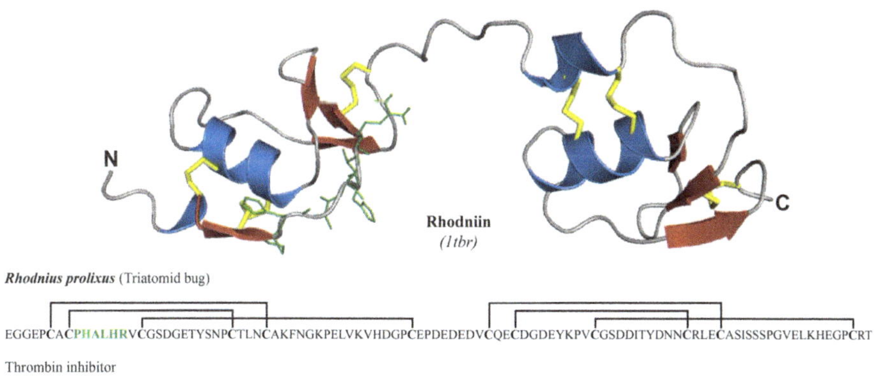

Figure 2.4 The Kazal domains of rhodniin from the assassin bug *Rhodnius prolixus*. 3D structure of rhodniin. The N- and C-termini are labelled and the PDB accession number is indicated in parentheses. The primary structure of the peptide along with the source organism and molecular target and shown below the 3D structure.

structure of a rhodniin/bovine α-thrombin complex revealed the dominant role of the disc-shaped N-terminal rhodniin domain in the interaction with thrombin.[82] The N-terminal domain residues Pro9, His10, Ala11, Leu12, His13 and Arg14 interact with the active-site cleft of thrombin, whereas the C-terminal domain interacts with the fibrinogen recognition exosite mainly through electrostatic interactions.

2.2.2.4 WAP Domain

The four-disulfide whey acidic protein (WAP)-type core domain (also called WAP-4-DSC or WFDC)[83] is particularly conserved in the subclass of small-secreted cysteine-rich mini-proteins like elafin or the secretory leukocyte protease inhibitor SLPI, for instance, that play important roles in host defence as antimicrobials and immunomodulators.[84,85] Two snake venom peptides belonging to the waprin family, nawaprin from the black-necked spitting cobra *Naja nigricollis* (Figure 2.5A) and omwaprin-a from the inland taipan *Oxyuranus microlepidotus* (Figure 2.5B), contain a single WAP domain and have similar functions.[86,87] These two peptides are of similar length (51 and 50 residues, respectively, which is characteristic of WAP domains), and they have the same eight-cysteine scaffold (C–C–C–C–CC–C–C pattern with disulfide connectivity C^{I}–C^{VI}, C^{II}–C^{VII}, C^{III}–C^{V}, C^{IV}–C^{VIII}), as well as identical

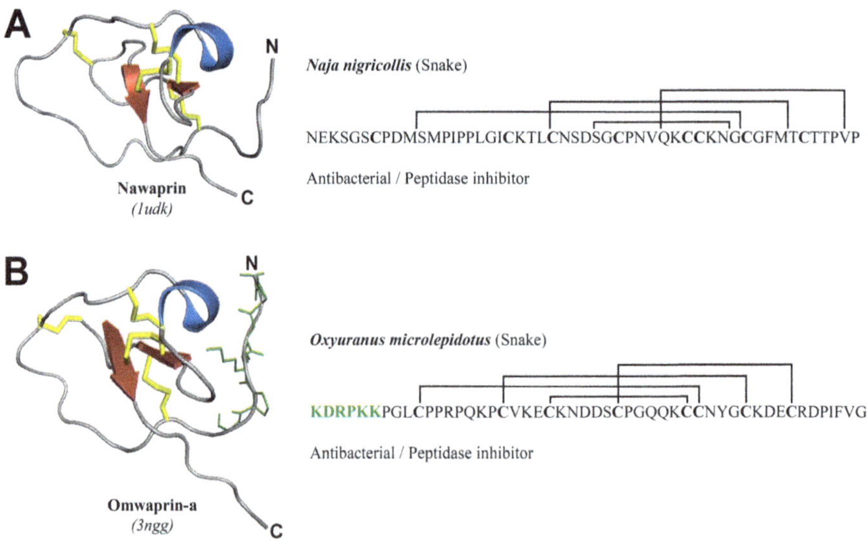

Figure 2.5 WAP domains of two members of the waprin family. Left panels show the 3D structures of (A) nawaprin determined using NMR spectroscopy, and (B) omwaprin-a determined using X-ray crystallography. The N- and C-termini are labelled and PDB accession numbers are indicated in parentheses. Right panels show the primary structure and disulfide framework of these peptides and list the source organism and molecular function.

secondary structures and tertiary folds. Nawaprin and omwaprin-a possess an overall flat shape with a spiral backbone organized in one outer and one inner segment stabilized by the four disulfide bridges.[86] The outer part contains only a small 3_{10} helix connected to the inner segment that comprises an antiparallel β-sheet. Interestingly, whereas only hypothetical specificities of action have been formulated for nawaprin, it has been demonstrated that in contrast to elafin and SLPI, omwaprin-a selectively targets its bactericidal activity against certain species of Gram-positive bacteria (*Bacillus megaterium* and *Staphylococcus warneri*) by causing membrane disruption in a dose-dependent manner; in contrast, it is non-toxic to mice.[87] Moreover, the same study revealed that the structure and disulfide scaffold of omwaprin-a are essential for sustaining the function of the toxin, as witnessed by the complete loss of activity upon reduction and alkylation of the cysteine residues, even with its six N-terminal pharmacophoric residues still present.

2.2.2.5 SXC Motif

The six-cysteine SXC or ShKT motif is found in diverse organisms such as mammals,[88,89] the nematode *Caenorhabditis elegans*,[90] and cnidarians (sea anemones and jellyfish). The Kv channel blocker ShK from the Caribbean sea anemone *Stichodactyla helianthus* (Figure 2.6A) and aurelin from the moon jellyfish *Aurelia aurita* (Figure 2.6B) illustrate this diversity. Both peptide toxins contain a unique SXC motif comprising six cysteines arranged in a C–C–C–C–C–C pattern with disulfide connectivity C^{I}–C^{VI}, C^{II}–C^{IV}, C^{III}–C^{V}.[91–93] The disulfide-rich SXC domain is an all-α motif with two short α-helices linked to portions of the peptide backbone *via* two disulfide bonds.[93–95] The third disulfide bond (C^{I}–C^{VI}) links the N- and C-terminal regions of the toxin. The 40-residue aurelin peptide has moderate antibacterial activity against Gram-positive *Listeria monocytogenes* and Gram-negative *Escherichia coli*.[93] In contrast, ShK (35 residues, 4.05 kDa) potently inhibits human Kv1.3 (IC_{50} ≈ 10–100 pM). This channel is a therapeutic target for autoimmune disease because of its upregulation in effector memory T cells. ShK-186, a derivative of ShK with improved stability and selectivity for Kv1.3, recently passed Phase Ia clinical trials for treatment of multiple sclerosis (see Chapter 10 for further details).

2.2.2.6 PLA_2

PLA_2 enzymes are found in almost all living entities, from bacteria to mammals. They are essential hydrolytic enzymes involved in various physiological processes, such as lipid metabolism,[96] signal transduction,[97] and host defence.[98] In the world of venomous animals, secreted Ca^{2+}-dependent PLA_2s are particularly abundant in snake, bee and wasp venoms. PLA_2s are at the edge between polypeptides and proteins, being larger than most venom peptides, and functionally viable as monomers or homo- or heterodimers.

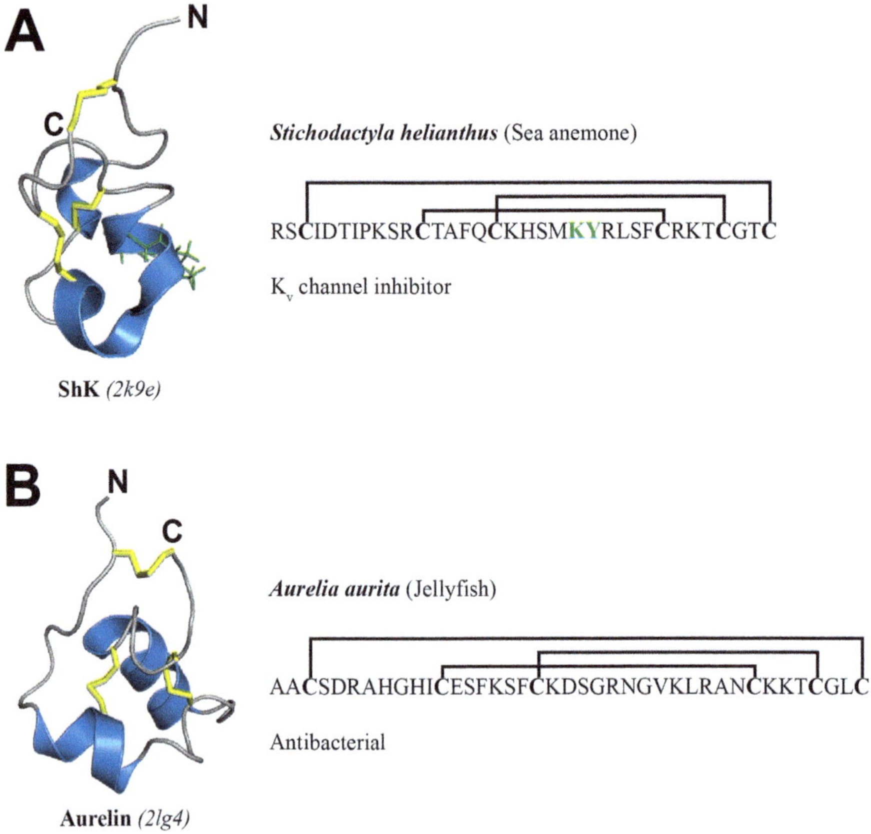

Figure 2.6 The SXC folds of ShK and aurelin. Left panels show the 3D structures of (A) ShK and (B) aurelin. The N- and C-termini are labelled and PDB accession numbers are indicated in parentheses. Right panels show the primary structure and disulfide framework of these peptides and list the source organism and molecular function.

We have chosen to illustrate a subtype of secretory PLA_2s using three different snake venom mini-proteins: the acidic PLA_2 5 isolated from the venom of the cobra *Naja sagittifera* (Figure 2.7A), the basic PLA_2 notexin from the common tiger snake (*Notechis scutatus scutatus*) (Figure 2.7B), and ammodytoxin A from the Western sand viper (*Vipera ammodytes ammodytes*) (Figure 2.7C). These toxins are 118–122 residues in length, and they have different cysteine frameworks. PLA_2 5 from *N. sagittifera* contains 16 cysteine residues disposed in a C–C–C–C–CC–C–C–C–C–C–C–C–C–C pattern with disulfide connectivity C^{I}–C^{X}, C^{II}–C^{XVI}, C^{III}–C^{VI}, C^{IV}–C^{VII}, C^{V}–C^{XV}, C^{VIII}–C^{XIV}, C^{IX}–C^{XII}, C^{XI}–C^{XIII}.[99] Notexin has 14 cysteine residues (C–C–C–CC–C–C–C–C–C–C–C–C–C) that form seven disulfide bridges (C^{I}–C^{VIII}, C^{II}–C^{XIV}, C^{III}–C^{V}, C^{IV}–C^{XIII}, C^{VI}–C^{XII}, C^{VII}–C^{X}, C^{IX}–C^{XI}).[100] Ammodytoxin A also has 14 cysteine residues, although their disposition (C–C–CC–CC–C–C–C–C–C–C–C–C) and

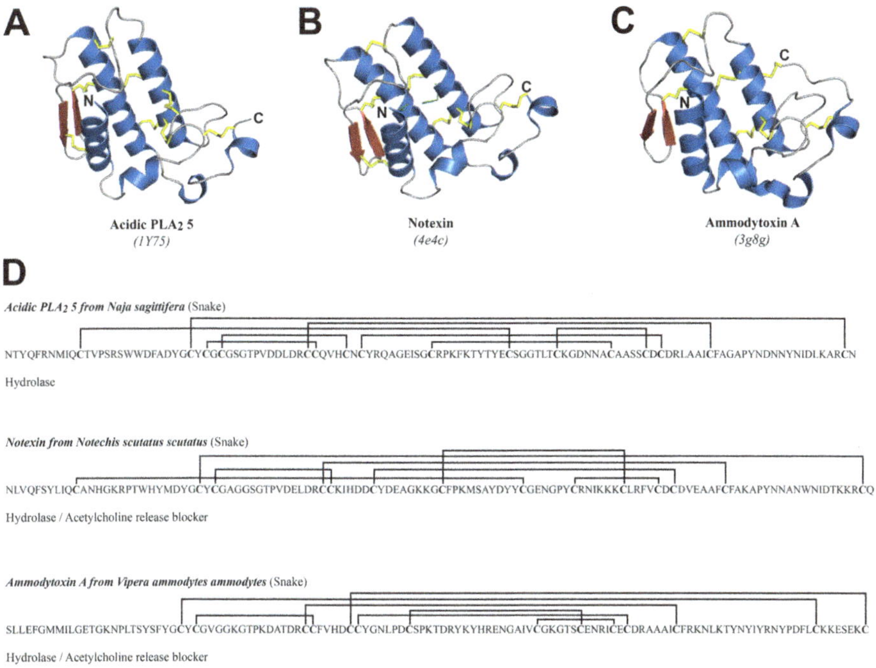

Figure 2.7 Structure of PLA_2 enzymes from snake venoms. The 3D structures of (A) acidic PLA_2 5, (B) notexin, and (C) ammodytoxin A are shown at the top of the figure. The N- and C-termini are labelled and PDB accession numbers are indicated in parentheses. (D) Primary structure and disulfide framework of these PLA_2 enzymes. The source organism and molecular function are indicated.

disulfide connectivity (C^{I}–C^{XIII}, C^{II}–C^{IV}, C^{III}–C^{XII}, C^{V}–C^{XIV}, C^{VI}–C^{XI}, C^{VII}–C^{IX}, C^{VIII}–C^{X}) are not the same as notexin.[101] Despite their different primary structure (~28% identity) and cysteine frameworks, these three members of the PLA_2 family share identical secondary elements and very similar overall 3D architectures (see Figure 2.7A). They all display a central core comprised of three long α-helices flanked by a small β-hairpin on the side, as well as two short helical turns in the C-terminal region, with the numerous disulfide bridges stabilizing the compact structure.[99,101,102] Apart from exerting classical PLA_2 hydrolytic activities on the 2-acyl groups in 3-sn-phosphoglycerides, notexin and ammodytoxin A are also able to increase the release of acetylcholine (ACh) at neuromusclar junctions via a mechanism that remains to be precisely elucidated.[102–107]

2.2.3 Taxon-Specific Disulfide Frameworks

In the second part of this chapter, we focus on unique disulfide frameworks that are only found in specific animal taxa, and describe the 3D architecture of these venom peptides.

2.2.3.1 *Neurotoxin III Fold in Sea Anemones (Order Actinaria)*

The first taxon-specific disulfide framework we will discuss is found in neurotoxin III (ATX III), a 27-residue peptide isolated from the Mediterranean snakelocks sea anemone (*Anemonia sulcata*). ATX III contains six cysteine residues that form a CC–C–C–C–C pattern and are organized into three disulfide bonds (C^{I}–C^{V}, C^{II}–C^{IV}, C^{III}–C^{VI}; Figure 2.8).[108–110] ATX III is devoid of any regular secondary structure; rather, the 3D structure determined using nuclear magnetic resonance (NMR) spectroscopy[111] reveals a compact architecture containing two distorted type I β-turns (Cys6–Gly9 and Trp8–Cys11), two inverse γ-turns (Pro12–Gly14 and Gln15–Cys17), and two other chain reversals that enable ATX III to adopt an overall globular and twisted fold. ATX III slows inactivation of Na_V channels in crayfish giant axons.[112] However, there have been no investigations into structure–activity relationships of the toxin.

2.2.3.2 *Snakes (Order Squamates)*

2.2.3.2.1 Snake Three-Finger Toxins

The three-finger toxins (3FTxs) are a snake-specific disulfide-rich peptide family characterized by a conserved three-dimensional fold and disulfide framework.[113] All 3FTxs contain four conserved disulfide bridges in their core region, but extra disulfide bonds are sometimes present.[114] The typical fold of 3FTxs is comprised of three β-stranded loops, symbolized as fingers, which stretch and diverge from a single globular and hydrophobic cysteine-rich core region comprising the four conserved disulfide bonds.[115,116] Despite their common three-dimensional topology, 3FTxs have diverse pharmacology with toxins from this family capable of targeting nicotinic acetylcholine receptors (nAChRs), muscarinic acetylcholine receptors (mAChRs), acetylcholinesterase, Ca_V channels, adrenergic receptors, and ASICs.[113,117,118] 3FTxs

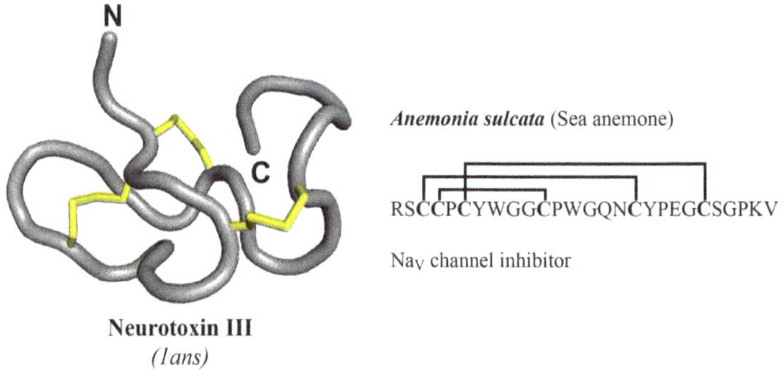

Figure 2.8 Structure of the sea anemone peptide neurotoxin III. Left panel shows the 3D structure of neurotoxin III determined using NMR spectroscopy. The N- and C-termini are labelled and the PDB accession number is indicated in parentheses. Right panel shows the primary structure and disulfide framework of neurotoxin III and lists the source organism and molecular function. Note the lack of regular secondary structure.

have been used extensively as molecular probes for studying the pharmacology and tissue/cellular distribution of nAChRs.[119–121]

α-Elapitoxin-Nk2a (Figure 2.9A) is a typical 3FTx isolated from the venom of the monocled cobra (*Naja kaouthia*). This 71-residue peptide contains 10 cysteine residues (C–C–C–C–C–C–C–CC–C pattern) that form five disulfide bonds with connectivity (C^{I}–C^{III}, C^{II}–C^{VI}, C^{IV}–C^{V}, C^{VII}–C^{VIII}, C^{IX}–C^{X}).[122–124]

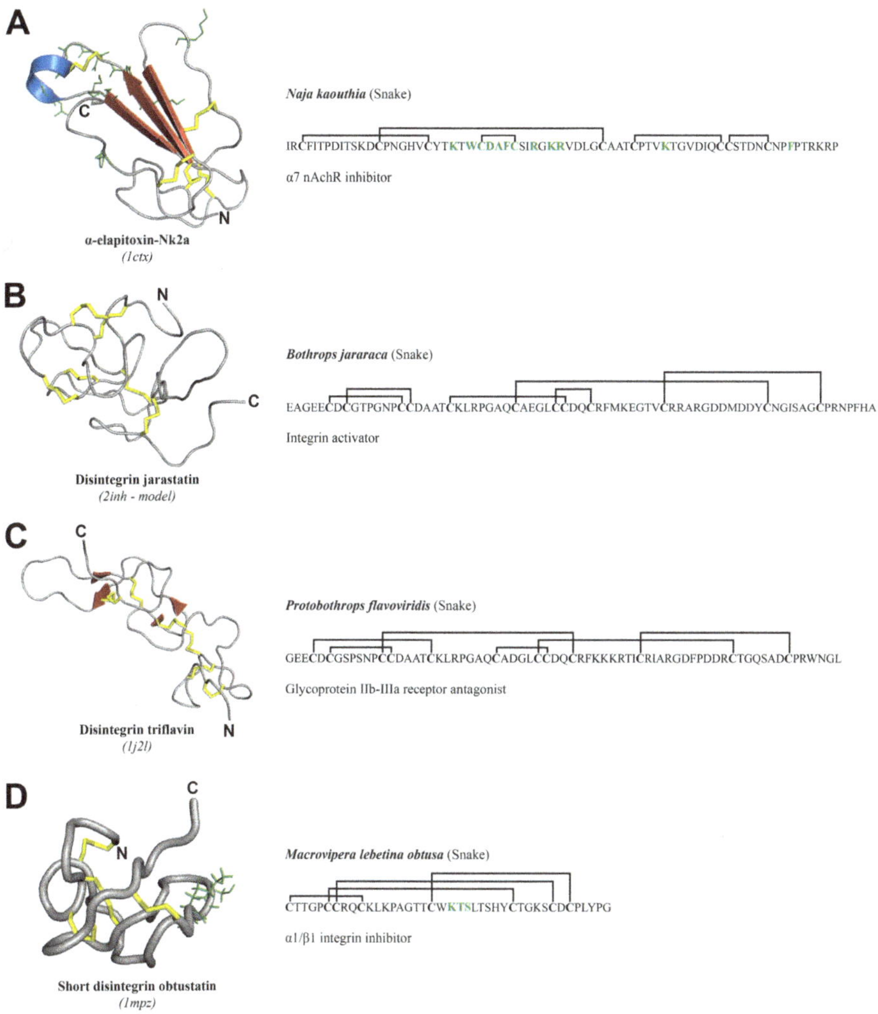

Figure 2.9 3FTs and disintegrins from snake venom. Left panels show the 3D structures of (A) α elapitoxin-Nk2a (X-ray; PDB 1ctx), (B) jarastatin (model; PDB 2inh), (C) triflavin (X-ray; PDB 1j2l), and (D) obtustatin (NMR; PDB 1mpz). The N- and C-termini are labelled and PDB accession numbers are indicated in parentheses. Right panels show the primary structure and disulfide framework of these peptides and indicate the source organism and molecular function.

α-Elapitoxin-Nk2a contains the three loops typically found in 3FTxs plus two distorted right-handed helical turns at the tip of loop II, stabilized by the fifth disulfide bond C^{IV}–C^{V}.[123] The monomeric form of the toxin binds selectively and with high affinity to the α7 nAChR, thereby triggering paralysis in envenomated prey.[125–127] Loop II of α-elapitoxin-Nk2a, which is stabilized by the extra disulfide bond and contains the helical turns, is crucial for high affinity interaction of the toxin with the α7 nAChR.[128–130]

2.2.3.2.2 Snake Disintegrins

The 3FTs display diverse pharmacology on a conserved structural scaffold. In contrast, members of the viper-specific disintegrin protein family share a similar function, namely inhibition of platelet aggregation and integrin-dependent cell adhesion, but they encompass a wider range of folds than the 3FTs.[131,132] To illustrate this structural diversity we discuss three disintegrins that exhibit different disulfide scaffolds and 3D folds: jarastatin from the South American pit viper *Bothrops jararaca* (Figure 2.9B), triflavin from the venom of *Protobothrops flavoviridis* (Figure 2.9C), a viper endemic to the Ryukyu islands of Japan, and obstutatin from the Asian blunt-nosed viper (*Macrovipera lebetina obtusa*) (Figure 2.9D). Jarastatin and triflavin are medium-sized disintegrins, with 73 and 70 residues, respectively, and they contain six disulfide bonds.[133–135] Obstustatin is smaller (41 residues) and it contains only four disulfide bridges.[136,137] Jarastatin and triflavin have identical cysteine patterns (C–C–CC–C–C–CC–C–C–C–C) but different cysteine connectivities (C^{I}–C^{III}, C^{II}–C^{IV}, C^{V}–C^{VIII}, C^{VI}–C^{XI}, C^{VII}–C^{IX}, C^{X}–C^{XII} and C^{I}–C^{V}, C^{II}–C^{IV}, C^{III}–C^{IX}, C^{VI}–C^{VIII}, C^{VII}–C^{XI}, C^{X}–C^{XII}, respectively). Obtustatin has only eight cysteine residues (C–CC–C–C–C–C–C pattern) with connectivity C^{I}–C^{IV}, C^{II}–C^{VI}, C^{III}–C^{VII}, C^{V}–C^{VIII}. The 3D structures of these toxins (note that the jarastatin structure is a theoretical model produced *in silico*) reveal different overall shapes and a lack of major secondary structure elements. Triflavin adopts an elongated and rigid structure characterized by a series of turns and two short regions composed of antiparallel β-strands.[138] This anti-aggregant agent inhibits the interaction between fibrinogen and platelets by binding to the glycoprotein IIb-IIIa receptor on the platelet surface.[133] Obtustatin has a compact globular shape comprised almost exclusively of turns without any regular secondary structure.[137] Obtustatin potently and selectively inhibits the α1/β1 integrin *via* a triad of loop residues (Lys21–Thr22–Ser23) that differs from the usual disintegrin Arg-Gly-Asp pharmacophore motif.[139]

2.2.3.3 Marine Cone Snails (Genus Conus*)*

Several toxins found in the venom of marine cone snails from the genus *Conus* display unique cysteine scaffolds. We exemplify such peptides with descriptions of ι-conotoxin RXIA, α-conotoxin PIVA, and ε-conotoxin TxVA.

2.2.3.3.1 ι-Conotoxins

ι-Conotoxin RXIA was first discovered in the venom of the piscivorous (fish-hunting) rayed cone snail *Conus radiatus*[140] (Figure 2.10A). ι-Conotoxin RXIA contains 46 residues with eight cysteines arranged in an atypical framework (C–C–CC–CC–C–C pattern with disulfide connectivity C^{I}–C^{IV}, C^{II}–C^{VI},

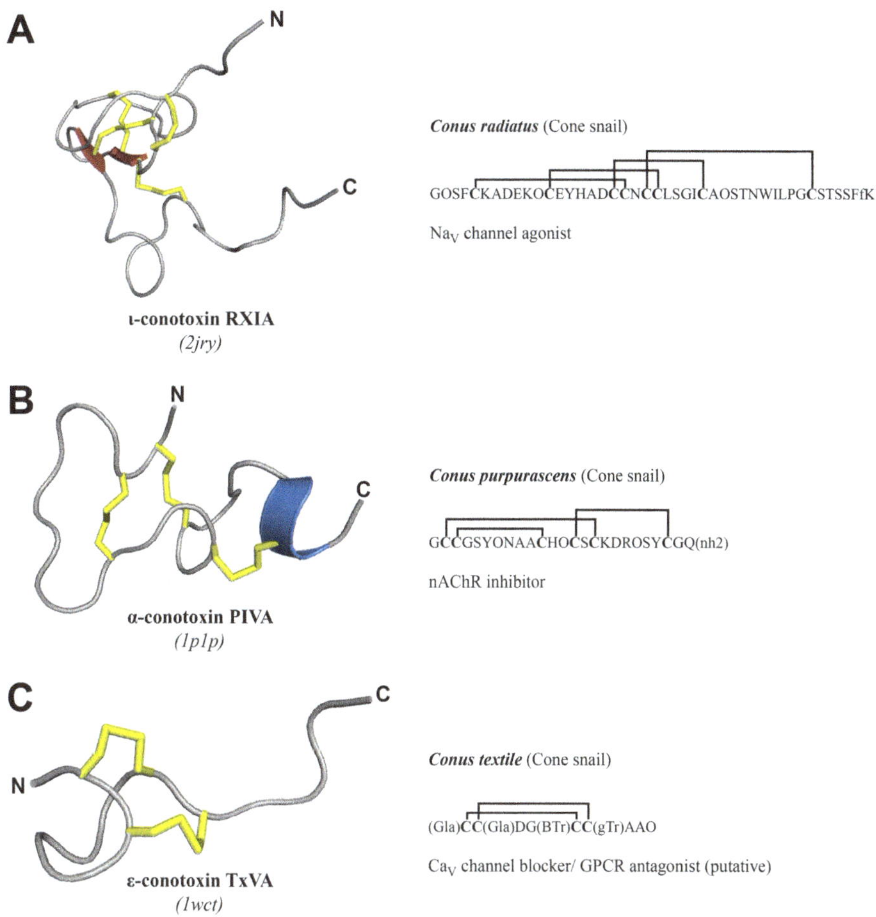

Figure 2.10 Taxon-specific disulfide frameworks from cone snail venom. Left panels show the 3D structures of (A) ι-conotoxin RXIA, (B) α-conotoxin PIVA, and (C) ε-conotoxin TxVA. The N- and C-termini are labelled and PDB accession numbers are indicated in parentheses. Right panels show the primary structure and disulfide framework of these peptides and indicate the source organism and molecular function. PTMs are indicated as follows O = hydroxyproline; f = D-Phe; (Gla) = γ-carboxyglutamate; (BTr) = bromotryptophan; (gTr) = glycosylated threonine; (nh2) = amidated C-terminus.

C^{III}–C^{VII}, C^{V}–C^{VIII}).[141] ι-conotoxin RXIA has numerous post-translational modifications (PTMs) in addition to the four disulfide bonds, including hydroxyproline at positions 2, 11 and 29 and an unusual D-Phe as the third last residue.[141,142] The 3D structure of ι-conotoxin RXIA comprises a series of turns and two antiparallel β-strands in a compact globular region stabilized by the four disulfide bonds, as well as a long flexible C-terminal tail.[141] Electrophysiology experiments revealed that ι-conotoxin RXIA is an excitatory peptide that acts as an agonist of frog and mouse $Na_V1.2$, $Na_V1.6$ and $Na_V1.7$ channels ($Na_V1.6 > Na_V1.2 > Na_V1.7$) by shifting the voltage-dependence of activation to more hyperpolarized potentials.[141,143] Mutation of the D-Phe residue to L-Phe caused a two-fold reduction in affinity towards $Na_V1.6$ and a two-fold faster off-rate, as well as a complete inactivity on $Na_V1.2$, indicating that this PTM contributes significantly to toxin reversibility and selectivity.[141,143]

2.2.3.3.2 α-Conotoxins

α-Conotoxin PIVA from the piscivorous purple cone *Conus purpurascens* is a typical α-conotoxin that displays a taxon-specific cysteine framework (Figure 2.10B). This short, 25-residue conopeptide contains six cysteine residues (CC–C–C–C–C) that form three disulfide bonds (C^{I}–C^{V}, C^{II}–C^{III}, C^{IV}–C^{VI}).[144] Like ι-conotoxin RXIA, it is also rich in PTMs, with hydroxyproline residues at positions 7, 13, and 20 as well as an amidated C-terminus. Han *et al.* described the overall shape of α-conotoxin PIVA as an "iron" in which the external and highly charged hydrophilic Ser15–Arg19 segment is the "handle", while the rest of the toxin forms the "bottom plate".[145] α-Conotoxin PIVA does not contain significant secondary structure, with the exception of a single turn of 3_{10} helix formed by residues Ser21–Gly24. Like most α-conotoxins, α-conotoxin PIVA inhibits postsynaptic nAChRs, particularly the α1/β1/γ/δ subtype.[146]

2.2.3.3.3 ε-Conotoxins

ε-Conotoxin TxVA from the venom of the molluscivorus cone snail *Conus textile* (Figure 2.10C) is a very short 13-residue peptide with a unique cysteine pattern composed of two sets of tandem cysteines (CC–CC) with disulfide connectivity C^{I}–C^{III}, C^{II}–C^{IV}. TxVA is replete with PTMs, including γ-carboxyglutamate at positions 1 and 4, bromotryptophan at position 7, a glycosylated threonine at position 10, and hydroxyproline at the C-terminus.[147–149] It is the prototypic member of the T-superfamily of conotoxins that is defined by the presence of two sets of tandem cysteines separated by four to seven residues.[150] The side chains of the two γ-carboxyglutamate residues create an electronegative patch extending outward from a deep cleft.[148] On the C-terminal face of the peptide, the glycosylated Thr10 and Hyp13 confine a cluster of hydrophobic residues around the brominated Trp7 in the constrained inter-cysteine region.[148] It has been suggested that TxVA might target presynaptic Ca_V channels or GPCRs based on the fact that it causes a reduction of calcium influx and neurotransmitter release in *Aplysia* cholinergic synapses.[148] However, no unequivocal target has been defined for any member of the T-superfamily with the exception of τ-CnVA, from the piscivorous cone snail *Conus consors*, which was recently shown to be a micromolar antagonist of the somatostatin sst3 receptor.[150]

2.2.3.4 CSα/β Toxins in Scorpions (Order Scorpiones)

Scorpions and centipedes are the oldest terrestrial venomous taxa, with the oldest scorpion fossils dating back 430 million years to the Silurian period.[2] Although they inhabit a diverse range of habitats, scorpions (~1750 extant species) are far less speciose than their closest venomous relatives, the spiders (~45 000 extant species).[2] Scorpion venoms are dominated by so-called cysteine stabilized α/β (CSα/β) toxins that are evolutionarily derived from CSα/β defensins, innate-immunity-related antimicrobial peptides that are found in plants, fungi, nematodes, and arthropods.[151,152] It was recently demonstrated that only a modest number of mutations are required to convert a CSα/β defensin into a neurotoxin.[153] Most CSα/β scorpion toxins target Kv or Na_V channels.[152] The core cysteine framework of CSα/β scorpion toxins is the C–C–C–C–C–C pattern found in CSα/β defensins, but many toxins are elaborated with additional disulfide bonds, as outlined in the examples below.

2.2.3.4.1 Butantoxin (α-KTx12.1)

The potassium channel toxin α-KTx12.1, also called butantoxin or TsTX-IV, was first isolated from the venom of the Brazilian yellow scorpion (*Tityus serrulatus*) (Figure 2.11A).[154] This 40-residue peptide contains a cysteine framework composed of eight cysteine residues (C–C–C–C–C–C–C–C) with disulfide connectivity (C^{I}–C^{II}, C^{III}–C^{VI}, C^{IV}–C^{VII}, C^{V}–C^{VIII}).[155,156] The C^{III}–C^{VI}, C^{IV}–C^{VII}, and C^{V}–C^{VIII} disulfide bonds constitute the core disulfide framework found in CSα/β defensins, with the N-terminal C^{I}–C^{II} disulfide being an elaboration of the core fold. Butantoxin is a typical CSα/β toxin in which an α-helix is nestled on the face of a β-hairpin, with this core fold stabilized by the C^{III}–C^{VI}, C^{IV}–C^{VII} and C^{V}–C^{VIII} disulfide bonds.[155,157,158] The remaining N-terminal C^{I}–C^{II} disulfide bridge is unique to this toxin and does not appear to confer additional stability.[155] Butantoxin blocks high-conductance K_{Ca} channels and inhibits Shaker Kv channels with low affinity.[154] Preliminary SAR studies revealed that His28 is important for interaction with its molecular targets.[157]

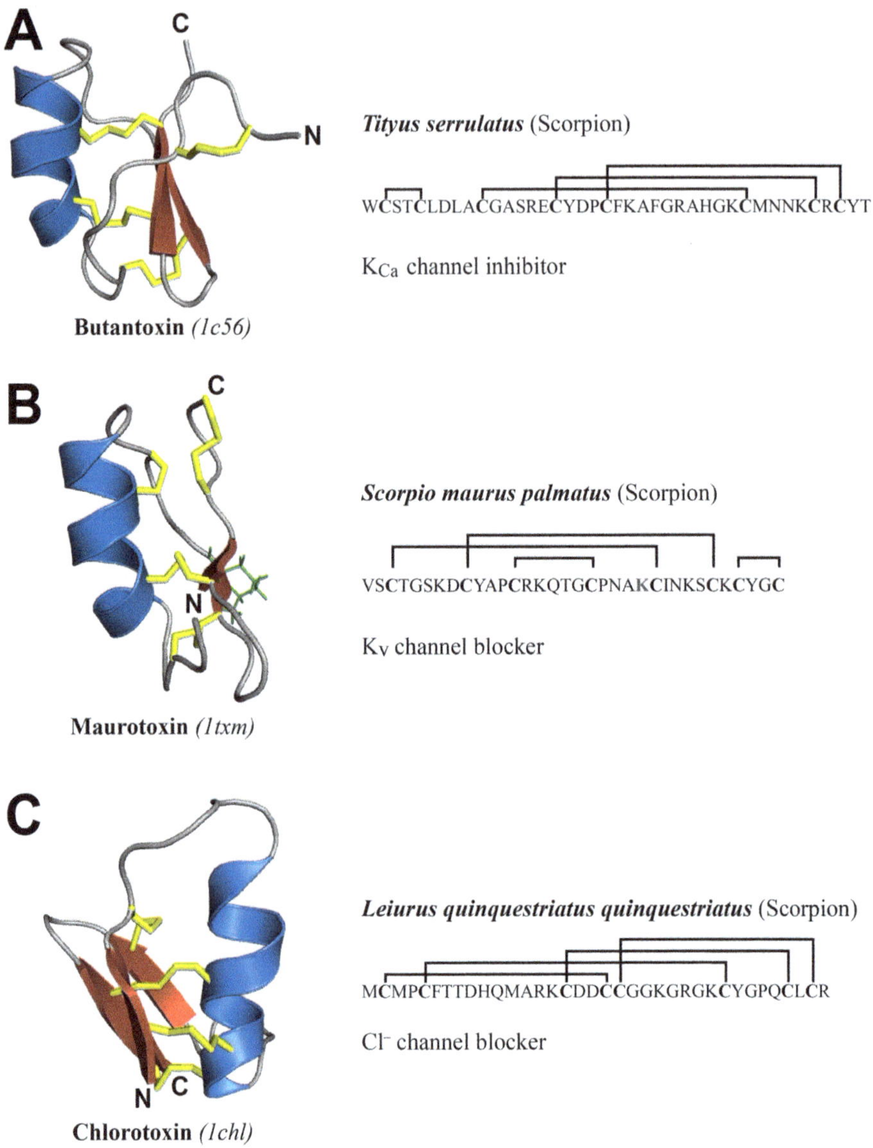

Figure 2.11 The scorpion-specific CSα/β framework. Left panels show the 3D structures of (A) butantoxin, (B) maurotoxin, and (C) chlorotoxin. The N- and C-termini are labelled and PDB accession numbers are indicated in parentheses. Right panels show the primary structure and disulfide framework of these peptides and indicate the source organism and molecular function. Note that all three peptides exhibit a core CSα/β defensin fold consisting of an α-helix on one face of a two- or three-stranded antiparallel β-sheet.

2.2.3.4.2 Maurotoxin (α-KTx6.2)

α-KTx6.2, also known as maurotoxin (Figure 2.11B), from the venom of the chactoid scorpion *Scorpio maurus palmatus* is a 34-residue peptide comprising an eight-cysteine scaffold (C–C–C–C–C–C–C–C) with disulfide connectivity C^{I}–C^{V}, C^{II}–C^{VI}, C^{III}–C^{IV}, C^{VII}–C^{VIII}.[159–163] The C^{I}–C^{V}, C^{II}–C^{VI}, and C^{III}–C^{IV} disulfide bonds constitute the core disulfide framework found in CSα/β defensins, with the C-terminal C^{VII}–C^{VIII} disulfide being an elaboration of the core fold. The core CSα/β fold is comprised of a single α-helix (Ser6–Gln16) abutting a two-stranded antiparallel β-sheet (β-strands Lys23–Asn26 and Ser28–Cys31).[163] Maurotoxin can block the Shaker B channel as well as Kv1.1 (IC_{50} = 45 nM), Kv1.2 (IC_{50} = 0.8 nM), and Kv1.3 (IC_{50} = 180 nM) channels with high affinity.[160,161,164]

2.2.3.4.3 Chlorotoxin

Chlorotoxin is a 36-residue CSα/β toxin isolated from the venom of the Egyptian scorpion *Leiurus quinquestriatus quinquestriatus*. It contains eight cysteine residues distributed in a C–C–C–CC–C–C–C pattern with disulfide connectivity C^{I}–C^{IV}, C^{II}–C^{VI}, C^{III}–C^{VII}, C^{V}–C^{VIII}.[165,166] In this case, the C^{II}–C^{VI}, C^{III}–C^{VII}, and C^{V}–C^{VIII} disulfide bonds constitute the core disulfide framework found in CSα/β defensins, with the N-terminal C^{I}–C^{IV} disulfide being an elaboration of the core fold. Relative to the classic CSα/β defensin fold, chlorotoxin contains an additional N-terminal β-strand, which leads to the formation of a three-stranded antiparallel β-sheet; the extra disulfide bond links this additional β-strand to the α-helix (Figure 2.11C).[166] Chlorotoxin induces paralysis in crayfish and cockroaches and inhibits small conductance chloride channels isolated from rat epithelia and brain.[165,167] Chlorotoxin selectively binds to gliomas (a type of malignant brain cancer) and inhibits glioma cell invasion by virtue of interactions with annexin A2 and matrix metalloproteinase-2 on the cell surface.[168,169] Moreover, it was recently reported that chlorotoxin is capable of permeating the human blood–brain barrier.[170] A chlorotoxin conjugate is currently in clinical trials as an imaging agent for intraoperative visualization of cancer foci.[171–173]

2.3 Discussion

In this chapter we have provided an overview of the broad diversity of structures encountered in secreted venom peptides. In the majority of these small proteins, the 3D architecture is maintained by precise and often elaborate disulfide-rich scaffolds. As well as providing stability and in many cases resistance to proteases, these rigid disulfide frameworks allow pharmacophoric amino acid side chains to be displayed in precise spatial orientations, thereby optimizing the interaction between these venom peptides and their molecular targets. Moreover, the disulfide bonds often direct the 3D architecture of these peptides to such an extent that the inter-cystine loops are highly permissive to mutations, thereby facilitating the evolution of new toxic functions on the same 3D scaffold. The consequence of this is that a single scaffold, such as the ICK motif, can support a wide range of pharmacological activities that aid the process of envenomation.

In summary, the disulfide-rich scaffolds found in venom peptides provide stability and resistance to proteases, and they facilitate the evolution of high affinity binding and diverse pharmacology, traits that make these peptides ideal candidates for drug discovery programmes. We found that 86 disulfide frameworks account for the 2022 secreted cysteine-rich peptides and polypeptides reported to date. Although some of these disulfide scaffolds are found in taxonomically diverse venomous animals (*e.g.* the ICK, Kunitz, Kazal, WAP, SXC and PLA_2 folds), other scaffolds such as snake 3FTxs, scorpion CSα/β toxins, and α-conotoxins are found only in specific taxa. The diversity of cysteine-rich venom peptides is certain to exceed what is currently known, as only a very small percentage of venomous taxa (<1%) has been examined to date. Advances in transcriptomic and proteomic analyses of venoms, which are discussed in Chapter 3, are rapidly transforming the field and are likely to lead to the discovery of many new disulfide-rich scaffolds in hitherto unstudied venomous animals.

Acknowledgements

V.L. acknowledges the provision of an Institute for Molecular Bioscience (IMB) Postgraduate Award and support from Program Grant 569927 from the Australian National Health and Medical Research Council (NHMRC). G.F.K. acknowledges support from the NHMRC (Principal Research Fellowship; Project Grants APP1012338 and APP1034958) and the Australian Research Council (ARC) (Discovery Grants DP130103813 and DP110103129). M.M. acknowledges support from the ARC (Future Fellowship FT110100925 and Discovery Grant DP140101098).

References

1. R. J. Lewis and M. L. Garcia, *Nat. Rev. Drug Discov.*, 2003, **2**, 790–802.
2. G. F. King, *Expert Opin. Biol. Ther.*, 2011, **11**, 1469–1484.
3. V. Lavergne, R. J. Taft and P. F. Alewood, *Curr. Top. Med. Chem.*, 2012, **12**, 1514–1533.
4. Q. Kaas, R. Yu, A. H. Jin, S. Dutertre and D. J. Craik, *Nucl. Acids Res.*, 2012, **40**, D325–D330.

5. V. Herzig, D. L. Wood, F. Newell, P. A. Chaumeil, Q. Kaas, G. J. Binford, G. M. Nicholson, D. Gorse and G. F. King, *Nucl. Acids Res.,* 2011, **39**, D653–D657.
6. M. Kita, D. S. Black, O. Ohno, K. Yamada, H. Kigoshi and D. Uemura, *J. Am. Chem. Soc.,* 2009, **131**, 18038–18039.
7. C. M. Whittington, A. T. Papenfuss, D. P. Locke, E. R. Mardis, R. K. Wilson, S. Abubucker, M. Mitreva, E. S. Wong, A. L. Hsu, P. W. Kuchel, K. Belov and W. C. Warren, *Genome Biol.,* 2010, **11**, R95.
8. A. G. Craig, E. C. Jimenez, J. Dykert, D. B. Nielsen, J. Gulyas, F. C. Abogadie, J. Porter, J. E. Rivier, L. J. Cruz, B. M. Olivera and J. M. McIntosh, *J. Biol. Chem.,* 1997, **272**, 4689–4698.
9. T. S. Kang, D. Georgieva, N. Genov, M. T. Murakami, M. Sinha, R. P. Kumar, P. Kaur, S. Kumar, S. Dey, S. Sharma, A. Vrielink, C. Betzel, S. Takeda, R. K. Arni, T. P. Singh and R. M. Kini, *FEBS J.,* 2011, **278**, 4544–4576.
10. D. Fass, *Annu. Rev. Biophys.,* 2012, **41**, 63–79.
11. K. Nareoja and J. Nasman, *Acta Physiol.,* 2012, **204**, 186–201.
12. D. J. Adams, B. Callaghan and G. Berecki, *Br. J. Pharmacol.,* 2012, **166**, 486–500.
13. G. M. Nicholson, M. J. Little and L. C. Birinyi-Strachan, *Toxicon,* 2004, **43**, 587–599.
14. S. Mouhat, N. Andreotti, B. Jouirou and J. M. Sabatier, *Curr. Pharm. Des.,* 2008, **14**, 2503–2518.
15. R. S. Norton and S. I. McDonough, *Curr. Pharm. Des.,* 2008, **14**, 2480–2491.
16. G. F. King and M. C. Hardy, *Annu. Rev. Entomol.,* 2013, **58**, 475–496.
17. R. J. Lewis, *IUBMB Life,* 2004, **56**, 89–93.
18. R. de O. Beleboni, A. B. Pizzo, A. C. Fontana, O. G. C. R. de, J. Coutinho-Netto and W. F. Dos Santos, *Eur. J. Pharmacol.,* 2004, **493**, 1–17.
19. R. J. Lewis, *Toxicon,* 2012, **59**, 524–528.
20. M. A. Hayashi and A. C. Camargo, *Toxicon,* 2005, **45**, 1163–1170.
21. N. Marsh and V. Williams, *Toxicon,* 2005, **45**, 1171–1181.
22. A. C. Camargo, D. Ianzer, J. R. Guerreiro and S. M. Serrano, *Toxicon,* 2012, **59**, 516–523.
23. A. Gomes, P. Bhattacharjee, R. Mishra, A. K. Biswas, S. C. Dasgupta and B. Giri, *Indian J. Exp. Biol.,* 2010, **48**, 93–103.
24. I. Vetter, J. L. Davis, L. D. Rash, R. Anangi, M. Mobli, P. F. Alewood, R. J. Lewis and G. F. King, *Amino Acids,* 2011, **40**, 15–28.
25. M. Magrane and UniProt Consortium, *Database (Oxford),* 2011, **2011**, bar009.
26. P. K. Pallaghy, K. J. Nielsen, D. J. Craik and R. S. Norton, *Protein Sci.,* 1994, **3**, 1833–1839.
27. J. Gracy, D. Le-Nguyen, J. C. Gelly, Q. Kaas, A. Heitz and L. Chiche, *Nucl. Acids Res.,* 2008, **36**, D314–D319.

28. D. Chung, S. Gaur, J. R. Bell, J. Ramachandran and L. Nadasdi, *Int. J. Pept. Protein Res.,* 1995, **46**, 320–325.
29. N. J. Saez, S. Senff, J. E. Jensen, S. Y. Er, V. Herzig, L. D. Rash and G. F. King, *Toxins,* 2010, **2**, 2851–2871.
30. B. Gao, P. J. Harvey, D. J. Craik, M. Ronjat, M. De Waard and S. Zhu, *Biosci. Rep.,* 2013, **33**, e00047.
31. N. L. Daly and D. J. Craik, *Curr. Opin. Chem. Biol.,* 2011, **15**, 362–368.
32. G. P. Miljanich, *Curr. Med. Chem.,* 2004, **11**, 3029–3040.
33. V. J. Basus, L. Nadasdi, J. Ramachandran and G. P. Miljanich, *FEBS Lett.,* 1995, **370**, 163–169.
34. K. J. Nielsen, L. Thomas, R. J. Lewis, P. F. Alewood and D. J. Craik, *J. Mol. Biol.,* 1996, **263**, 297–310.
35. R. J. Lewis, S. Dutertre, I. Vetter and M. J. Christie, *Pharmacol. Rev.,* 2012, **64**, 259–298.
36. K. J. Nielsen, D. Adams, L. Thomas, T. Bond, P. F. Alewood, D. J. Craik and R. J. Lewis, *J. Mol. Biol.,* 1999, **289**, 1405–1421.
37. K. Peng, Q. Shu, Z. Liu and S. Liang, *J. Biol. Chem.,* 2002, **277**, 47564–47571.
38. J. J. Smith, T. R. Cummins, S. Alphy and K. M. Blumenthal, *J. Biol. Chem.,* 2007, **282**, 12687–12697.
39. Y. Xiao, J. P. Bingham, W. Zhu, E. Moczydlowski, S. Liang and T. R. Cummins, *J. Biol. Chem.,* 2008, **283**, 27300–27313.
40. J. J. Cox, F. Reimann, A. K. Nicholas, G. Thornton, E. Roberts, K. Springell, G. Karbani, H. Jafri, J. Mannan, Y. Raashid, L. Al-Gazali, H. Hamamy, E. M. Valente, S. Gorman, R. Williams, D. P. McHale, J. N. Wood, F. M. Gribble and C. G. Woods, *Nature,* 2006, **444**, 894–898.
41. J. Weiss, M. Pyrski, E. Jacobi, B. Bufe, V. Willnecker, B. Schick, P. Zizzari, S. J. Gossage, C. A. Greer, T. Leinders-Zufall, C. G. Woods, J. N. Wood and F. Zufall, *Nature,* 2011, **472**, 186–190.
42. D. B. Rupasinghe, O. Knapp, L. V. Blomster, A. B. Schmid, D. J. Adams, G. F. King and M. J. Ruitenberg, *Channels,* 2012, **6**, 103–110.
43. S. England and M. J. de Groot, *Br. J. Pharmacol.,* 2009, **158**, 1413–1425.
44. J. J. Clare, *Expert Opin. Investig. Drugs,* 2010, **19**, 45–62.
45. S. D. Dib-Hajj, Y. Yang, J. A. Black and S. G. Waxman, *Nat. Rev. Neurosci.,* 2013, **14**, 49–62.
46. J. K. Klint, S. Senff, D. B. Rupasinghe, S. Y. Er, V. Herzig, G. M. Nicholson and G. F. King, *Toxicon,* 2012, **60**, 478–491.
47. J. D. Revell, P. E. Lund, J. E. Linley, J. Metcalfe, N. Burmeister, S. Sridharan, C. Jones, L. Jermutus and M. A. Bednarek, *Peptides,* 2013, **44**, 40–46.
48. P. Escoubas, J. R. De Weille, A. Lecoq, S. Diochot, R. Waldmann, G. Champigny, D. Moinier, A. Menez and M. Lazdunski, *J. Biol. Chem.,* 2000, **275**, 25116–25121.
49. N. J. Saez, M. Mobli, M. Bieri, I. R. Chassagnon, A. K. Malde, R. Gamsjaeger, A. E. Mark, P. R. Gooley, L. D. Rash and G. F. King, *Mol. Pharmacol.,* 2011, **80**, 796–808.

50. M. Mazzuca, C. Heurteaux, A. Alloui, S. Diochot, A. Baron, N. Voilley, N. Blondeau, P. Escoubas, A. Gelot, A. Cupo, A. Zimmer, A. M. Zimmer, A. Eschalier and M. Lazdunski, *Nat. Neurosci.,* 2007, **10**, 943–945.
51. G. Pignataro, R. P. Simon and Z. G. Xiong, *Brain,* 2007, **130**, 151–158.
52. J. I. Fletcher, B. E. Chapman, J. P. Mackay, M. E. Howden and G. F. King, *Structure,* 1997, **5**, 1525–1535.
53. M. R. Brown, D. D. Sheumack, M. I. Tyler and M. E. Howden, *Biochem. J.,* 1988, **250**, 401–405.
54. N. S. Bende, E. Kang, V. Herzig, F. Bosmans, G. M. Nicholson, M. Mobli and G. F. King, *Biochem. Pharmacol.,* 2013, **85**, 1542–1554.
55. J. I. Kim, S. Konishi, H. Iwai, T. Kohno, H. Gouda, I. Shimada, K. Sato and Y. Arata, *J. Mol. Biol.,* 1995, **250**, 659–671.
56. G. Corzo, P. Escoubas, M. Stankiewicz, M. Pelhate, C. P. Kristensen and T. Nakajima, *Eur. J. Biochem.,* 2000, **267**, 5783–5795.
57. X.-H. Wang, M. Connor, R. Smith, M. W. Maciejewski, M. E. H. Howden, G. M. Nicholson, M. J. Christie and G. F. King, *Nat. Struct. Biol.,* 2000, **7**, 505–513.
58. S. J. Gunning, F. Maggio, M. J. Windley, S. M. Valenzuela, G. F. King and G. M. Nicholson, *FEBS J.,* 2008, **275**, 4045–4059.
59. A. D. de Araujo, V. Herzig, M. J. Windley, S. Dziemborowicz, M. Mobli, G. M. Nicholson, P. F. Alewood and G. F. King, *Antioxid. Redox Signal.,* 2013, **19**, 1976–1980.
60. S. Reiss, M. Sieber, V. Oberle, A. Wentzel, P. Spangenberg, R. Claus, H. Kolmar and W. Losche, *Platelets,* 2006, **17**, 153–157.
61. M. Werle, K. Kafedjiiski, H. Kolmar and A. Bernkop-Schnurch, *Int. J. Pharm.,* 2007, **332**, 72–79.
62. Z. Miao, G. Ren, H. Liu, R. H. Kimura, L. Jiang, J. R. Cochran, S. S. Gambhir and Z. Cheng, *Bioconjug. Chem.,* 2009, **20**, 2342–2347.
63. J. J. Smith, J. M. Hill, M. J. Little, G. M. Nicholson, G. F. King and P. F. Alewood, *Proc. Natl. Acad. Sci. U.S.A.,* 2011, **108**, 10478–10483.
64. K. Sunagar, E. A. Undheim, A. H. Chan, I. Koludarov, S. A. Munoz-Gómez, A. Antunes and B. G. Fry, *Toxins,* 2013, **5**, 2456–2487.
65. C. J. Bohlen, A. Priel, S. Zhou, D. King, J. Siemens and D. Julius, *Cell,* 2010, **141**, 834–845.
66. M. Y. Sachkova, A. A. Slavokhotova, E. V. Grishin and A. A. Vassilevski, *Insect Mol. Biol.*, 2014, **23**, 527–538.
67. N. D. Rawlings, D. P. Tolle and A. J. Barrett, *Biochem. J.,* 2004, **378**, 705–716.
68. M. J. Pugia, R. Valdes, Jr. and S. A. Jortani, *Adv. Clin. Chem.,* 2007, **44**, 223–245.
69. L. Zhuo and K. Kimata, *Connect. Tissue Res.,* 2008, **49**, 311–320.
70. W. Antuch, K. D. Berndt, M. A. Chavez, J. Delfin and K. Wüthrich, *Eur. J. Biochem.,* 1993, **212**, 675–684.
71. R. Garcia-Fernandez, T. Pons, A. Meyer, M. Perbandt, Y. Gonzalez-Gonzalez, D. Gil, M. de los Angeles Chavez, C. Betzel and L. Redecke, *Acta Crystallogr. Sect. F,* 2012, **68**, 1289–1293.

72. J. Delfin, I. Martinez, W. Antuch, V. Morera, Y. Gonzalez, R. Rodriguez, M. Marquez, A. Saroyan, N. Larionova, J. Diaz, G. Padron and M. Chavez, *Toxicon,* 1996, **34**, 1367–1376.
73. K. Peng, Y. Lin and S. P. Liang, *Acta Biochim. Biophys. Sin.,* 2006, **38**, 457–466.
74. H. Schweitz, C. Heurteaux, P. Bois, D. Moinier, G. Romey and M. Lazdunski, *Proc. Natl. Acad. Sci. U.S.A.,* 1994, **91**, 878–882.
75. B. Gilquin, A. Lecoq, F. Desne, M. Guenneugues, S. Zinn-Justin and A. Menez, *Proteins,* 1999, **34**, 520–532.
76. R. Garcia-Fernandez, T. Pons, M. Perbandt, P. A. Valiente, A. Talavera, Y. Gonzalez-Gonzalez, D. Rehders, M. A. Chavez, C. Betzel and L. Redecke, *J. Struct. Biol.,* 2012, **180**, 271–279.
77. S. Liang, *Toxicon,* 2004, **43**, 575–585.
78. N. D. Rawlings, A. J. Barrett and A. Bateman, *Nucl. Acids Res.,* 2012, **40**, D343–D350.
79. L. A. Kazal, D. S. Spicer and R. A. Brahinsky, *J. Am. Chem. Soc.,* 1948, **70**, 304–340.
80. B. Schlott, J. Wohnert, C. Icke, M. Hartmann, R. Ramachandran, K. H. Guhrs, E. Glusa, J. Flemming, M. Gorlach, F. Grosse and O. Ohlenschlager, *J. Mol. Biol.,* 2002, **318**, 533–546.
81. T. Friedrich, B. Kröger, S. Bialojan, H. G. Lemaire, H. W. Höffken, P. Reuschenbach, M. Otte and J. Dodt, *J. Biol. Chem.,* 1993, **268**, 16216–16222.
82. A. van de Locht, D. Lamba, M. Bauer, R. Huber, T. Friedrich, B. Kröger, W. Höffken and W. Bode, *EMBO J.,* 1995, **14**, 5149–5157.
83. S. Ranganathan, K. J. Simpson, D. C. Shaw and K. R. Nicholas, *J. Mol. Graph. Model.,* 1999, **17**, 106–113.
84. T. Moreau, K. Baranger, S. Dade, S. Dallet-Choisy, N. Guyot and M. L. Zani, *Biochimie,* 2008, **90**, 284–295.
85. T. S. Wilkinson, A. Roghanian, A. J. Simpson and J. M. Sallenave, *Biochem. Soc. Trans.,* 2011, **39**, 1409–1415.
86. A. M. Torres, H. Y. Wong, M. Desai, S. Moochhala, P. W. Kuchel and R. M. Kini, *J. Biol. Chem.,* 2003, **278**, 40097–40104.
87. D. G. Nair, B. G. Fry, P. Alewood, P. P. Kumar and R. M. Kini, *Biochem. J.,* 2007, **402**, 93–104.
88. B. C. Trask, T. Broekelmann, T. M. Ritty, T. M. Trask, C. Tisdale and R. P. Mecham, *Biochemistry,* 2001, **40**, 4372–4380.
89. A. Miyamoto, R. Lau, P. W. Hein, J. M. Shipley and G. Weinmaster, *J. Biol. Chem.,* 2006, **281**, 10089–10097.
90. S. W. Tsang, C. Q. Nguyen, D. H. Hall and K. L. Chow, *Dev. Biol.,* 2007, **312**, 353–366.
91. O. Castaneda, V. Sotolongo, A. M. Amor, R. Stocklin, A. J. Anderson, A. L. Harvey, A. Engstrom, C. Wernstedt and E. Karlsson, *Toxicon,* 1995, **33**, 603–613.

92. J. Pohl, F. Hubalek, M. Byrnes, K. Nielsen, A. Woods and M. Pennington, *Lett. Pept. Sci.,* 1994, **1**, 291–297.
93. T. V. Ovchinnikova, S. V. Balandin, G. M. Aleshina, A. A. Tagaev, Y. F. Leonova, E. D. Krasnodembsky, A. V. Men'shenin and V. N. Kokryakov, *Biochem. Biophys. Res. Commun.,* 2006, **348**, 514–523.
94. J. E. Tudor, P. K. Pallaghy, M. W. Pennington and R. S. Norton, *Nat. Struct. Biol.,* 1996, **3**, 317–320.
95. K. Kalman, M. W. Pennington, M. D. Lanigan, A. Nguyen, H. Rauer, V. Mahnir, K. Paschetto, W. R. Kem, S. Grissmer, G. A. Gutman, E. P. Christian, M. D. Cahalan, R. S. Norton and K. G. Chandy, *J. Biol. Chem.,* 1998, **273**, 32697–32707.
96. E. A. Dennis, *J. Biol. Chem.,* 1994, **269**, 13057–13060.
97. C. C. Leslie, *J. Biol. Chem.,* 1997, **272**, 16709–16712.
98. X. D. Qu and R. I. Lehrer, *Infect. Immun.,* 1998, **66**, 2791–2797.
99. T. Jabeen, N. Singh, R. K. Singh, J. Jasti, S. Sharma, P. Kaur, A. Srinivasan and T. P. Singh, *Proteins,* 2006, **62**, 329–337.
100. J. Halpert and D. Eaker, *J. Biol. Chem.,* 1975, **250**, 6990–6997.
101. F. A. Saul, P. Prijatelj-Znidarsic, B. Vulliez-le Normand, B. Villette, B. Raynal, J. Pungercar, I. Krizaj and G. Faure, *J. Struct. Biol.,* 2010, **169**, 360–369.
102. B. Westerlund, P. Nordlund, U. Uhlin, D. Eaker and H. Eklund, *FEBS Lett.,* 1992, **301**, 159–164.
103. J. Halpert, D. Eaker and E. Karlsson, *FEBS Lett.,* 1976, **61**, 72–76.
104. A. Ritonja and F. Gubensek, *Biochim. Biophys. Acta,* 1985, **828**, 306–312.
105. H. B. Fathi, E. G. Rowan and A. L. Harvey, *Toxicon,* 2001, **39**, 1871–1882.
106. U. Petrovic, J. Sribar, A. Paris, M. Rupnik, M. Krzan, N. Vardjan, F. Gubensek, R. Zorec and I. Krizaj, *Biochem. Biophys. Res. Commun.,* 2004, **324**, 981–985.
107. P. Prijatelj, N. Vardjan, E. G. Rowan, I. Krizaj and J. Pungercar, *Biochimie,* 2006, **88**, 1425–1433.
108. G. Martinez and C. Kopeyan, *FEBS Lett.,* 1977, **84**, 247–252.
109. L. Beress, G. Wunderer and E. Wachter, *Hoppe Seylers Z. Physiol. Chem.,* 1977, **358**, 985–988.
110. R. S. Norton, K. Cross, V. Braach-Maksvytis and E. Wachter, *Biochem. J.,* 1993, **293**, 545–551.
111. N. Manoleras and R. S. Norton, *Biochemistry,* 1994, **33**, 11051–11061.
112. A. Warashina, Z. Y. Jiang and T. Ogura, *Pflügers Arch.,* 1988, **411**, 88–93.
113. R. M. Kini and R. Doley, *Toxicon,* 2010, **56**, 855–867.
114. T. Endo and N. Tamiya, *Pharmacol. Ther.,* 1987, **34**, 403–451.
115. A. Menez, *Toxicon,* 1998, **36**, 1557–1572.
116. V. Tsetlin, *Eur. J. Biochem.,* 1999, **264**, 281–286.
117. S. Diochot, A. Baron, M. Salinas, D. Douguet, S. Scarzello, A. S. Dabert-Gay, D. Debayle, V. Friend, A. Alloui, M. Lazdunski and E. Lingueglia, *Nature,* 2012, **490**, 552–555.

118. C. I. Schroeder, L. D. Rash, X. Vila-Farres, K. J. Rosengren, M. Mobli, G. F. King, P. F. Alewood, D. J. Craik and T. Durek, *Angew. Chem. Int. Ed.*, 2014, **53**, 1017–1020.
119. E. Mulugeta, E. Karlsson, A. Islam, R. Kalaria, H. Mangat, B. Winblad and A. Adem, *Brain Res.*, 2003, **960**, 259–262.
120. S. Nirthanan and M. C. Gwee, *J. Pharmacol. Sci.*, 2004, **94**, 1–17.
121. V. Tsetlin, I. Shelukhina, E. Kryukova, G. Burbaeva, L. Starodubtseva, M. Skok, O. Volpina and Y. Utkin, *Life Sci.*, 2007, **80**, 2202–2205.
122. A. V. Osipov, I. E. Kasheverov, Y. V. Makarova, V. G. Starkov, O. V. Vorontsova, R. K. Ziganshin, T. V. Andreeva, M. V. Serebryakova, A. Benoit, R. C. Hogg, D. Bertrand, V. I. Tsetlin and Y. N. Utkin, *J. Biol. Chem.*, 2008, **283**, 14571–14580.
123. C. Betzel, G. Lange, G. P. Pal, K. S. Wilson, A. Maelicke and W. Saenger, *J. Biol. Chem.*, 1991, **266**, 21530–21536.
124. R. Le Goas, S. R. LaPlante, A. Mikou, M. A. Delsuc, E. Guittet, M. Robin, I. Charpentier and J. Y. Lallemand, *Biochemistry*, 1992, **31**, 4867–4875.
125. B. M. Martin, B. A. Chibber and A. Maelicke, *J. Biol. Chem.*, 1983, **258**, 8714–8722.
126. M. Alkondon and E. X. Albuquerque, *Eur. J. Pharmacol.*, 1990, **191**, 505–506.
127. A. Grozio, L. Paleari, A. Catassi, D. Servent, M. Cilli, F. Piccardi, M. Paganuzzi, A. Cesario, P. Granone, G. Mourier and P. Russo, *Int. J. Cancer*, 2008, **122**, 1911–1915.
128. D. Servent, V. Winckler-Dietrich, H. Y. Hu, P. Kessler, P. Drevet, D. Bertrand and A. Menéz, *J. Biol. Chem.*, 1997, **272**, 24279–24286.
129. S. Antil, D. Servent and A. Menéz, *J. Biol. Chem.*, 1999, **274**, 34851–34858.
130. S. Antil-Delbeke, C. Gaillard, T. Tamiya, P. J. Corringer, J. P. Changeux, D. Servent and A. Menéz, *J. Biol. Chem.*, 2000, **275**, 29594–29601.
131. M. A. McLane, C. Marcinkiewicz, S. Vijay-Kumar, I. Wierzbicka-Patynowski and S. Niewiarowski, *Proc. Soc. Exp. Biol. Med.*, 1998, **219**, 109–119.
132. M. A. McLane, E. E. Sanchez, A. Wong, C. Paquette-Straub and J. C. Perez, *Curr. Drug Targets Cardiovasc. Haematol. Disord.*, 2004, **4**, 327–355.
133. T. F. Huang, J. R. Sheu, C. M. Teng, S. W. Chen and C. S. Liu, *J. Biochem.*, 1991, **109**, 328–334.
134. A. L. Coelho, M. S. de Freitas, A. L. Oliveira-Carvalho, V. Moura-Neto, R. B. Zingali and C. Barja-Fidalgo, *Exp. Cell Res.*, 1999, **251**, 379–387.
135. D. A. Cidade, L. S. Wermelinger, G. Lobo-Hajdu, A. M. Davila, C. Bon, R. B. Zingali and R. M. Albano, *Toxicon*, 2006, **48**, 590–599.
136. M. P. Moreno-Murciano, D. Monleon, J. J. Calvete, B. Celda and C. Marcinkiewicz, *Protein Sci.*, 2003, **12**, 366–371.
137. M. Paz Moreno-Murciano, D. Monleon, C. Marcinkiewicz, J. J. Calvete and B. Celda, *J. Mol. Biol.*, 2003, **329**, 135–145.

138. Y. Fujii, D. Okuda, Z. Fujimoto, K. Horii, T. Morita and H. Mizuno, *J. Mol. Biol.*, 2003, **332**, 1115–1122.
139. D. Monleon, M. P. Moreno-Murciano, H. Kovacs, C. Marcinkiewicz, J. J. Calvete and B. Celda, *J. Biol. Chem.*, 2003, **278**, 45570–45576.
140. E. C. Jimenez, R. P. Shetty, M. Lirazan, J. Rivier, C. Walker, F. C. Abogadie, D. Yoshikami, L. J. Cruz and B. M. Olivera, *J. Neurochem.*, 2003, **85**, 610–621.
141. O. Buczek, D. Wei, J. J. Babon, X. Yang, B. Fiedler, P. Chen, D. Yoshikami, B. M. Olivera, G. Bulaj and R. S. Norton, *Biochemistry*, 2007, **46**, 9929–9940.
142. O. Buczek, D. Yoshikami, G. Bulaj, E. C. Jimenez and B. M. Olivera, *J. Biol. Chem.*, 2005, **280**, 4247–4253.
143. B. Fiedler, M. M. Zhang, O. Buczek, L. Azam, G. Bulaj, R. S. Norton, B. M. Olivera and D. Yoshikami, *Biochem. Pharmacol.*, 2008, **75**, 2334–2344.
144. C. Hopkins, M. Grilley, C. Miller, K. J. Shon, L. J. Cruz, W. R. Gray, J. Dykert, J. Rivier, D. Yoshikami and B. M. Olivera, *J. Biol. Chem.*, 1995, **270**, 22361–22367.
145. K. H. Han, K. J. Hwang, S. M. Kim, S. K. Kim, W. R. Gray, B. M. Olivera, J. Rivier and K. J. Shon, *Biochemistry*, 1997, **36**, 1669–1677.
146. R. W. Teichert, E. Lopez-Vera, J. Gulyas, M. Watkins, J. Rivier and B. M. Olivera, *Biochemistry*, 2006, **45**, 1304–1312.
147. C. S. Walker, D. Steel, R. B. Jacobsen, M. B. Lirazan, L. J. Cruz, D. Hooper, R. Shetty, R. C. DelaCruz, J. S. Nielsen, L. M. Zhou, P. Bandyopadhyay, A. G. Craig and B. M. Olivera, *J. Biol. Chem.*, 1999, **274**, 30664–30671.
148. A. C. Rigby, E. Lucas-Meunier, D. E. Kalume, E. Czerwiec, B. Hambe, I. Dahlqvist, P. Fossier, G. Baux, P. Roepstorff, J. D. Baleja, B. C. Furie, B. Furie and J. Stenflo, *Proc. Natl. Acad. Sci. U.S.A.*, 1999, **96**, 5758–5763.
149. D. E. Kalume, J. Stenflo, E. Czerwiec, B. Hambe, B. C. Furie, B. Furie and P. Roepstorff, *J. Mass Spectrom.*, 2000, **35**, 145–156.
150. C. Petrel, H. G. Hocking, M. Reynaud, G. Upert, P. Favreau, D. Biass, M. Paolini-Bertrand, S. Peigneur, J. Tytgat, N. Gilles, O. Hartley, R. Boelens, R. Stocklin and D. Servent, *Biochem. Pharmacol.*, 2013, **85**, 1663–1671.
151. B. G. Fry, K. Roelants, D. E. Champagne, H. Scheib, J. D. Tyndall, G. F. King, T. J. Nevalainen, J. A. Norman, R. J. Lewis, R. S. Norton, C. Renjifo and R. C. de la Vega, *Annu. Rev. Genomics Hum. Genet.*, 2009, **10**, 483–511.
152. J. J. Smith, V. Herzig, G. F. King and P. F. Alewood, *Cell. Mol. Life Sci.*, 2013, **70**, 3665–3693.
153. S. Zhu, S. Peigneur, B. Gao, Y. Umetsu, S. Ohki and J. Tytgat, *Mol. Biol. Evol.*, 2014, **31**, 546–559.
154. J. C. Novello, E. C. Arantes, W. A. Varanda, B. Oliveira, J. R. Giglio and S. Marangoni, *Toxicon*, 1999, **37**, 651–660.

155. S. K. Holaday, Jr., B. M. Martin, P. L. Fletcher, Jr. and N. R. Krishna, *Arch. Biochem. Biophys.,* 2000, **379**, 18–27.
156. A. M. Pimenta, P. Mansuelle, C. R. Diniz and M. F. Martin-Eauclaire, *J. Pept. Sci.,* 2003, **9**, 132–140.
157. S. Oyama, Jr., P. Pristovsek, L. Franzoni, T. A. Pertinhez, E. Schinina, C. Lucke, H. Ruterjans, E. C. Arantes and A. Spisni, *Protein Sci.,* 2005, **14**, 1025–1038.
158. S. M'Barek, B. Chagot, N. Andreotti, V. Visan, P. Mansuelle, S. Grissmer, M. Marrakchi, M. El Ayeb, F. Sampieri, H. Darbon, Z. Fajloun, M. De Waard and J. M. Sabatier, *Proteins,* 2005, **60**, 401–411.
159. H. Rochat, R. Kharrat, J. M. Sabatier, P. Mansuelle, M. Crest, M. F. Martin-Eauclaire, F. Sampieri, R. Oughideni, K. Mabrouk, G. Jacquet, J. Van Rietschoten and M. El Ayeb, *Toxicon,* 1998, **36**, 1609–1611.
160. R. Kharrat, K. Mabrouk, M. Crest, H. Darbon, R. Oughideni, M. F. Martin-Eauclaire, G. Jacquet, M. el Ayeb, J. Van Rietschoten, H. Rochat and J. M. Sabatier, *Eur. J. Biochem.,* 1996, **242**, 491–498.
161. R. Kharrat, P. Mansuelle, F. Sampieri, M. Crest, R. Oughideni, J. Van Rietschoten, M. F. Martin-Eauclaire, H. Rochat and M. El Ayeb, *FEBS Lett.,* 1997, **406**, 284–290.
162. Z. Fajloun, A. Mosbah, E. Carlier, P. Mansuelle, G. Sandoz, M. Fathallah, E. di Luccio, C. Devaux, H. Rochat, H. Darbon, M. De Waard and J. M. Sabatier, *J. Biol. Chem.,* 2000, **275**, 39394–393402.
163. E. Blanc, J. M. Sabatier, R. Kharrat, S. Meunier, M. el Ayeb, J. Van Rietschoten and H. Darbon, *Proteins,* 1997, **29**, 321–333.
164. I. Regaya, C. Beeton, G. Ferrat, N. Andreotti, H. Darbon, M. De Waard and J. M. Sabatier, *J. Biol. Chem.,* 2004, **279**, 55690–55696.
165. J. A. DeBin, J. E. Maggio and G. R. Strichartz, *Am. J. Physiol.,* 1993, **264**, C361–C369.
166. G. Lippens, J. Najib, S. J. Wodak and A. Tartar, *Biochemistry,* 1995, **34**, 13–21.
167. L. Soroceanu, Y. Gillespie, M. B. Khazaeli and H. Sontheimer, *Cancer Res.,* 1998, **58**, 4871–4879.
168. J. Deshane, C. C. Garner and H. Sontheimer, *J. Biol. Chem.,* 2003, **278**, 4135–4144.
169. K. Kesavan, J. Ratliff, E. W. Johnson, W. Dahlberg, J. M. Asara, P. Misra, J. V. Frangioni and D. B. Jacoby, *J. Biol. Chem.,* 2010, **285**, 4366–4374.
170. A. N. Mamelak and D. B. Jacoby, *Expert Opin. Drug. Deliv.,* 2007, **4**, 175–186.
171. M. Veiseh, P. Gabikian, S. B. Bahrami, O. Veiseh, M. Zhang, R. C. Hackman, A. C. Ravanpay, M. R. Stroud, Y. Kusuma, S. J. Hansen, D. Kwok, N. M. Munoz, R. W. Sze, W. M. Grady, N. M. Greenberg, R. G. Ellenbogen and J. M. Olson, *Cancer Res.,* 2007, **67**, 6882–6888.

172. M. R. Stroud, S. J. Hansen and J. M. Olson, *Curr. Pharm. Des.*, 2011, **17**, 4362–4371.
173. M. Akcan, M. R. Stroud, S. J. Hansen, R. J. Clark, N. L. Daly, D. J. Craik and J. M. Olson, *J. Med. Chem.*, 2011, **54**, 782–787.
174. J. I. Fletcher, R. Smith, S. I. O'Donoghue, M. Nilges, M. Connor, M. E. H. Howden, M. J. Christie and G. F. King, *Nat. Struct. Biol.*, 1997, **4**, 559–566.

CHAPTER 3

Venoms-Based Drug Discovery: Proteomic and Transcriptomic Approaches

SÉBASTIEN DUTERTRE*[a], EIVIND A. B. UNDHEIM[b,c], SANDY S. PINEDA[b], AI-HUA JIN[b], VINCENT LAVERGNE[b], BRYAN G. FRY[b], RICHARD J. LEWIS[b], PAUL F. ALEWOOD[b], AND GLENN F. KING[b]

[a]Institut des Biomolécules Max Mousseron, UMR 5247, Université Montpellier 2 - CNRS, Place Eugène Bataillon, 34095 Montpellier Cedex 5, France; [b]Institute for Molecular Bioscience, The University of Queensland, St Lucia, Queensland 4072, Australia; [c]Centre for Advanced Imaging, The University of Queensland, St Lucia, Queensland 4072, Australia
*E-mail: Sebastien.Dutertre@univ-montp2.fr

3.1 Introduction

A number of traditional medications have been derived from the venom of snakes, spiders and frogs.[1,2] In recent times, venom peptides have been subjected to intense scientific investigation, in part due to the recent re-orientation of biopharmaceutical companies towards "biologics". Peptides often show greater specificity, higher potency, and lower toxicity than traditional small-molecule drugs, and they have emerged as a commercially important class of therapeutics.[3] For example, peptides have become recognised treatments for numerous human diseases, including diabetes, HIV,

RSC Drug Discovery Series No. 42
Venoms to Drugs: Venom as a Source for the Development of Human Therapeutics
Edited by Glenn F. King

Published by the Royal Society of Chemistry, www.rsc.org

hepatitis, chronic pain, and cancer.[2,3] Furthermore, peptides remain invaluable research tools for dissecting the physiological functions of many human receptors and for unravelling the biological mechanisms underlying disease.

The traditional venom bioprospecting strategy is based on bioassay-guided fractionation (see Chapter 4 for further details). While this procedure has been, and still continues to be, successful (*e.g.*, see ref. 4), it is time-consuming and typically requires large amounts of venom. As a result, this strategy has introduced a bias towards the largest and the most abundant venomous species, which have been intensively studied. Smaller and rarer species are likely to be just as interesting from a drug development perspective, but to efficiently exploit these limited/restricted resources, improvements in the sensitivity of analytical techniques are required. The collection of large quantities of small/rare animals is often not only impractical and unsustainable, but also ethically questionable. Therefore, state-of-the-art discovery strategies are expected to deliver a maximum of information from limited samples. An integrated approach, whereby transcriptomic and proteomic data are combined together with dedicated bioinfomatic tools, has the potential to achieve this goal (Figure 3.1).

The recent development of second-generation sequencing technologies, initially fuelled by the Human Genome Project, allows rapid access to all sequences expressed in a venom gland. While several different platforms are available, the 454-pyrosequencing technology (Roche) provides the longest reads (>350 bp on average), which in the absence of a reference genome, remains a major advantage for *de novo* transcriptome assembly. Critically, post-translational modifications (PTMs) can be essential to the biological activity of a venom peptide, thus requiring integration of proteomic data. For more than a decade, mass spectrometry (MS) has been the method of choice for studying the complexity of venoms.[5] In particular, soft ionisation technologies such as matrix-assisted laser desorption-ionisation (MALDI) and electrospray ionisation (ESI) are now heavily utilised to unravel the composition of these proteinaceous mixtures.[6–8] While venoms can be studied as a whole sample (mass fingerprinting, profiling), they are usually pre-fractionated in order to achieve better resolution and higher coverage, and to minimise ion suppression effects. A liquid chromatography step can be carried out off-line (MALDI) or online (ESI), each method providing high quality, yet complementary data sets.[9] In addition, recent improvements in the sensitivity (dynamic range) and accuracy of mass spectrometers are allowing high-throughput analysis from minute samples.[10] For instance, using a combined Orbitrap-ETD with a targeted chemical derivatisation strategy, the full sequences of 31 peptide toxins were obtained from just 7% of the crude venom of a single marine cone snail.[11] In this chapter, we review current venoms-based drug discovery strategies, with an emphasis on the integration of proteomic and transcriptomic data.

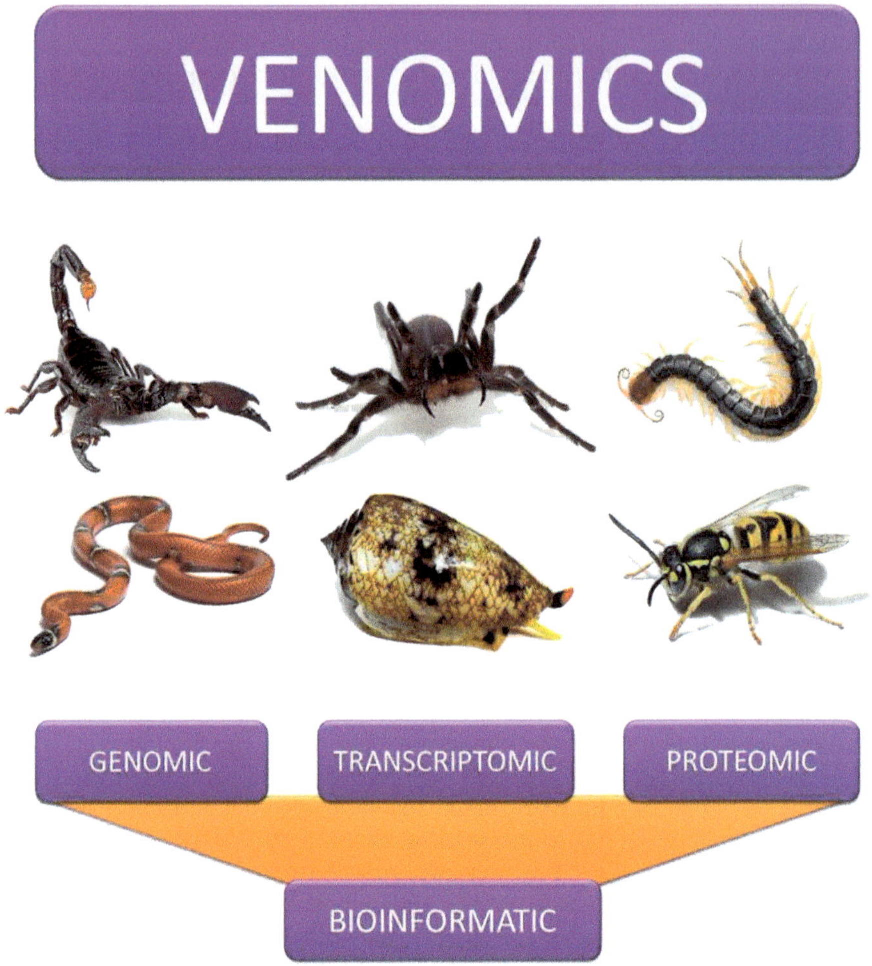

Figure 3.1 Integrated venomics-based biodiscovery strategy. Combined state-of-the-art technologies maximise scientific outcomes from limited material, even single specimens. Venomics is generally considered to involve a holistic study of venom composition using a combination of genomic, transcriptomic, and proteomic approaches.

3.2 Venom-Gland Transcriptomics

Throughout the course of evolution, venom proteins from both vertebrates and invertebrates have been optimised to target specific receptors with high affinity and often exquisite selectivity, making them excellent pharmacological tools and drug leads.[2,12–25] The number of venom-derived peptides that are in pre-clinical or clinical trials has increased significantly in the last 15 years.[2]

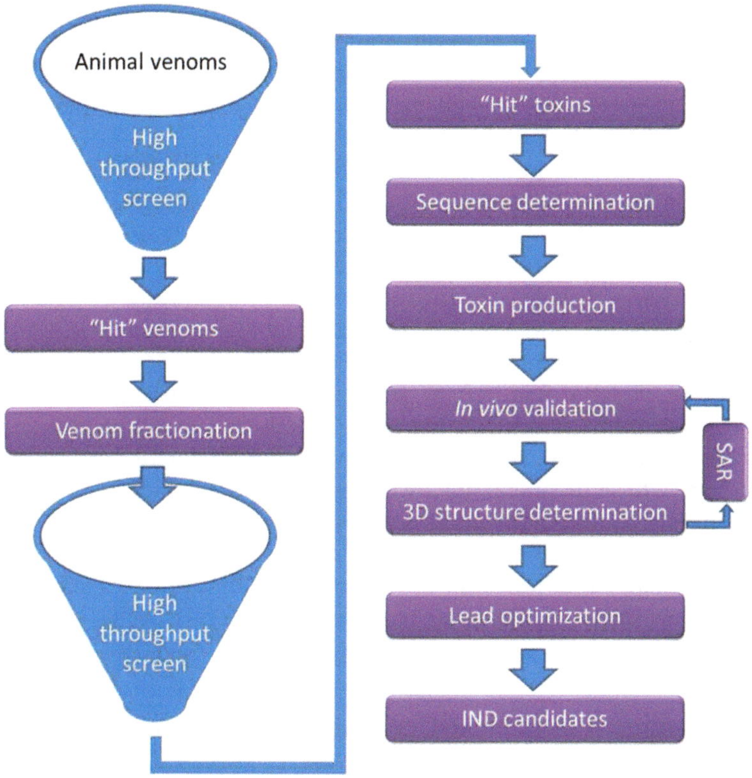

Figure 3.2 Key elements of a venoms-based drug discovery pipeline. A robust high-throughput screen is essential for rapid identification of "hit" venoms and subsequent isolation of bioactive peptides. An efficient peptide production system is required to produce sufficient material for functional and structural characterisation and to examine structure–activity relationships. Adapted from ref. 12.

Until recently, the study of toxic peptides from venomous animals was mostly limited to the isolation and biochemical characterisation of toxins of medical importance. Little or no attention was paid to the genes, cellular machinery, and other important processes involved in assembly of the final product expressed in the venom. The main approach utilised for isolation of venom peptides has been bioassay-guided fractionation (Figure 3.2). Animal venoms were generally screened in medium- to high-throughput assays against targets of therapeutic interest, then "hit venoms" were chromatographically fractionated and individual fractions re-screened in order to isolate peptides responsible for bioactivity. These peptides were then sequenced *via* a combination of Edman degradation, tandem mass spectrometry (MS/MS) techniques, and venom-gland transcriptomics. In some cases, incomplete sequence information acquired *via* MS/MS and/or Edman degradation has been used to design primers to amplify transcripts encoding the toxin of interest from a venom-gland cDNA library. This has the advantage of providing useful information about the signal and propeptide regions of the toxin precursor as well as the sequence of transcripts encoding paralogues (and orthologues in related species).[26]

Although these methods have proved useful for acquiring information about single families of toxins (*i.e.*, families of paralogues/orthologues) they are inefficient, time-consuming, and low throughput and therefore not suitable for providing an overview of the full complexity of a venom. Thus, after more than a decade of trying to isolate venom peptides using a targeted approach, there was a move away from sequencing specific toxin transcripts to random amplification of venom-gland transcripts; this became possible as the amount of RNA required to generate cDNA libraries decreased from several micrograms to nanograms due to the introduction of new enzymes, adaptors, vectors and, most importantly, improvements in sequencing technology. These technical advances resulted in a considerable increase in the number of toxin-encoding expressed sequence tags (ESTs) being sequenced using Sanger sequencing, which significantly increased the rate at which new toxin superfamilies were discovered. This approach also led to the first overview of the entire toxin repertoire of several species of venomous animals.

3.2.1 Sanger *versus* Next-Generation Sequencing

Since it was first reported in 1977, the rapid determination of DNA sequences using the method described by Frederick Sanger and Alan Coulson transformed the field of biology, as it provided a tool for acquiring entire gene sequences and subsequently entire genomes.[27–29] For more than 30 years, this approach was regarded as the gold standard in sequencing. Sanger sequencing remained the method of choice for some time following the introduction of second-generation sequencing methods such as pyrosequencing technology in 1985, despite the fact that new technologies promised enormous potential and higher throughput. Even today, the majority of toxin precursor sequences described in the literature (>95%) were obtained using Sanger sequencing.

Over the past decade, the advent of second-generation or next-generation sequencing has transformed the way DNA sequencing is performed. The majority of second-generation sequencing platforms rely in principle on a multi-step process that requires the conversion of the template, so that it can be attached to the surface of a bead, for the amplification of that particular template to occur prior to the sequencing reaction—a process that differs quite markedly from the traditional capillary electrophoresis principle applied in Sanger sequencing.[27–29] Second-generation techniques resulted in an exponential increase in the number of reads that could be acquired in a single sequencing run (*i.e.*, from thousands to millions) due to massive parallelisation and miniaturisation of the reactions.

Second-generation sequencers can be divided into two main types, based on the read length and throughput. The longest read lengths can be achieved using 454 pyrosequencing, with an average length of 450–800 bp. In contrast, platforms such as Hi Seq200, miSeq, 5500xl SOLiD system and HeliScope provide shorter read lengths, ranging from 25 to 150 bp, but yield much higher throughput.[30] Read length is a critical factor when sequencing venom-gland cDNA libraries for two reasons. First, for most venomous animals, there are no reference genomes that can be used as templates for assembly. Second, many venoms tend to be dominated by large multigene families containing numerous members with extensive sequence overlap, making bioinformatic reconstruction difficult for short read-length platforms. Thus, due to the longer reads it provides, 454 pyrosequencing has to date been the platform of choice to investigate venom-gland preparations from a wide array of venomous species, including bats,[31] centipedes,[32] cone snails,[33–37] cuttlefish, squid and octopus,[38] scorpions,[39] snakes,[40–42] and spiders.[43,44] In contrast, only a few venom-gland transcriptomes have been analysed using HiSeq or Illumina.[45–47]

In some cases, the raw read lengths available using the 454 platform (typically >350 bp) are sufficiently long to cover the full length of venom-peptide precursors, thus obviating the need for bioinformatic reconstruction. We recently took advantage of this fact to unravel the mechanisms by which marine cone snails produce highly complex venoms.[37] The applicability of this approach, however, is obviously dictated by the length of the toxin precursors expressed in the venom gland. At present, marine cone snails are unique in that the majority of toxin diversity is encoded by transcripts that are shorter than the average read length of commonly available high-throughput sequencing platforms. In general, until third-generation sequencing technologies become commonly available, successful bioinformatic reconstruction of venom-gland transcriptomes sequenced using second-generation platforms will be vital for meaningful data analyses.

3.2.2 Bioinformatic Analysis of cDNA Libraries without a Reference Genome

De novo assembly of reads generated *via* Sanger or next-generation sequencing using non-model organisms has always been regarded as a challenge. Currently available bioinformatic tools provide better support when assemblies are generated using a reference genome, especially when short- to medium-sized reads have been acquired. This is reflected in the fact that the majority of transcriptomic studies that have been published since the introduction of the 454, Illumina, ABI/SOLiD and or HeliScope platforms are mainly from organisms for which extensive genomic or Sanger EST sequences are available. To overcome this hurdle, several bioinformatic tools have been developed and tested for a range of assembly projects with reads obtained by a single technology or hybrids (*e.g.*, a combination of Sanger and second-generation sequencing). Overwhelmingly, the high throughput of second-generation sequencing projects creates a bioinformatic bottleneck that is hard to solve in the absence of adequate tools for non-model organisms. Currently there are only a few online tools and programs that can be used for *de novo* assembly, all with their own inherent advantages and disadvantages. Therefore, custom-made programs often need to be specifically designed to access information related to venom toxins, including toxin-gene superfamilies, relative levels of expression (reads per sequence), likely post-translational modifications, and protease cleavage sites in the toxin precursors (see Section 3.5).

3.3 Venom Proteomics

The traditional proteomic approach to obtain the amino acid sequence of peptide drug leads has largely relied on the use of automated Edman degradation and amino acid composition analysis followed by confirmation of molecular weight *via* MS (*e.g.*, the discovery and characterisation of exendin-4[48] and ShK[49]). While this approach enables confident assignment of peptide sequences, it suffers from both low throughput and high sample requirement. Furthermore, sequencing by Edman degradation can be prohibitively expensive for large numbers of peptides. However, in recent years, the development of highly sensitive, high-resolution MS instruments that provide novel fragmentation techniques has provided a way around these issues. As a result, in the field of venomics, "proteomics" has become near synonymous with MS analyses.

3.3.1 The Mass Spectrometer

A simplified version of a modern mass spectrometer contains three parts: the ionisation source produces multiple different gas-phase ions, the mass filter/analyser separates the ions, and the ion detector amplifies and transfers the signals. Of the large number of ionisation modes historically available, only two are commonly used today for the analysis of proteins and peptides, namely ESI and MALDI. Both techniques can be used to analyse large proteins and protein complexes,[50,51] but they have gained popularity in venoms-based drug discovery laboratories primarily because they are ideally suited to the ionisation of small proteins and peptides.

In ESI, ions are formed directly from solution, which means it is amenable to online high-throughput systems. In addition, multiply charged ions are dominant, and this allows both a wide range of molecular weights to be accurately measured within a narrow *m*/*z* range and the optimal utilisation of some fragmentation techniques. Although the details of the ion formation mechanisms in MALDI is still a matter of debate, the overarching principle is fairly simple: a sample co-crystallised with a matrix is irradiated with an ultraviolet (UV) or infra-red (IR) laser; this causes desorption of charged analytes into the gas phase, and these are then accelerated through a force field and into the mass analyser. While the co-crystallisation step renders MALDI unsuitable for high-throughput sample analysis, the importance of the matrix in the ionisation of analytes also means MALDI is a very versatile source of ionisation. An MS application illustrating this is the use of MALDI in MS imaging, or the ionisation of analytes directly from tissue sections,[52] which has recently been applied to venom glands.[53]

Along with the type of ionisation source, the mass analyser and detector dictate the utility of a mass spectrometer. To achieve the best performance in speed, sensitivity, resolution and accuracy, most modern MS instruments have hybrid mass analysers that combine different performance characteristics offered by various types of analysers (*e.g.*, Time Of Flight [TOF], Quadropole [Q], Trap, Fourier Transform types [FT]).[54] These combinations also greatly affect the type of fragmentation, and hence sequence information, that can be obtained from a sample (see below). Ultimately, the range of combinations of sources and mass analysers available means MS is ideally suited to the wide spectrum of analytical challenges encountered in toxinology and venoms-based drug discovery.

3.3.2 Liquid Chromatography–Mass Spectrometry

Due to the relatively high complexity of most venoms, an efficient separation by liquid chromatography (LC) is important before sample introduction to the ionisation source and it therefore forms an integral part of most MS-based analyses of venoms. For MALDI-MS this can be conveniently performed using an online autospotter, although the throughput is limited by the additional analysis time of the spotted samples. In contrast, for ESI-MS, online LC separation enables high-throughput venom analysis that can be reproducibly achieved in a very short time frame. LC separation of intact venom components is usually performed using reverse-phase (RP) high-performance liquid chromatography (HPLC) with large pore (300 Å) C_{18} columns. Nano-LC-ESI-MS has been favoured, as it provides high sensitivity for many peptides and proteins and the extremely low flow rates ($\sim$200 nL min^{-1}) provide enormous improvement to the efficiency of ionisation. It is, however, limited by the separation efficiency of current capillary column technology and the very limited amount of sample loading (maximum 2 μg of a linear dynamic sample) capacity. Therefore, Nano-LC-ESI-MS is not ideal when components are unevenly distributed in complex venoms, unless another dimension of separation is applied first. Conventional columns provide maximum sample loading capacity, which can enhance discovery of peptides/proteins expressed at low levels in venom. In addition, the newly developed uHPLC system provides comparable sensitivity and efficiency, but with a much faster gradient, thereby yielding shorter run times. Figure 3.3 compares the venom profile of the predatory marine cone snail *Conus marmoreus* obtained using conventional and uHPLC columns.

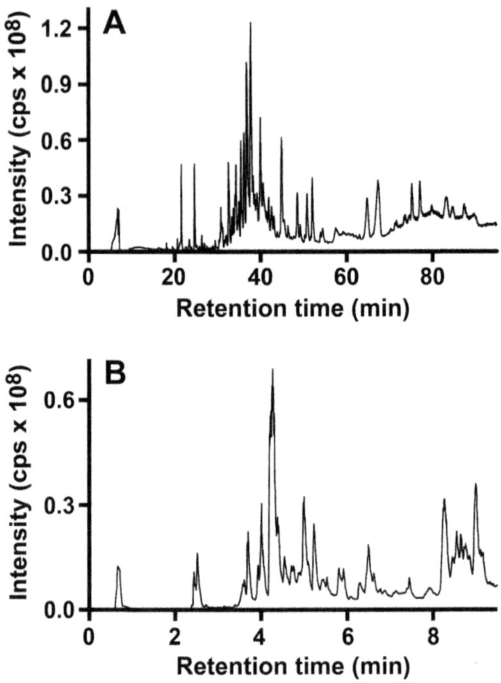

Figure 3.3 HPLC chromatograms showing fractionation of venom from the cone snail *C. marmoreus*. (A) Venom chromatogram obtained using a Thermo C_{18} column (4.6 × 150 mm, 5 μm, 300 Å) with elution using a linear gradient of 1% solvent B (90% acetonitrile/0.1% formic acid) per minute at a flow rate of 0.25 ml min^{-1} over 80 min, which revealed 6254 peptides. (B) Venom chromatogram obtained using an Agilent uHPLC column (2.1 × 100 mm, 1.8 μm, 300 Å) with elution using a linear gradient of 10% solvent B (90% acetonitrile/0.1% formic acid) per minute at a flow rate of 0.25 ml min^{-1} over 8 min, which revealed 930 peptides. Mass detection was by ESI-MS using an AB Sciex TripleTOF 5600 System, a hybrid quadruple TOF MS equipped with a DuoSpray ionisation source.

3.3.3 MS Sequencing

Sequencing by mass spectrometry can be roughly divided into two general types: true *de novo* sequencing and search-based sequencing. Traditional *de novo* sequencing efforts have focused on sequencing either one or a few functionally characterised proteins and peptides, and they have often been combined with Edman degradation. Incomplete peptide sequence information, especially due to ambiguous *de novo* fragmentation interpretations or incomplete C-terminal sequences from Edman degradation, make the two techniques complementary. Although these approaches have made invaluable contributions to venom peptide/protein discovery, they are generally low throughput and time-consuming. However, technological and bioinformatic advancements means *de novo* shotgun sequencing of whole venoms is now possible.[55] Similarly, full-length peptide toxins can now routinely be sequenced *de novo* on certain mass spectrometers.[11]

While technological developments have enabled rapid, true *de novo* sequencing of whole venoms and peptides, this still remains a daunting task on most mass spectrometers. Search-based MS sequencing is therefore a much more commonly applied approach. Using this technique, observed fragment spectra are searched against theoretical fragment spectra based on a sequence template using bioinformatic tools such as Mascot (Matrix Science), Spectrum Mill (Agilent), or Protein Pilot (ABSciex). Although the nature of the approach means that it is not by itself sufficient for full sequence characterisation of novel toxins, in combination with a venom gland transcriptome it enables both accurate and high-throughput characterisation of novel venoms.[31,37,45,56]

Central to both *de novo* and search-based MS sequencing is the generation of fragments of a precursor peptide or protein to obtain sequence information. The most common peptide fragmentation techniques can be grouped into two types based on the underlying mechanism of cleavage of the peptide backbone, namely thermal- and radical-driven fragmentation. Due to the popularity of Q/TOF, triple-Q, and TOF/TOF mass analysers, thermally mediated fragmentation techniques are the most frequently employed and include post-source decay (PSD), collision-induced dissociation (CID), and higher collision energy dissociation (HCD).[57] During these tandem MS experiments (MS/MS, or MS^n for higher order MS experiments), cleavage, primarily of the amide bond of the peptide backbone, is caused either by dissociation during flight in a unimolecular process (PSD), or by collision into molecules of an inert gas (*e.g.*, N_2 or Ar) at high energies (CID and HCD). Although both fast and sensitive, these techniques tend also to "knock off" labile PTMs, leaving them undetected. They are also largely restricted to relatively small peptides (<3 kDa) for the generation of informative spectra, and use of proteolytic enzymes is therefore usually necessary. This approach is often termed "bottom-up" proteomics, where fragments rather than entire proteins are analysed by MS, and it represents the most common approach for proteomic characterisation of venoms by MS, particularly using CID-based MS/MS.[57]

The other main form of peptide fragmentation that is increasingly being used to acquire sequence information from peptides and proteins is radical-driven fragmentation. In this approach, an electron capture (electron capture dissociation (ECD)), loss (electron detachment dissociation (EDD)), or transfer (electron transfer dissociation (ETD) and negative electron transfer dissociation (NETD)) results in the formation of a radical, which then induces dissociation at the N-Cα (ECD, ETD) or Cα-CO (EDD, NETD) bond.[58,59] In contrast to thermally mediated fragmentation, radical-driven fragmentation such as ETD is not necessarily restricted in terms of precursor size and is therefore used in both bottom-up and top-down (intact proteins) proteomics.[11,60,61] Furthermore, even labile PTMs are usually retained, allowing for their identification. While the aforementioned fragmentation events are induced in the mass analyser, fragmentation methods analogous to both ETD and NETD can also be employed in-source in MALDI-MS by using certain matrices such as 1,5-diaminonapthalene (ETD-like) or 5-nitro-salicylic acid (NETD-like, but in positive ion mode).[62–65] Although in-source decay (ISD) normally requires pure sample for interpretation of spectra and does not represent a high-throughput strategy, the ability to induce multiple, complementary types of fragmentation using the same source and mass analyser illustrates the versatility of MALDI-MS.

3.3.4 Optimising Coverage

While each mode of fragmentation represents a powerful tool for the characterisation of venoms and toxins of interest, their complementary nature can be used to maximise coverage and number of high confidence protein identifications.[60] Additionally, results obtained using MALDI-MS and ESI-MS have been shown to contain a substantial proportion of non-overlap and can be further used to maximise sequencing depth.[37] Furthermore, orthogonal separation of enzymatic fragments, venom components, or enriching for components of interest (*e.g.*, in antivenomics[66]) can greatly enhance the sequencing depth. These methods can be used to compensate for the lack of orthogonal fragmentation methods available to many venoms-based drug discovery laboratories, particularly when complemented with transcriptomic data.

3.4 Bioinformatic Processing of Large-Scale Sequencing Data

Recent advances in next-generation sequencing technologies used at the three levels of the omics cascade (genomics, transcriptomics, proteomics) provide large amounts of data that need to be efficiently analysed using high-throughput and purpose-built bioinformatics tools. In the case of the complementary transcriptomic and proteomic approaches employed in venoms-based drug discovery, the treatment of these data usually follows the following step-wise process: (i) extraction of the precursor protein of interest from *in silico* translated cDNA sequences obtained from transcriptome sequencing platforms; (ii) isolation of the mature peptide fragment from the parent precursor sequence supported by MS experiments. Once the toxin sequence has been deciphered, the peptide can be produced by synthetic or recombinant methods and its activity analysed.

Mature venom peptides vary in size depending on the source organism. For instance, venom peptides from cone snails are typically less than 40 residues in length (see Chapter 6 for more details), whereas snake venoms contain larger peptides and small enzymes such as phospholipase A_2 that can comprise up to 160 amino acid residues (see Chapter 5 for more details). As a consequence, the size of the transcript encoding the precursor form of these toxins will also vary greatly, which begs the question of how best to perform an assembly of the reads into longer contigs. Assembly is much easier when a reference genome is available, which is not the case for the vast majority of venomous animals. Thus, *de novo* transcriptome assembly has to be considered. Although this step can be relatively laborious and time-consuming, several *de novo* assemblers are available, such as Trinity,[67] SOAPdenovo-Trans,[68] and Oases.[69] Today, most next-generation sequence assemblers use de Bruijn graphs but still lack efficiency when given highly repetitive sequences. The graph resolution is often aborted after a number of cycles, and no sequences (or chimeras) are generated. This issue is currently a main focus of the scientific community.

After the cDNA sequences have been assembled (or if it has been decided to use raw reads), they can be translated *in silico* in order to obtain the corresponding mature toxin sequences. However, since the cDNA library results from upstream amplification of mRNA, it is often of interest to study the sequence of the entire toxin precursor (delimited by start and stop codons) instead of the mature peptide sequence. The precursor of secreted proteins such as those found in venoms contain an N-terminal signal sequence responsible for trafficking the protein to the endoplasmic reticulum.[70,71] Although the signal peptide and propeptide are enzymatically trimmed from the precursor protein to produce the active mature toxin, they can be useful for understanding the biology of the source organism and for toxin classification. For example, the signal peptide sequences in conotoxin precursors have been conserved much better than that the mature toxin sequences during the course of evolution, and consequently the signal peptide sequences are used to classify conopeptides into gene superfamilies and provide hints about the structure and pharmacological activity of the mature toxins[72] (see Chapter 6 for more details). Nevertheless, the recognition of these signature motifs among the thousands to millions of sequences generated by transcriptome sequencing platforms cannot be achieved manually with high accuracy.

In addition to the time and effort that needs to be invested, this approach raises the question of the phylogenetic specificities of venomous animal taxa, and the ability of global bioinformatics programs to accurately recognise and classify toxins from diverse organisms. Depending on the objective of the study, the size of the query data set, and the nature of the endogenous precursor toxin (*e.g.*, presence of propeptide regions, length of mature peptide), various processing pipelines can be adopted. Despite being particularly unsuited to large-scale proteomic analyses (time-consuming, computationally intense, limited specificity), multiple sequence alignment programs, such as ClustalW[73] or MUSCLE[74] for instance, have been widely used and are probably the simplest approach to set up. For greater accuracy, model-based programs made from precise training sets should be used instead. Moreover, because different venomous animals produce a variety of peptide toxins with unique sequence features, "tailor-made" models must be designed for specific taxa in order to efficiently detect toxin from non-toxin sequences. An example is ConoSorter, a high-throughput program for large-scale analysis of cone snail transcriptomes and proteomes.[75] The ConoSorter algorithm uses two complementary searching approaches based on regular expressions (a "rigid" Boolean search) and profile Hidden Markov Models[76] (a more flexible stochastic searching strategy for distantly related sequences) to specifically detect, with a specificity ≥99%, conopeptide sequences and to classify them into gene superfamilies by automatically detecting their signal, propeptide and mature toxin signatures. ConoSorter also enables the user to identify new conopeptide families by calculating a set of sequence characteristics, and by automatically searching the ConoServer database[72] for known annotated toxins.

The second challenge of using transcriptomic data for toxin screening is to deconvolute the mature peptide sequence from the precursor protein sequence. While the signal peptide can be reliably inferred with programs based on the recognition of consensus eukaryotic cleavage sites, such as SignalP for instance,[77] the presence of other potential regions trimmed from the parent protein, such as the propeptide, provides additional challenges in identifying the mature peptide region. In some cases, specific bioinformatic tools have been developed for this purpose, such as SpiderP, which is specifically designed to detect propeptide regions in spider-toxin precursors.[43] Thus, it is often necessary to perform complementary MS experiments to obtain sequence information on mature toxins isolated directly from venom. Matching of the peptide fragments to their parent precursor proteins can then be achieved with proprietary MS software such as ABSCIEX ProteinPilot™ (Paragon algorithm), which also predicts post-translational modifications of the peptide fragments.

3.5 Summary

Our ancestral fear of snakes, scorpions and spiders is directly linked to their venomous nature and ability to cause human fatalities. Nevertheless, despite their lethal potential, the therapeutic value of venoms has also been recognised since ancient times. For example, snake venom has been used in Ayurvedic medicine since the 7th century BCE to prolong life and treat arthritis and gastrointestinal ailments, while Chinese traditional medicine has a long history of using venomous decoctions to treat anything from opium addiction or arthritis to heart diseases and cancer.[2,78]

The modern era of venoms-based drug discovery began in the 1970s with the development of the blockbuster antihypertensive drug captopril from an inhibitor of angiotensin-converting enzyme discovered in the venom of the Brazilian viper *Bothrops jaracaca*[79] (see Chapter 6 for more details). Subsequently, five more venom-derived drugs have been approved by the United States Food and Drug Administration (FDA), the most recent of which is Byetta®, a glucagon-like peptide-1 agonist isolated from the saliva of the Gila monster, which is used for the treatment of type 2 diabetes mellitus.[2] Despite these successes, the traditional approach of venoms-based drug discovery *via* activity-guided fractionation has proved highly inefficient. Since there are likely to be more than 200 000 venomous animals on Earth, each with a unique venom comprising a minimum of 50–200 peptide toxins, it can be estimated that more than 10 million venom peptides await evaluation. In this chapter, we have shown how integration of second-generation sequencing technologies with state-of-the-art proteomics can greatly accelerate the discovery of bioactive venom peptides, a strategy known as venomics. We predict greatly increased use of this integrated discovery strategy over the next 2–5 years, and especially exciting will be its application to novel and previously difficult-to-study small venomous species. With the emergence of third-generation sequencing platforms, which promise improvements in fidelity and ease of analysis, venomics will become more affordable and applicable to an even wider range of venomous animals, thus providing an even greater opportunity to discover therapeutic leads from animal venoms.

References

1. A. L. Harvey, *Trends Pharmacol. Sci.,* 2002, **23**, 201–203.
2. G. F. King, *Expert Opin. Biol. Ther.,* 2011, **11**, 1469–1484.
3. V. M. Ahrens, K. Bellmann-Sickert and A. G. Beck-Sickinger, *Future Med. Chem.,* 2012, **4**, 1567–1586.
4. S. Yang, Y. Xiao, D. Kang, J. Liu, Y. Li, E. A. Undheim, J. K. Klint, M. Rong, R. Lai and G. F. King, *Proc. Natl. Acad. Sci. U.S.A.,* 2013, **110**, 17534–17539.
5. P. Escoubas, *Toxicon,* 2006, **47**, 609–613.
6. P. Escoubas, B. L. Sollod and G. F. King, *Toxicon,* 2006, **47**, 650–663.
7. P. Escoubas, L. Quinton and G. M. Nicholson, *J. Mass Spectrom.,* 2008, **43**, 279–295.
8. J. Davis, A. Jones and R. J. Lewis, *Peptides,* 2009, **30**, 1222–1227.
9. D. Biass, S. Dutertre, A. Gerbault, J. L. Menou, R. Offord, P. Favreau and R. Stocklin, *J. Proteomics,* 2009, **72**, 210–218.
10. T. Nilsson, M. Mann, R. Aebersold, J. R. Yates, 3rd, A. Bairoch and J. J. Bergeron, *Nat. Methods,* 2010, 7, 681–685.
11. B. M. Ueberheide, D. Fenyo, P. F. Alewood and B. T. Chait, *Proc. Natl. Acad. Sci. U.S.A.,* 2009, **106**, 6910–6915.
12. B. M. Olivera, G. P. Miljanich, J. Ramachandran and M. E. Adams, *Annu. Rev. Biochem.,* 1994, **63**, 823–867.
13. J. M. McIntosh, A. D. Santos and B. M. Olivera, *Annu. Rev. Biochem.,* 1999, **68**, 59–88.
14. J. M. McIntosh, B. M. Olivera and L. J. Cruz, *Methods Enzymol.,* 1999, **294**, 605–624.
15. L. D. Possani, B. Becerril, M. Delepierre and J. Tytgat, *Eur. J. Biochem.,* 1999, **264**, 287–300.
16. R. J. Lewis and M. L. Garcia, *Nat. Rev. Drug Discovery,* 2003, **2**, 790–802.
17. C. J. Armishaw and P. F. Alewood, *Curr. Protein Pept. Sci.,* 2005, **6**, 221–240.

18. A. M. Pimenta and M. E. De Lima, *J. Pept. Sci.*, 2005, **11**, 670–676.
19. G. Estrada, E. Villegas and G. Corzo, *Nat. Prod. Rep.*, 2007, **24**, 145–161.
20. P. Escoubas and F. Bosmans, *Expert Opin. Drug Discovery,* 2007, **2**, 823–835.
21. J. W. Fox and S. M. T. Serrano, *Curr. Pharm. Des.*, 2007, **13**, 2927–2934.
22. P. Escoubas and G. F. King, *Expert Rev. Proteomics,* 2009, **6**, 221–224.
23. S. Dutertre and R. J. Lewis, *J. Biol. Chem.*, 2010, **285**, 13315–13320.
24. N. J. Saez, S. Senff, J. E. Jensen, S. Y. Er, V. Herzig, L. D. Rash and G. F. King, *Toxins,* 2010, **2**, 2851–2871.
25. C. Y. Koh and R. M. Kini, *Toxicon,* 2012, **59**, 497–506.
26. B. L. Sollod, D. Wilson, O. Zhaxybayeva, J. P. Gogarten, R. Drinkwater and G. F. King, *Peptides,* 2005, **26**, 131–139.
27. Y.-H. Rogers and J. C. Venter, *Nature,* 2005, **437**, 326–327.
28. S. C. Schuster, *Nat. Methods,* 2008, **5**, 16–18.
29. A. J. Wilhelm, *New Biotechnol.*, 2009, **25**, 195–203.
30. T. P. Niedringhaus, D. Milanova, M. B. Kerby, M. P. Snyder and A. E. Barron, *Anal. Chem.*, 2011, **83**, 4327–4341.
31. D. H. W. Low, K. Sunagar, E. A. B. Undheim, S. A. Ali, A. C. Alagon, T. Ruder, T. N. W. Jackson, S. P. Gonzalez, G. F. King, A. Jones, A. Antunes and B. G. Fry, *J. Proteomics,* 2013, **89**, 95–111.
32. E. A. B. Undheim, A. Jones, J. W. Holland, K. Clauser, S. S. Pineda, G. F. King and B. G. Fry, *Mol. Biol. Evol.*, 2014, **31**, 2124–2148.
33. H. Hu, P. Bandyopadhyay, B. Olivera and M. Yandell, *BMC Genomics,* 2011, **12**, 60.
34. A. Violette, D. Biass, S. Dutertre, D. Koua, D. Piquemal, F. Pierrat, R. Stöcklin and P. Favreau, *J. Proteomics,* 2012, **75**, 5215–5225.
35. Y. Terrat, D. Biass, S. Dutertre, P. Favreau, M. Remm, R. Stöcklin, D. Piquemal and F. Ducancel, *Toxicon,* 2012, **59**, 34–46.
36. H. Hu, P. Bandyopadhyay, B. Olivera and M. Yandell, *BMC Genomics,* 2012, **13**, 284.
37. S. Dutertre, A.-h. Jin, Q. Kaas, A. Jones, P. F. Alewood and R. J. Lewis, *Mol. Cell. Proteomics,* 2013, **12**, 312–329.
38. T. Ruder, K. Sunagar, E. A. B. Undheim, S. A. Ali, T.-C. Wai, D. H. W. Low, T. N. W. Jackson, G. F. King, A. Antunes and B. G. Fry, *J. Mol. Evol.,* 2013, **76**, 192–204.
39. M. Rendón-Anaya, L. Delaye, L. D. Possani and A. Herrera-Estrella, *PLoS ONE,* 2012, 7, e43331.
40. D. R. Rokyta, K. P. Wray, A. R. Lemmon, E. M. Lemmon and S. B. Caudle, *Toxicon,* 2011, **57**, 657–671.
41. J. Durban, P. Juarez, Y. Angulo, B. Lomonte, M. Flores-Diaz, A. Alape-Giron, M. Sasa, L. Sanz, J. Gutierrez, J. Dopazo, A. Conesa and J. Calvete, *BMC Genomics,* 2011, **12**, 259.
42. J. Durban, A. Perez, L. Sanz, A. Gomez, F. Bonilla, S. Rodriguez, D. Chacon, M. Sasa, Y. Angulo, J. Gutierrez and J. Calvete, *BMC Genomics,* 2013, **14**, 234.

43. E. S. Wong, M. C. Hardy, D. Wood, T. Bailey and G. F. King, *PLoS One,* 2013, **8**, e66279.
44. E. A. Undheim, K. Sunagar, V. Herzig, L. Kely, D. H. Low, T. N. Jackson, A. Jones, N. Kurniawan, G. F. King, S. A. Ali, A. Antunes, T. Ruder and B. G. Fry, *Toxins,* 2013, **5**, 2488–2503.
45. E. S. W. Wong, D. Morgenstern, E. Mofiz, S. Gombert, K. M. Morris, P. Temple-Smith, M. B. Renfree, C. M. Whittington, G. F. King, W. C. Warren, A. T. Papenfuss and K. Belov, *Mol. Cell. Proteomics,* 2012, **11**, 1354–1364.
46. D. Rokyta, K. Wray and M. Margres, *BMC Genomics,* 2013, **14**, 394.
47. I. M. Francischetti, T. C. Assumpcao, D. Ma, Y. Li, E. C. Vicente, W. Uieda and J. M. Ribeiro, *J. Proteomics,* 2013, **82**, 288–319.
48. J. Eng, W. A. Kleinman, L. Singh, G. Singh and J. P. Raufman, *J. Biol. Chem.,* 1992, **267**, 7402–7405.
49. O. Castañeda, V. Sotolongo, A. M. Amor, R. Stocklin, A. J. Anderson, A. L. Harvey, A. Engstrom, C. Wernstedt and E. Karlsson, *Toxicon,* 1995, **33**, 603–613.
50. R. H. van den Heuvel, E. van Duijn, H. Mazon, S. A. Synowsky, K. Lorenzen, C. Versluis, S. J. Brouns, D. Langridge, J. van der Oost, J. Hoyes and A. J. Heck, *Anal. Chem.,* 2006, **78**, 7473–7483.
51. I. A. Kaltashov, C. E. Bobst and R. R. Abzalimov, *Protein Sci.,* 2013, **22**, 530–544.
52. R. M. Caprioli, T. B. Farmer and J. Gile, *Anal. Chem.,* 1997, **69**, 4751–4760.
53. E. A. Undheim, K. Sunagar, B. R. Hamilton, A. Jones, D. J. Venter, B. G. Fry and G. F. King, *J. Proteomics,* 2014, **102**, 1–10.
54. G. L. Glish and D. J. Burinsky, *J. Am. Soc. Mass Spectrom.,* 2008, **19**, 161–172.
55. N. Bandeira, K. R. Clauser and P. A. Pevzner, *Mol. Cell. Proteomics,* 2007, **6**, 1123–1134.
56. B. G. Fry, E. A. Undheim, S. A. Ali, T. N. Jackson, J. Debono, H. Scheib, T. Ruder, D. Morgenstern, L. Cadwallader, D. Whitehead, R. Nabuurs, L. van der Weerd, N. Vidal, K. Roelants, I. Hendrikx, S. P. Gonzalez, I. Koludarov, A. Jones, G. F. King, A. Antunes and K. Sunagar, *Mol. Cell. Proteomics,* 2013, **12**, 1881–1899.
57. J. V. Olsen, B. Macek, O. Lange, A. Makarov, S. Horning and M. Mann, *Nat. Methods,* 2007, **4**, 709–712.
58. I. Anusiewicz, M. Jasionowski, P. Skurski and J. Simons, *J. Phys. Chem. A,* 2005, **109**, 11332–11337.
59. M. S. Kim and A. Pandey, *Proteomics,* 2012, **12**, 530–542.
60. A. Guthals, K. R. Clauser, A. M. Frank and N. Bandeira, *J. Proteome Res.,* 2013, **12**, 2846–2857.
61. L. Fornelli, J. Parra, R. Hartmer, C. Stoermer, M. Lubeck and Y. O. Tsybin, *Anal. Bioanal. Chem.,* 2013, **405**, 8505–8514.
62. M. Takayama, *J. Am. Soc. Mass Spectrom.,* 2001, **12**, 1044–1049.
63. R. S. Brown and J. J. Lennon, *Anal. Chem.,* 1995, **67**, 3990–3999.

64. D. Asakawa, N. Smargiasso, L. Quinton and E. De Pauw, *J. Mass Spectrom.*, 2013, **48**, 352–360.
65. D. Asakawa and M. Takayama, *J. Am. Soc. Mass Spectrom.*, 2011, **22**, 1224–1233.
66. J. J. Calvete, L. Sanz, Y. Angulo, B. Lomonte and J. M. Gutierrez, *FEBS Lett.*, 2009, **583**, 1736–1743.
67. M. G. Grabherr, B. J. Haas, M. Yassour, J. Z. Levin, D. A. Thompson, I. Amit, X. Adiconis, L. Fan, R. Raychowdhury, Q. Zeng, Z. Chen, E. Mauceli, N. Hacohen, A. Gnirke, N. Rhind, F. di Palma, B. W. Birren, C. Nusbaum, K. Lindblad-Toh, N. Friedman and A. Regev, *Nat. Biotechnol.*, 2011, **29**, 644–652.
68. Y. Xie, G. Wu, J. Tang, R. Luo, J. Patterson, S. Liu, W. Huang, G. He, S. Gu, S. Li, X. Zhou, T. W. Lam, Y. Li, X. Xu, G. K. Wong and J. Wang, *Bioinformatics*, 2014, **30**, 1660–1666.
69. M. H. Schulz, D. R. Zerbino, M. Vingron and E. Birney, *Bioinformatics*, 2012, **28**, 1086–1092.
70. S. L. Rusch and D. A. Kendall, *Mol. Membr. Biol.*, 1995, **12**, 295–307.
71. G. Schatz and B. Dobberstein, *Science*, 1996, **271**, 1519–1526.
72. Q. Kaas, R. Yu, A. H. Jin, S. Dutertre and D. J. Craik, *Nucleic Acids Res.*, 2012, **40**, D325–D330.
73. M. A. Larkin, G. Blackshields, N. P. Brown, R. Chenna, P. A. McGettigan, H. McWilliam, F. Valentin, I. M. Wallace, A. Wilm, R. Lopez, J. D. Thompson, T. J. Gibson and D. G. Higgins, *Bioinformatics*, 2007, **23**, 2947–2948.
74. R. C. Edgar, *Nucleic Acids Res.*, 2004, **32**, 1792–1797.
75. V. Lavergne, S. Dutertre, A. H. Jin, R. J. Lewis, R. J. Taft and P. F. Alewood, *BMC Genomics*, 2013, **14**, 708.
76. S. R. Eddy, *Bioinformatics*, 1998, **14**, 755–763.
77. J. D. Bendtsen, H. Nielsen, G. von Heijne and S. Brunak, *J. Mol. Biol.*, 2004, **340**, 783–795.
78. G. F. King, *Aust. Biochem.*, 2013, **44**, 13–15.
79. D. W. Cushman and M. A. Ondetti, *Hypertension*, 1991, **17**, 589–592.

CHAPTER 4

Venoms-Based Drug Discovery: Bioassays, Electrophysiology, High-Throughput Screens and Target Identification

IRINA VETTER[a,b], WAYNE C. HODGSON[c], DAVID J. ADAMS[d], AND PETER McINTYRE*[d]

[a]School of Pharmacy, The University of Queensland, St Lucia 4072, Australia; [b]Institute for Molecular Bioscience, The University of Queensland, St Lucia 4072, Australia; [c]Department of Pharmacology, Monash University, Clayton 3168, Australia; [d]Health Innovations Research Institute and School of Medical Sciences, RMIT University, Bundoora 3083, Australia
*E-mail: peter.mcintyre@rmit.edu.au

4.1 Introduction

Venoms are complex mixtures of active components, ranging from simple organic molecules to polymeric molecules such as acylpolyamines and protein toxins. The effects of envenomation can be manifold, and understanding the role of specific components may be obscured by this complexity. Therefore, venom components must be fractionated, and possibly concentrated, before their activities can be properly characterised. Physical fractionation of venom components is a labour-intensive activity. Alternatively, protein components can be isolated by molecular cloning from the mRNA of the venom gland, which offers a sensitive method to isolate pure protein components, if suitable

RSC Drug Discovery Series No. 42
Venoms to Drugs: Venom as a Source for the Development of Human Therapeutics
Edited by Glenn F. King

Published by the Royal Society of Chemistry, www.rsc.org

methods of expression that faithfully recapitulate native synthesis can be identified. Neither approach is high-throughput, so identifying venom components takes considerable time and effort. Despite this restriction, convenient and useful screening technologies and assays are available to identify and characterise molecules with novel biological activities, once isolated. Such methods are used in modern drug discovery and involve automated procedures and information-rich readouts from cellular assay systems.

In general, the likely biological effects and molecular targets of venom components will be deduced from the effects of envenomation on whole animals or experimental exposure in tissue-based assay systems. In some cases, the molecular mechanism will remain a mystery. Initial methods for fractionating and characterising venoms often used *in vivo* assays on experimental animals or tissues isolated from them. Nowadays, fractionation procedures are more likely to be monitored by functional assays in cellular and cell-free systems, and are a fast and efficient way to identify the protein targets of venom toxins. Alternatively, physical methods can be used, including mass spectrometry, protein sequencing, nuclear magnetic resonance (NMR) and other biophysical approaches. The approach taken to screen venom components using cellular assays depends on the functional role of the venom components.

In this chapter, we will describe *in vivo* methods of assaying venoms and venom components, and then describe approaches that use more defined assay systems. Many venom components activate or inhibit ion channels or cellular signalling components. These components range from agonists or antagonists of ion channels (*e.g.* voltage-gated calcium, sodium and potassium channels) or signalling receptors (*e.g.* acetylcholine, histamine, 5-hydroxytryptamine and kinin receptors), to protein toxins and hydrolytic enzymes. As such, they are a rich source of potential drugs.

A range of assay technologies is in place to isolate and characterise venom components. The first step is to isolate a single component, either physically by purification or genetically by molecular cloning. The second step is to develop an assay with which to measure the activity of the component of interest. This assay step can be broadly divided into binding assays and functional assays, although molecular homology can provide clues to an unknown function of an uncharacterised molecule. Binding assays identify distinct molecular interactions, but do not tell us about the ability of the interaction to activate or inhibit the target protein's biological function, whereas functional assays tell us about agonism and antagonism. Consequently, functional assays are of more value in early stages of drug discovery, so we will concentrate on them in this review.

4.2 Isolation of Venom Components

Venom characterisation traditionally used electrophoresis or liquid chromatography of crude venom to separate and isolate individual venom components.[1] Components of interest are identified by functional or immunological assay and isolated by column chromatography and other techniques of classical biochemical purification. These methods are best suited to identifying proteinaceous components larger than 10 kDa.

Peptides and non-protein venom constituents are more difficult to isolate and identify, and are usually purified using liquid chromatography. A well-known example of this is the bradykinin potentiating peptides (BPP), which are isolated from snake venom and were precursors of the blockbuster antihypertensive drug captopril, an angiotensin-converting enzyme (ACE) inhibitor.[2]

Non-proteinaceous substances, such as lipid metabolites, are not well detected by the usual ultraviolet (UV) absorbance methods used to monitor chromatographic columns, so they are less well studied and characterised.

Another means of isolating and purifying venom proteins is by molecular cloning. The mRNA that encodes venom proteins and their precursors can be isolated from venom glands and efficiently made into representative cDNA libraries of all of the venom components. These components can be isolated as single clones encoding full-length proteins and used to synthesise the venom precursor proteins in bacterial or eukaryotic expression systems. Applying these proteomic, mass spectrometric and transcriptomic methods to help identify toxins from scorpion venom has been reviewed[3] (also see Chapter 3).

What is not so straightforward is recapitulating the wide range of post-translational modifications (PTMs) that can modify the original precursor peptide to produce functional toxins. PTMs can include disulfide bond formation (often in complex patterns), lipid and/or carbohydrate conjugation, proteolytic cleavage, C-terminal amidation, and hydroxylation of prolines. Toxins can also contain more exotic PTMs, such as L-to-D epimerisation of an amino acid, carboxylation of glutamate, hydroxylation of valine and lysine, bromination of tryptophan, cyclisation of an N-terminal glutamine to pyroglutamate, and sulfation of tyrosine.[4] The functional integrity of the mature proteins can depend entirely on making sure these modifications are made correctly.

Despite these caveats, many toxin proteins have been identified by molecular cloning, expressed and purified from such expression systems, and then shown to be active in functional assay (see below). In some cases, it is not possible to make toxins by recombinant DNA approaches and peptide synthesis is the method of choice for production. In particular, conotoxins are more amenable to peptide synthesis. A good example of this is ω-conotoxin MVIIA, a venom peptide from the marine cone snail *Conus magus*, which has been successfully synthesised at sufficient scale to enable its registration as the drug Prialt®, a treatment for refractory chronic pain (see ref. 5 and Chapter 12 for excellent reviews of the issues surrounding chemical synthesis of venom peptides).

4.3 Assay of Venom Components

There is no single assay with which to characterise all venom components, because there are many different functions and targets they may affect. Rather, there are several general approaches that may be adapted to screening for particular functional effects. A number of cellular expression systems are suitable for expressing isolated, cloned receptors, ion channels or enzyme targets so the effects of venom components can be closely examined. These include *Xenopus laevis* oocytes, yeast, baculovirus-infected insect cells and mammalian cell lines that either do not have differentiated characteristics (*e.g.* HEK293 cells) or do have differentiated characteristics of specialised cells, like the neuronal receptors in SH-SY5Y neuroblastoma cells. Many venom proteins are rich in disulfide bonds and other PTMs, making it difficult to accurately synthesise the mature protein so that it is identical to the native protein. The expression system or chemical synthetic scheme used to make the protein must be highly optimised to obtain structural fidelity. As an example, many conotoxins from marine snails have up to three disulfide bonds that must form in a complex pattern that is essential for correct biological function. Apart from the pattern of these bonds, these conotoxins share few distinct features.[6]

4.4 Biological Systems: Tissues, Primary and Immortalised Cells

Cell surface receptors and ion channels are important drug targets, since they are the means by which cells sense their environments and are 'sensory gateways to signalling' for cells. Therefore, developing assays suitable for characterising these targets is a major activity in the discovery of new pharmaceutical agents. Intracellular targets, such as enzymes, may also be targets and must be analysed with specific assays to measure their activity.

In addition to native tissues, venoms that target receptors and ion channels can be assayed in cell types that endogenously express the target of interest. For example, painful or analgesic venom components with an unknown mechanism of action can be screened in cellular assays using cultured primary sensory neurons from the dorsal root ganglion or trigeminal ganglion of rats or mice.[7] These cells are a heterogeneous population containing a subset of pain-sensing neurons. Alternatively, immortalised neuronal cell lines are a more homogeneous cell system that may express targets of interest. For example, the human neuroblastoma cell line SH-SY5Y expresses a range of human receptors and ion channels, such as voltage-gated sodium channels, and so have been used for screening compounds that act at these targets.[8] The problem with this approach is that the cells used may not express the target of interest or have the intracellular signalling mechanisms necessary for a response that resembles that seen *in vivo*. Another potential caveat is that such cells may express more than one potential target, which could confound the interpretation of results.

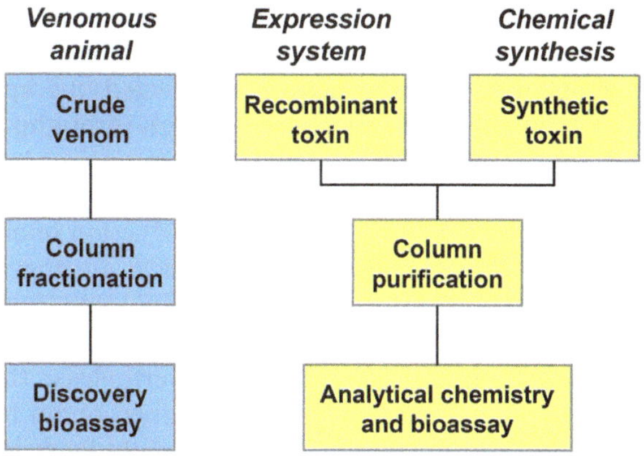

Figure 4.1 Approaches to venom component identification.

Assays that use cultured native cells from heterogeneous populations must be performed using a method that enables identification and analysis of cells with appropriate responses. Therefore, methods that sample single cells, such as microscopy or electrophysiology, must be used.

Another common approach is to develop assays using cloned ion channels expressed in mammalian cells, such as HEK293 or Chinese hamster ovary (CHO) cells, *Xenopus* oocytes, or other heterologous expression systems. These cells are then assayed using either electrophysiological techniques or fluorometric assays based on ion-selective or membrane potential-sensitive dyes (discussed further below). Figure 4.1 outlines a simple schema of approaches to isolating and identifying venom components.

4.5 Bioassay

A bioassay of venoms and venom components on animals or tissues is a powerful way to detect complex, integrated biological responses—something that is considered in detail later in this chapter—and analyse fractionated venom components. A bioassay is a rather specialised technique that is difficult to perform in parallel, so it is limited to examining relatively small sample numbers. Early *in vivo* studies of venom actions focused on determining the pharmacological and physiological effects of venoms and venom components after they were administered to animals. However, due to the difference in expression, function and structure of pharmacological targets between the natural prey of venomous animals and mammals, delineating pharmacological activity using *in vivo* bioassays has some pitfalls. For example, intraperitoneal injection of several peptides from *Conus geographus* elicited no overt behavioural or physiological changes in mice. However, when administered by intracerebroventricular injection, a shaking phenotype was observed, which led to the discovery that conotoxin GVIA is a highly selective $Ca_V2.2$ antagonist.[9–12] Despite the significant insight *in vivo* or *ex vivo* bioassays can provide, these approaches also suffer from ethical limitations and are poorly suited to high-throughput identification of venom components with specific action at selected therapeutic targets. Therefore, to accelerate drug discovery efforts, high-throughput approaches are increasingly being applied to isolate bioactive components from venoms.

Historically, *in vitro* and *in vivo* pharmacological techniques have played a vital role in the delineation of the mechanism of action of a wide range of animal venoms and toxins. However, technological advances have meant that more rapid and sensitive approaches to 'screen' toxins with potential to be drug leads are now available. Consequently, traditional techniques in drug discovery have become less popular. Nevertheless, the amount of information that can be gained from a well-designed organ bath or whole-animal experiment should not be underestimated. Indeed, in many cases, 'proof of function' cannot be determined or confirmed without such experiments. Using electrically stimulated preparations enables the examination of venoms/toxins with processes crucial for transmitter release, as well as binding of the transmitter at the effector tissue. Non-electrically stimulated preparations give valuable insight into the ability of venoms/toxins to contract or relax a variety of tissues that are key targets of important toxin classes. These include skeletal muscle, vascular smooth muscle, cardiac muscle and gastrointestinal smooth muscle.

Interpretation of data from *in vitro* and *in vivo* experiments is facilitated by:

- well-defined receptor populations and knowledge of the associated second messenger pathways
- availability and use of appropriate selective receptor antagonists/inhibitors
- inclusion of 'positive' and 'negative' controls to confirm the efficacy and selectivity of receptor antagonists/inhibitors.

4.5.1 Skeletal Muscle Preparations

Many animal venoms contain potent neurotoxins that have evolved to inhibit nerve–muscle transmission at the skeletal neuromuscular junction. This aids in the capture of prey by paralysing skeletal muscles involved in respiration and movement. Isolated skeletal muscle preparations have been instrumental in providing insight into the mode of action of these important venom components and examining the efficacy of antivenoms against these toxins. In particular, rat (or mouse) phrenic-nerve diaphragm,[13] toad rectus abdominis and chick biventer cervicis nerve–muscle[14] preparations have been highly valuable tools used in the pharmacological characterisation of snake venom neurotoxins.

The main difference between these preparations is the nature of the innervation of the skeletal muscle fibres. The avian preparation contains focally and multiply-innervated muscle fibres, while the mammalian and amphibian preparations contain only focally-innervated and multiply-innervated muscle fibres, respectively. Electrical stimulation of focally-innervated muscle fibres induces a transient twitch response due to the coordinated release of acetylcholine from the nerve terminal. Multiply-innervated fibres mediate a more prolonged contracture in response to exogenous nicotinic acetylcholine receptor (nAChR) agonists, such as carbachol, acetylcholine or suxamethonium.[15] In both mammalian and avian preparations, the competitive nAChR antagonist tubocurarine abolishes the effects of indirect twitches, confirming motor nerve stimulation. Importantly, the presence of both types of muscle fibres in the avian preparation enables identification of the site of action of neurotoxins as either post- or pre-synaptic. However, when testing crude venoms, the presence of presynaptically acting toxins is difficult to confirm because the more rapidly acting post-synaptic neurotoxins mask their activity.

The potency of venom neurotoxins can be compared using either mammalian or avian preparations by determining the time taken to inhibit indirect twitches by a defined amount. Historically, 50% and 90% inhibition have been regularly used as convenient ‘end points’ to determine t_{50} or t_{90} values, respectively. The magnitude of these values depend on the kinetics of toxin–receptor binding and, as such, often result in a rank order of ‘potency’ for venoms/toxins that differs from that obtained by murine LD_{50} studies. In contrast, murine LD_{50} studies indicate a concentration of venom/toxin lethal to 50% of the treated animals, but do not give any information on the time course of action of the venom/toxin apart from the fact that the animal is dead within the defined period of the study. LD_{50} studies are normally used for studying whole venoms. As the ‘lethality’ of venoms is likely to be due to the presence of multiple toxin types, this test provides little or no information on the level of toxicity associated with each toxin type in the venom.

Schild plot analysis can be used to determine the pA_2 value of an antagonist. This method depends on the competitive nature of the interaction between receptor and ligand. However, a depression of the maximum tissue response often occurs as a result of the pseudo-irreversible antagonism displayed by most of these neurotoxins. Hence, the pA_2 value of these toxins cannot be calculated using Schild plot analysis. The modified Lew–Angus method offers an alternative method for obtaining a reasonable estimate of the pA_2 value.[16,17] The pseudo-irreversible nature of many of these toxins highlights the necessity to exercise caution when referring to them as ‘curare-mimetic toxins’.[18]

The inhibitory effects of post-synaptic toxins (*e.g.* α-neurotoxins) on the electrically evoked twitches of a skeletal muscle preparation are normally characterised by a gradual decline in twitch height that starts shortly after the toxin/venom is added. In contrast, a lag period is often observed in *in vitro* skeletal muscle experiments after a presynaptic neurotoxin (*e.g.* β-neurotoxin) is added, before a characteristic triphasic effect on twitch height. *In vitro*, the rate of neurotoxicity onset depends on the bathing medium temperature and nerve stimulation frequency. Lowering both parameters delays neurotoxicity onset.[19] The triphasic effect is characterised by an initial decrease in twitch height followed by a transient increase in twitch height, then complete twitch inhibition. However, this effect is often only unmasked in a low Ca^{2+} or high Mg^{2+} bathing solution.[20] It has been reported that different species have different nerve–muscle preparation sensitivity to presynaptic neurotoxins.[21,22] The mouse phrenic-nerve diaphragm is more sensitive to paradoxin, taipoxin, textilotoxin and notexin than the chick biventer cervicis nerve-muscle preparation.[23] However, the chick biventer cervicis is more sensitive to crotoxin and β-bungarotoxin than the mouse phrenic nerve preparation.[22]

Skeletal muscle preparations were also instrumental in the discovery of dendrotoxins, which are widely used to examine K^+ channel function.[24] Unlike the neurotoxins described above, dendrotoxins potentiate indirect twitches in these preparations, an activity that was discovered when studying the effects of venom from the Eastern green mamba snake (*Dendroaspis angusticeps*).[25,26]

4.5.2 Smooth Muscle Preparations

4.5.2.1 Guinea Pig Ileum/Rabbit Jejunum

Guinea pig ileum and rabbit jejunum are gastrointestinal smooth muscle preparations with a broad range of receptors that aid the study of venoms/toxins. These include histaminergic, serotonergic and cholinergic receptors. The guinea pig preparation is quiescent, unless electrically stimulated. In contrast, the rabbit jejunum displays regular rhythmic spontaneous contractions. Both preparations contract to a range of autacoids/neurotransmitters, or compounds that activate these receptors. Identification of BPPs and development of therapeutically useful ACE inhibitors[27] owes much to the observation that contractile responses to bradykinin were potentiated in the presence of venom from the South American pit viper *Bothrops jararaca*.[2,28] BPPs have been identified in a range of snake venoms[29,30] and scorpion venoms,[31,32] using gastrointestinal smooth muscle as a bioassay.

4.5.2.2 Vas Deferens and Prostate

The vas deferens is an excellent preparation for *in vitro* examination of the interaction between venoms/toxins and the sympathetic nervous system. Contractions induced by electrical stimulation of the sympathetic nerve fibres are mediated by noradrenaline and ATP release. ATP is responsible for the rapid phase of the contraction, while noradrenaline produces a slower contractile effect.[33–35]

Cone snail (*Conus* spp.) venom peptides have been shown to act at a diverse range of molecular targets, and they have been investigated for their therapeutic potential for a variety of diseases and chronic conditions.[36]

Experiments using the rat vas deferens were instrumental in identifying the activities of ρ-TIA from *C. tulipa* venom and χ-Mr1A from *C. marmoreus* venom. Interestingly, these peptides had opposing effects on the second phase (*i.e.* the noradrenaline-mediated component) of the contractile response to electrical stimulation of the prostatic segment of the vas deferens. ρ-TIA caused a concentration-dependent inhibition of the contractile response, whereas χ-Mr1A caused a concentration-dependent increase of the same phase.[37] This led to the discovery of cone snail peptides that are selective, non-competitive inhibitors of the noradrenaline transporter (*i.e.* χ-Mr1A) and α_1-adrenoceptor (*i.e.* ρ-TIA).

Recently, venom from the Eastern green mamba was shown to contain two novel peptides (ρ-Da1a and ρ-Da1b) that have similar sequences to muscarinic toxins and act against adrenoceptors.[38] ρ-Da1a concentration-dependently reduced phenylephrine-induced rabbit isolated prostatic tissue tone and was identified as a non-competitive antagonist of the α_{1a}-adrenoceptors. The authors proposed that ρ-Da1a could be used to treat benign prostatic hyperplasia given its selectivity (*i.e.* uroselective), lack of side effects compared with currently available drugs and mode of action.[38] The cardiovascular effects of the novel compound were examined in the anaesthetised rat preparation (see below).

4.5.3 Isolated Blood Vessels

Isolated blood vessels (*e.g.* rat aorta and mesenteric artery) are excellent preparations for examining venoms/toxins that contract or relax vascular smooth muscle, and the interaction of venoms/toxins with adrenoceptors. These vascular effects can be caused by a direct action on the underlying smooth muscle, involving the release of contracting or relaxing factors from the endothelial cells that line the lumen of these vessels or a combination of both actions. When required, the endothelium can be denuded and the presence or absence of endothelial cells confirmed by the ability of acetylcholine, or another endothelium-dependent vasodilator, to induce relaxation of pre-contracted tissues.

This technique has been used to identify natriuretic peptides in the venoms of a number of animals, including the Eastern green mamba,[39] the inland taipan (*Oxyuranus microlepidotus*[40]), the platypus (*Ornithorhynchus anatinus*[41]) and, interestingly, the lace monitor (*Varanus varius*), a venomous lizard.[42] The potential for these peptides to be used as drug leads continues to be explored.[27,43]

The effects of cone snail toxins have also been widely examined in a variety of blood vessel preparations. Wang *et al.* examined the ability of synthetic ω-conotoxin analogues to prevent electrically evoked noradrenaline release of superfused segments of rat tail artery to determine their selectivity for $Ca_V2.2$ voltage-gated calcium channels.[44] The pharmacology of ω-conotoxins, and many other cone snail peptides, is now well established.[45]

4.5.4 Anaesthetised Whole Animals

The use of whole animals enables the effects of venoms/toxins on a number of parameters to be simultaneously determined under more physiological conditions than using *in vitro* techniques. The rat is the most common species used for these experiments, although historically other species, including dogs, rabbits, cats and pigs, have been used. Depending on the recording apparatus used and cannulae placement (*e.g.* carotid artery, femoral artery, jugular vein or trachea), the researcher can monitor systemic blood pressure, heart rate and respiratory rate, and obtain an electrocardiogram recording. Blood and urine samples can also be collected for further biochemical analysis. However, identifying baroreceptor 'reflex' effects or their contribution to venom/toxin activity is difficult to assess, because these may be blunted by anaesthesia.

The activity of the natriuretic peptide PNP isolated from the venom of the Iranian viper (*Pseudocerastes persicus*) was confirmed in the anaesthetised rat preparation by measuring the peptide's effects on blood pressure, diuresis and sodium excretion.[46] To help identify drugs with the potential to lower high blood pressure, spontaneously hypertensive rats (SHR) can be used instead of normotensive animals. This approach has been used successfully to examine the hypotensive effects of BPPs.[47,48] Venomous marine animals (*e.g.* stonefish, lionfish, soldierfish and jellyfish), which often have profound effects on the blood pressure of envenomed humans, have also been extensively examined in the anaesthetised rat preparation.[49–53]

4.6 Isolated Cell Studies of Venom Components

Observations using whole animal or organ studies highlighted that toxins and venom components can profoundly affect cell excitability, including resting membrane potential, membrane receptors and ionic channels, and second messenger signalling. Consequently, studies began to assess venom and venom component effects at the single cell level. The electrophysiological techniques developed by Cole[54] and Marmont,[55] and refined by Hodgkin and Huxley,[56] rapidly found application in the study of the effects of venom on the giant squid axon[57,58] and neurons from other animal species soon after.[59–61] These electrophysiological recording techniques provide real-time information on neuronal transmission rather than measuring static end points. Therefore, they are a true continuous bioassay of signalling events.

Accordingly, electrophysiological studies have given significant insight into the pharmacological mechanisms of venoms and venom components, and enabled detailed characterisation of the effects of venoms on cell membrane potential, action potential threshold, ionic conductance and ion channel gating.

4.6.1 Electrophysiological Recording Techniques

The electrical activity of cells can be monitored through either an extracellular recording electrode placed outside the cell membrane (extracellular recording) or a sharp glass microelectrode that penetrates the cell membrane (intracellular recoding).

Extracellular recordings provide information on the electrical activity of cells and changes in rates of activity. Single-unit recording through extracellular glass electrodes measures the potentials generated by extracellular current and can be used to monitor the discharge activity, but not subthreshold events, of a single neuron.

Intracellular recordings enable measurement of, and control over, the transmembrane voltage or membrane potential. Membrane potentials can be measured either under current-clamp conditions, where changes in membrane potential and input resistance are measured, or voltage-clamp conditions, where the current flowing through the intracellular electrode is measured when the membrane potential is held at a fixed value. Accordingly, intracellular recording techniques have become the methods of choice for assessing the effects of toxins and venom components on membrane potential and conductances. In particular, data from intracellular recordings can give valuable information about the mechanism of action of toxins on excitable cells. For example, recordings under current-clamp conditions showed that Pacific ciguatoxin-1 (P-CTX-1) depolarised the membrane of rat sensory neurons and induced spontaneous action potential firing (Figure 4.2). Figure 4.3 shows representative data for the effect of P-CTX-1 on the voltage-dependence of activation of tetrodotoxin (TTX)-sensitive (Figure 4.3A) and TTX-resistant voltage-gated sodium channels recorded under voltage-clamp conditions (Figure 4.3B).[62]

4.6.1.1 Intracellular Microelectrode Recordings

Early studies using intracellular recordings were performed with small wire electrodes that impaled the relatively large squid axon.[54–56] The advent of pulled-glass microelectrodes[63] enabled the voltage-clamp technique to be applied to smaller cells. These sharp intracellular electrodes are typically thin glass pipettes, filled with an electrolyte solution such as KCl, which can penetrate cell membranes. The fine tip of these intracellular recording microelectrodes leads to relatively higher resistance and generally a higher level of noise and non-linearities.[64] In addition, the seal between sharp intracellular electrodes and the membrane is generally poor, giving rise to leak currents and recording artefacts.[65] However, sharp intracellular microelectrodes allow recording with minimal effects on the ionic concentrations of the cytoplasm and allow signalling to be assessed while the cell remains in its normal relationship with neighbouring cells, without the need for enzymatic dissociation.

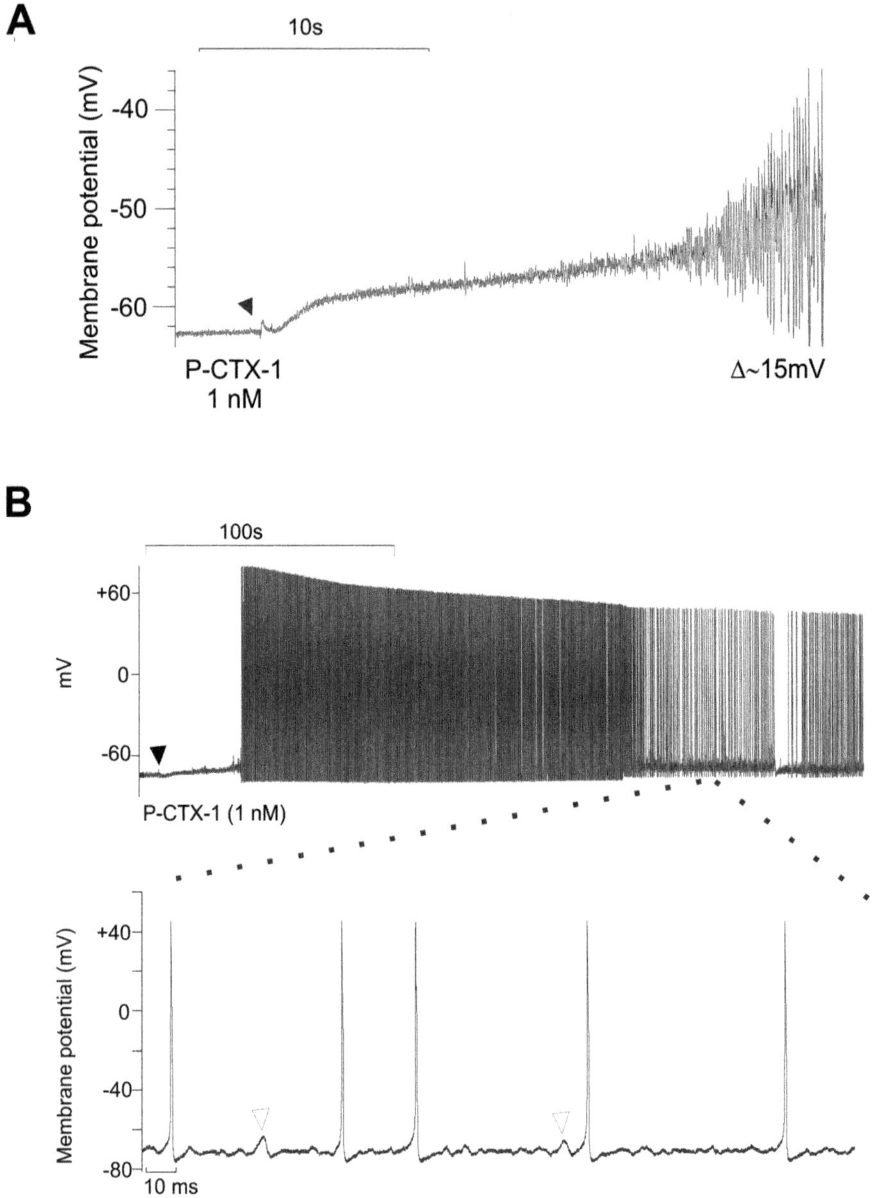

Figure 4.2 Effect of P-CTX-1 on resting membrane potential and action potential firing in DRG neurons. (A) Application of 1 nM P-CTX-1 (black arrow) caused depolarisation of the membrane potential recorded in current-clamp mode by ~10 mV on average. (B) P-CTX-1 causes spontaneous action potential firing in cultured DRG neurons. Upper panel: membrane depolarisation induced by P-CTX-1 (1 nM) rapidly leads to a series of action potentials. Detail expanded in lower panel: P-CTX-1 (1 nM) induced membrane oscillations, which were frequently followed by action potentials. Figure adapted from Vetter *et al.*[62]

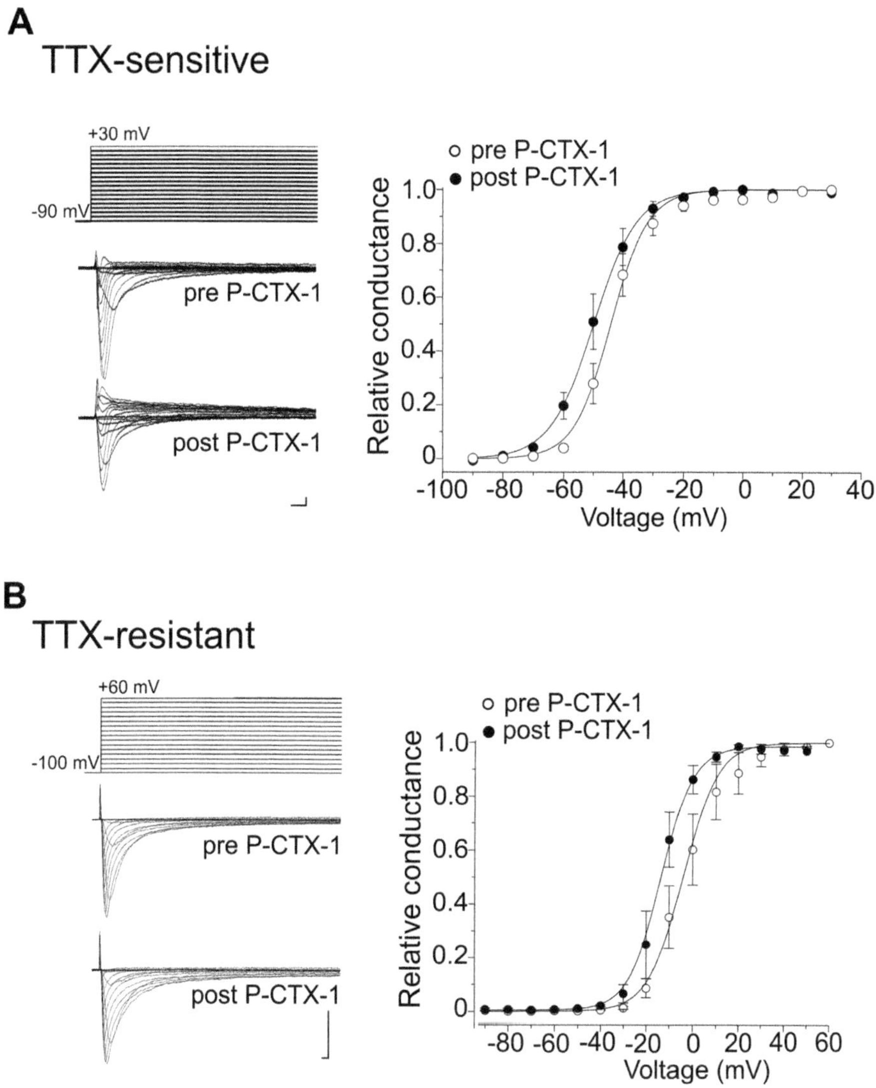

Figure 4.3 Effect of P-CTX-1 on the voltage-conduction relationship of TTX-sensitive and TTX-resistant Na_V currents. (A) TTX-sensitive: representative current traces for TTX-sensitive Na_V currents recorded from large-diameter (42.9 ± 1.4 μm) mouse DRG neurons. Upper panel: voltage protocol. Lower panels: current traces before (pre) and after (post) perfusion with 1 nM P-CTX-1. P-CTX-1 shifted the voltage-dependence of activation of TTX-sensitive Na_V channels to more negative potentials and induced significant channel activation at −60 mV. (B) TTX-resistant: representative recording of current traces recorded from ND7/23 cells heterologously expressing $Na_v1.8$. Upper panel: voltage protocol. Lower panels: current traces before (pre) and after (post) perfusion with 1 nM P-CTX-1. P-CTX-1 also elicited a 10 mV hyperpolarising shift of the voltage dependence of activation of $Na_v1.8$. Scale bars represent 1 ms and 1 nA. Figure adapted from Vetter *et al.*[62]

Two-electrode voltage clamping remains essential for recording from large cells, such as *Xenopus* oocytes. This technique has been used for studying the effects of toxins and venom peptides on a range of targets, including nAChRs; voltage-gated sodium (Na_V), calcium (Ca_V) and potassium (K_v) channel subtypes; acid-sensing ion channels (ASIC); and transient receptor potential (TRP) channels.[66–71]

4.6.1.2 Microelectrode Current-Clamp Technique

The single-microelectrode bridge technique has been used for the electrophysiological investigation of many neuronal preparations, primarily because most mammalian neurons are too small and inaccessible to allow penetration by two independent, membrane potential recording and current injection microelectrodes. Engel *et al.*[72] presented a theoretical treatment analysing the possible errors in resistance measurements introduced by the use of single microelectrode bridge technique. A major problem is the spatial inhomogeneity of the injected current and membrane potential in the neighbourhood of the impalement. Although Jack *et al.*[73] found only approximately 0.2% of the membrane area is subject to this error, even this small fraction led to an incorrect adjustment of the bridge circuit and consequential errors in measurements of membrane resistance and threshold levels for action potential generation.

Recordings from small cells require high-resistance glass microelectrodes. These electrodes are prone to large changes in resistance when passing large currents, and the bridge circuit needs to be carefully and frequently balanced. The discontinuous current clamp rapidly switches between membrane potential sampling and current passing modes to alleviate this problem.[74] Accurate measurement of membrane properties is possible with the discontinuous current clamp, provided that the membrane time constant is longer than the microelectrode time constant.

Intracellular microelectrode recording allows selective labelling of individual neurons from which electrical recordings have been made. The injection of intracellular dye (*e.g.* Lucifer Yellow, Cascade Blue) or enzyme (*e.g.* horseradish peroxidase) is particularly useful to probe structure–function relationships, such as neuronal morphology or cell–cell coupling of functionally characterised neurons. For example, this combined approach has been used to examine the correlation between the electrophysiological properties and neuropeptide content of mammalian autonomic neurons.[75,76]

4.6.1.3 Microelectrode Voltage-Clamp Technique

The voltage-clamp technique, introduced to determine the ionic conductances underlying the action potential in squid axon,[77] is a powerful tool for the investigation of electric current flow across cell membranes. It directly measures membrane current, which in turn allows membrane ionic conductances and permeabilities to be determined. Voltage clamping of neurons *via* intracellular microelectrodes has to overcome the problem of spatial inhomogeneity inherent in a non-spherical shape.[78] Neurons with an extensive dendritic arborisation are unlikely to be 'space' clamped, which may lead to quantitative errors and even propagated action potentials. Responses evoked adjacent to the location of the microelectrodes can, in principle, be adequately space clamped. Few studies, however, have directly tested this supposition.

Until the 1980s, voltage clamping was achieved using two microelectrodes: a membrane-potential-sensing electrode and a current-passing microelectrode. This limited the technique's application in small (<20 μm) mammalian central and peripheral neurons. The problems of access and cell size were reduced with the design of the single-electrode voltage clamp.[79] The principles of operation and application of this variant of the voltage-clamp technique have been described in detail.[80]

4.6.1.4 Patch-Clamp Technique

Developed in 1976 by Neher and Sakman[81] and further refined in 1980 by Sigworth and Neher,[82] the patch-clamp technique was designed to use a single electrode rather than two intracellular electrodes to voltage clamp mammalian cells. In this technique, the tip of a fire-polished glass micropipette (tip diameter <3 μm) is pressed against a 'naked' cell membrane. Suction applied to the pipette helps form a high resistance seal (typically >1 GΩ) between the pipette rim and membrane, which electrically isolates a cell-attached patch and permits currents flowing through single ionic channels to be recorded.[83] The high resistance 'gigaseal' permits voltage clamping with the same electrode and low noise recordings.[65] Single-channel currents as small as 0.5 pA can be resolved in the membrane under the electrode.

There are several procedures that permit recording from whole-cell or cell-free membrane patches after a gigaseal seal is formed.[84] The whole-cell recording configuration, in which the cell membrane is ruptured to allow access to the intracellular compartment, enables intracellular recordings in a similar manner to conventional sharp microelectrodes, except with the advantage of lower access resistance. This approach can be used to investigate the macroscopic response of the functional complement of ion channels in a cell. Most of the high-throughput electrophysiology platforms described later in this chapter use a version of this approach, the planar patch configuration, so recordings can be made in high throughput by automating whole-cell configuration on chips with a micro-structured aperture or opening. In these instruments, the cell is positioned over the opening using suction and the whole-cell configuration is achieved by applying further suction/voltage protocols.

The whole-cell recording configuration is of particular value in investigating the properties of very small cells (<20 μm diameter), which are not readily accessible to conventional microelectrode techniques. It allows rapid temporal and improved spatial voltage-clamp control of the cell and is better suited to the study of fast voltage-gated Na^+ and K^+ conductances in excitable cells than microelectrode techniques. The whole-cell configuration has been applied to electrophysiological studies of mammalian neurons; however, the need for a 'clean' membrane has restricted studies to acutely dissociated or cultured neurons. Sources of errors in patch-clamp measurements can arise from uncompensated liquid junction potentials and, in small cells whose input resistance (R_{in}) is in the order of the seal resistance, a significant fraction of the recorded current can actually flow through the seal resistance. Procedures for estimating junction potentials and analysing the effects of series and seal resistances on membrane current and reversal potential measurements have been described.[85,86]

A disadvantage of patch clamping is its technically challenging nature and the high level of expertise required to use it. The relatively wide tip opening can also lead to the pipette electrolyte solution diffusing into the cell, causing cell dialysis. Consequently, cellular mechanisms that require a diffusible intracellular second messenger or endogenous Ca^{2+} buffering mechanisms are impaired.[87]

To preserve intracellular regulatory systems and reduce 'rundown' or loss of membrane currents through cell dialysis, the perforated patch whole-cell recording configuration is an alternative to obtain electrical access to the cell.[88] Here, the membrane is perforated chemically to give electrical access, but restrict dialysis. Patch pipette tips are briefly dipped into the normal pipette solution and then back filled with a solution that also contains either a polyene antibiotic, nystatin or amphotericin-B. After the gigaseal forms between the recording pipette and cell membrane, a low-resistance pathway develops as the antibiotic diffuses into the cell membrane and forms pores that are selective for small monovalent ions. This typically takes 5–10 min and is indicated by the appearance of a slow capacitative transient and decrease in series resistance (R_s). However, R_s is usually three times more than that achieved by rupturing the membrane, which limits the method to currents that are neither too large in size nor too fast in kinetics. Due to incomplete access to the intracellular compartment, the perforated patch technique suffers from increased access resistance and recording noise. Perforated patch recordings of voltage-dependent Ca^{2+} currents and Ca^{2+}-dependent currents are routinely stable for >1 h, and the method can be combined with simultaneous measurements of intracellular [Ca^{2+}] through membrane-permeant, fluorescent indicator dyes.

4.6.1.5 *Single Cell Electrophysiological Recordings to Assess the Effect of Toxins on Excitable Cells*

Many venoms have evolved to target membrane receptors (*e.g.* nAChRs and G-protein-coupled receptors [GPCRs]), ion channels (*e.g.* voltage-gated ion channels such as Na_V, Ca_V and K_v), transporters (*e.g.* noradrenaline transporter) and enzymes (*e.g.* neuronal nitric oxide synthase) involved in cellular signalling and neuronal excitability.[22,37,89–91] Electrophysiological recordings have been instrumental in determining the varied pharmacological effects of toxins and venom components on cellular function, and have provided fundamental insight into ion channel structure and function. For example, nAChRs were isolated by virtue of high affinity binding of α-bungarotoxin from snake venom,[92–94] whereas the pore region of the K^+ channel was first defined through electrophysiological characterisation of mutants with altered binding of charybdotoxin from scorpion venom.[92,95] Similarly, binding of *Tityus* gamma toxin and TTX to Na_V channels revealed a TTX-binding site associated with the selectivity filter, and a toxin-binding site associated with the gating component of the Na_V channel.[96]

In recent years, electrophysiological studies of toxin-mediated effects have provided insight into the pharmacology of other membrane receptors and ion channels, such as the ryanodine receptor and TRP vanilloid 1 channel (TRPV1).[71,97,98] Vanillotoxins VaTx1, 2 and 3, and the double knot toxin DkTx from the venom of the Trinidad chevron tarantula (*Psalmopoeus cambridgei*) and Chinese bird spider (*Haplopelma schmidti*) have been reported to interact with TRPV1 channels.[71,97] Interestingly, DkTx caused persistent TRPV1 activation by trapping the ion channel in the open state *via* association with the pore-region of the channel. This suggests the pore-forming domain has a critical role in TRP channel gating.[71,97]

In combination with molecular biology approaches that allow specific ion channel isoform expression, electrophysiological recording techniques have also provided unique insight into the mechanism of action of toxins and the at times exquisite subtype selectivity of venom-derived ion channel modulators. For example, the ribbon form (Cys^1–Cys^4, Cys^2–Cys^3 disulfide framework) of α-conotoxin AuIB showed stoichiometry-dependent preference for α3β4 nAChR inhibition, with inhibition abolished when the α3 and β4 subunits were expressed in a ratio of 1 : 10.[99]

Electrophysiological recordings from cells heterologously expressing Na_V, Ca_V and Kv channel isoforms have also confirmed the high selectivity of venom peptides and toxins for specific ion channel subtypes. For example, ProTx-II from the venom of the green velvet tarantula (*Thrixopelma pruriens*) has attracted attention due to its potency at $Na_v1.7$, a subtype of the Na_V channel that is expressed preferentially in pain-sensing primary sensory neurons and has become an attractive analgesic target because of its loss-of-function mutations that confer congenital insensitivity to pain.[100,101] Similarly, the scorpion venom peptide margatoxin was recognised early as a selective $K_v1.3$ channel inhibitor, and ω-conotoxins, such as CVID, are highly selective for $Ca_v2.2$ channels over other Ca_V channel subtypes.[102,103]

4.6.2 High-Throughput Electrophysiology Assays

Ion channels permeable to calcium, sodium and potassium ions are important targets for many venom-derived compounds due to their prominent role in essential physiological processes such as neuronal signalling and muscle contractility. Given the crucial functional roles of ion channels, it is not surprising that many venomous animals have evolved peptides that target ion channels with remarkable selectivity and potency. Such venom peptides have contributed significantly to our understanding of the structure, function and pharmacology of a range of ion channels.

Electrophysiological measurement of potential difference or current flow across biological membranes is the benchmark method for studying ion channel characteristics. However, these methods need an operator with a high level of technical expertise, who studies one cell at a time. They do not permit substantial throughput because it is not possible to test many samples simultaneously.

The two approaches that are most commonly used for heterologous expression of cloned ion channels involve either gene transfection into small cells grown in tissue culture or injection of genetic material (cDNA or cRNA) into larger cells. The standard large cells used for cloned cDNA or RNA expression are the oocytes of the African frog *Xenopus laevis*.

The problem of manual operation and only recording from one cell at a time has been somewhat lessened by parallel screening approaches using RNA expression in *X. laevis* oocytes or by using mammalian cells that stably or transiently express the ion channel of interest. These developments have also greatly enhanced the usefulness of these methods. In addition, commercially available automated electrophysiology platforms, such as the OpusXpress (Axon Instruments) and Robocyte/HiClamp (MultiChannel Systems GmbH) for oocyte recording, and PatchXpress (Axon Instruments), Patchliner (Nanion Technologies GmbH), IonWorks Barracuda and Quattro (Molecular Devices), QPatch (Sophion) and Flyscreen 8500 (Flyion GmbH) for isolated cell recording, have increased throughput.[104–107] However, most systems still cannot rival plate-based, high-throughput screening assays. The advantages of these automated electrophysiology platforms, which can assay eight, 16 or more cells in parallel, have often come at the expense of assay flexibility and significant cost. In addition, the need for robust, high-quality cells with stable expression of the ion channel of interest and ideal patch-clamping characteristics (such as membrane seal and stability) remains.[105]

4.6.2.1 Automated Electrophysiology Platforms

The OpusXpress (Axon Instruments) was the first system developed to make high-throughput electrophysiology recordings to aid study of membrane receptors and ion channels expressed in *Xenopus* oocytes. The OpusXpress workstation is a two-electrode voltage-clamp system in which voltage control, data acquisition, fluid delivery and real-time analysis are automated. These functions are performed in parallel, enabling a single operator to simultaneously record from up to eight oocytes. Use of the OpusXpress for high-throughput screening of toxins and drugs has been described in detail.[108]

The IonWorks Barracuda and IonWorks Quattro (Molecular Devices) are automated planar-array electrophysiology systems based on 384-well Patch-Plates.[105,109] Both of these platforms are capable of population patch clamp, which can help mitigate data consistency problems that arise as a result of between-cell variability in the ion channel expression level, seal formation and cell viability.[110–112] However, as a result of their high-throughput capabilities, these systems suffer from decreased assay flexibility with respect to the addition protocols of test compounds and their ability to accurately control voltage and compensate for technical artefacts.[105,109]

Compared with the IonWorks Barracuda and Quattro, the PatchXpress 7000A and Patchliner have only moderate throughput, using a $SealChip_{16}$ and NPC-16 planar electrode, respectively, to establish gigaohm seals. These platforms feature rapid, automated fluid exchange, enabling parallel recordings from ligand-gated and voltage-gated ion channels with sophisticated control over voltage and pulse protocols.[109,112–119] Therefore, the PatchXpress and Patchliner are well suited for detailed pharmacological and biophysical characterisation of venom components that are active at ion channels.[120]

The QPatch series (Sophion) is based on multi-channel planar electrodes that enable recordings from 8–48 cells in parallel.[112,121–124] The latest member of the family, the Qube, significantly increases throughput, with 384 parallel measurement sites. Like the IonWorks Barracuda and Quattro, both single- and multi-hole (population) recordings are possible with the Qube. It also features gigaohm seals and sophisticated integrated fluidics that enable rapid fluid exchange, and produces high-quality recordings from voltage-gated and ligand-gated ion channels.[109] In addition, a glass-coated microfluidics channel has been incorporated into the Qube, reducing the inadvertent loss of compounds, and IC_{50} values obtained using this technology are generally in good agreement with literature values obtained using conventional electrophysiology.[121,125–127]

The Flyscreen 8500 (FlyIon) uses 'flip-the-tip' technology, an adaption of traditional electrophysiology in which suction pulses applied to a cell inside a glass pipette result in true whole-cell configuration, gigaohm seals and resultant high-quality electrophysiology data.[105,128] This platform is fully automated and requires comparatively low volumes of compound. However, with only six electrodes, throughput is low,[105] making the FlyScreen particularly suited to pharmacological characterisation of venom components in limited supply.

4.6.2.2 Challenges with Electrophysiology Assays

Electrophysiological assays of venoms and venom components remain crucial for detailed characterisation of the pharmacology and mechanism of action of venom peptides, as they undoubtedly provide the highest quality data and highest information content relating to ion channel function. Together with high temporal resolution, these assays can provide information on the state-dependency of ion channel modulation, which is increasingly recognised as important.[105,109,112] However, these assays are not usually well suited for the high-throughput primary identification and isolation of novel ion channel modulators from venoms.

The introduction of automated electrophysiology platforms has provided significant advances, enabling detailed electrophysiological characterisation in moderate or high-throughput format. However, major challenges relating to ion channel expression levels and cell quality remain. Notably, cell confluence, passage number, cell background, harvesting techniques and general cell handling can influence the seal formation, patch success and data quality.[105,109] Overall, relatively modest improvements in throughput have often been achieved at the expense of assay flexibility, data quality and information content. Therefore, the aim of developing high-throughput electrophysiology assays suitable for large-scale venom and venom component screening has not yet been realised. Nonetheless, electrophysiological techniques remain indispensable for detailed pharmacological characterisation, including mode of action and subtype selectivity studies, as well as secondary or safety screening of bioactive compounds isolated from venoms.

4.6.3 Biochemical Signalling

Venom components can be assayed effectively in a variety of biochemical assays that measure cell signalling molecules, such as intracellular calcium, inositol trisphosphate and cyclic AMP.

These assays systematically isolate and/or characterise bioactive molecules from the complex mixture of peptides, proteins and small molecules in venoms, and require at least high assay sensitivity, accuracy, robustness and reproducibility, reflected by a high Z′ score.[129] These types of assays can be performed in multiwell plates with high throughput if necessary. The ability to easily adapt them to 96-, 384- or even 1536-well format enables miniaturisation, which is a significant advantage if crude venom material is in limited supply. Thus, biochemical signalling assays performed in high-throughput format may permit more detailed characterisation of venoms from small cone snails, spiders, centipedes, ticks and scorpions, which often only provide microgram amounts of venom from a single milking or dissection, and have not been extensively studied using more traditional *ex vivo* or *in vivo* approaches.

Multiwell-plate-based assays have the advantage of enabling multiple assay points per condition, allowing pharmacological characteristics, such as concentrations, which enable half-maximal inhibition or activation of a receptor to be determined with more confidence.

Plate-based screening assays are of great use in this type of analysis, particularly if the primary target of the venom component is known. In parallel with the ever-increasing number of receptors, ligand- and voltage-gated ion channels, enzymes and transporters targeted by venoms and venom components, the repertoire of plate-based assays used to define pharmacological activity now expands on more traditional approaches, such as electrophysiological recordings and radioligand binding studies. In recent years, in addition to absorbance/fluorescence-based assays and enzyme-linked immunosorbent assays (ELISA), label-free technologies and assays based on the detection of bioluminescence, fluorescence polarisation, fluorescence-resonance energy transfer (FRET), bioluminescence resonance energy transfer (BRET) and scintillation proximity have become available and are increasingly used for pharmacological characterisation and assay-guided isolation of bioactive venom components.

In the following sections, we discuss the advantages and disadvantages of some of these techniques in the context of venoms-based drug discovery programs.

4.6.3.1 Binding Assays

Binding assays are designed to detect the interaction of a compound with receptors, enzymes, transporters and biological targets of interest by displacing a ligand incorporating a fluorescent or radioactive label. Importantly, binding assays can be performed on tissues, whole cells or recombinantly expressed protein, so they are very versatile. These assays are generally performed in moderate- to high-throughput plate-based formats and are relatively low cost. However, additional safety precautions and specialised training are needed to handle radioligands.

The major disadvantage with binding assays is that a labelled ligand will only be displaced if the test compound either binds to the same or an allosterically linked site.[112,130,131] Given the large number of binding sites that have been described for a range of receptors and ion channels targeted by toxins and venom components (*e.g.* six defined toxin binding sites on Na_V channels), this predisposes binding assays to a particularly high rate of false negatives. For example, the spider venom peptide ProTx-I, a Na_V1.7 inhibitor, couldn't displace radiolabelled batrachotoxin at Na_V1.7 heterologously expressed in HEK293 cells.[132] In addition, while labelled ligands are generally assumed to be structurally identical to their unlabelled analogues, this may not necessarily be the case, especially in the case of large fluorescent tags, and any structural differences could introduce additional errors.

Since binding studies only detect the physical interaction of ligands with the target, they cannot provide information on the pharmacological nature of such interactions, or distinguish between agonists and antagonists.[112,130,131] Similarly, high-affinity binding does not always translate to high-functional affinity. Therefore, while binding assays can be used in 'pull-down' assays to isolate an unknown pharmacological target of toxins and venom peptides,[133] functional assays have been increasingly used for pharmacological characterisation in recent years.

4.6.3.2 Fluorescence-Based Assays

Many ion channels are permeable to ions that can be visualised with specific fluorescent dyes, such as Fura2 and Fluo3 for calcium, or sodium green and SBFI for sodium. Fluorescent dyes are also available to measure chloride and potassium. Alternatively, membrane depolarisation can be measured with oxonol- or carbocyanin-based membrane potential-sensitive dyes. Membrane potential dyes are suitable to detect the activity of any receptor or ion channel that leads to a change in membrane voltage. However, these dyes tend to be prone to fluorescence artefacts and generally have a poorer signal-to-noise-ratio than fluorescence assays assessing accumulation of Ca^{2+} ions (Figure 4.4).

All of these dyes are available from Molecular Probes Ltd., along with detailed technical literature on their use. Assays based on these dyes may be performed using a suitable fluorescence microscope or in multiwell-plate-based systems, which rely on detecting fluorescence with photomultipliers, cameras or flow cytometers.

Based on their capability to measure fluorescence from 96 or more wells simultaneously, platforms such as the FLIPR^{TETRA} (Molecular Devices), Cell Lux (Perkin Elmer) and FDSS 6000 (Hamamatsu) enable true high-throughput analysis of venoms and venom components. They have also been used in recent years for the discovery, isolation and characterisation of bioactive peptides from snakes, spiders and cone snails.[134–136]

Due to the ubiquitous nature of the Ca^{2+} signal, the large Ca^{2+} gradient across cell membranes and superior spectrophotometric properties of Ca^{2+} dyes,[137] high-throughput Ca^{2+}assays are particularly attractive for isolating, identifying and characterising venom components with activity at a range of voltage- and ligand-gated ion channels (*e.g.* TRP channels and nAChRs) and GPCRs. Ca^{2+} flux assays provide kinetic rather than simple end-point information and often permit detection of agonists and antagonists in the same assay. Such functional information can also provide important insight into the molecular identity of the primary pharmacological target of the venom component, if it is unknown.

However, lack of activity in these assays does not necessarily imply lack of biological activity *per se*. Interesting pharmacological effects that might have been detected using *ex vivo* or *in vivo* assays can easily be missed using bioassays assessing specific signalling pathways. In addition, non-specific or off-target effects may be erroneously interpreted as activity at the pharmacological target of interest. For example, venom components that permeabilise cell membranes, affect membrane integrity or lead to pore formation can lead to Ca^{2+} signals that may be misinterpreted as Ca^{2+} influx through specific ion channels (Figure 4.5).

Similarly, while it is possible to screen whole crude venom, non-specific effects from salts, small molecules or neurotransmitters found in crude venom may elicit false positive responses or obscure activity at the target of interest. Therefore, separating these venom components, for example using reversed-phase high-performance liquid chromatography (HPLC), is often needed before screening, as discussed above. While some fluorescence assays can tolerate small amounts of organic solvents, such as methanol and acetonitrile, from these purification steps, most of the time these need to be removed before bioactivity testing.

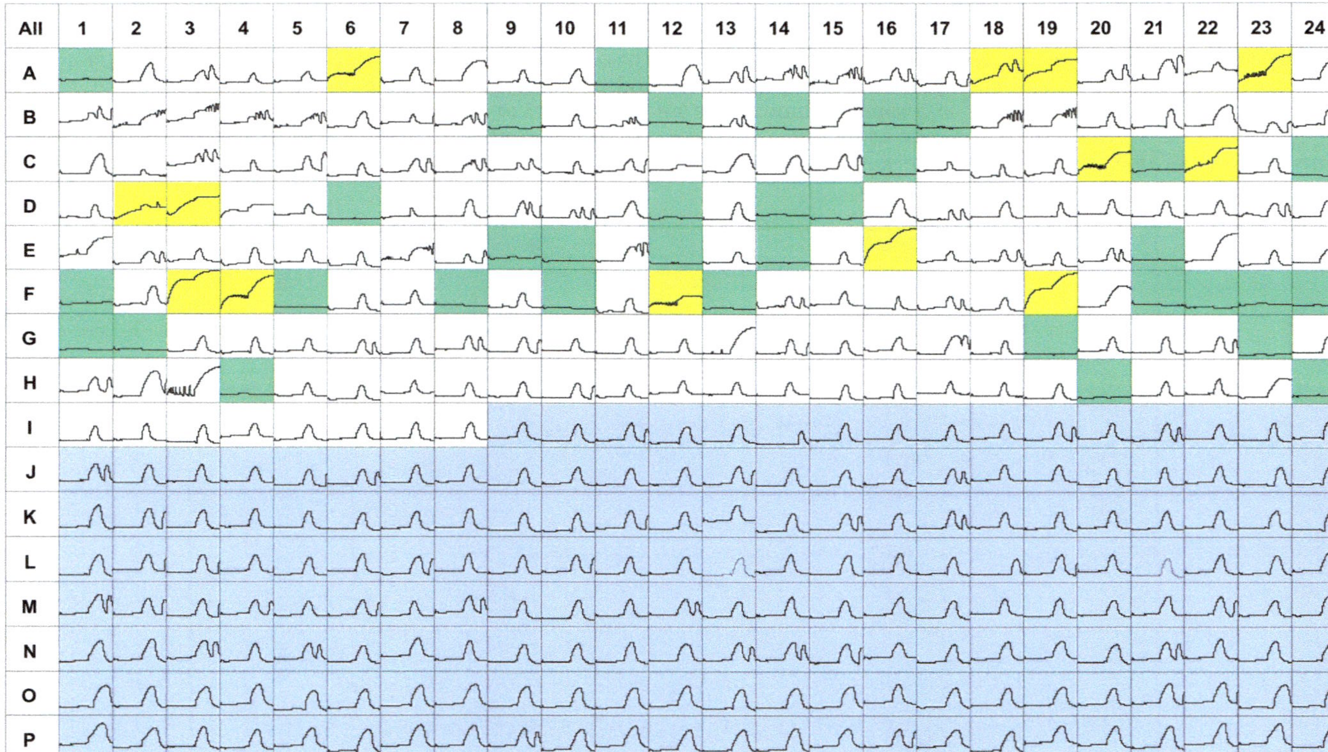

Figure 4.4 Example of a high-throughput FLIPR-based membrane potential assay. The effect of venoms and venom peptides on veratridine-induced membrane potential changes were assessed in CHO cells heterologously expressing $Na_V1.3$. A two-addition protocol was used, consisting of initial addition of venoms and venom components, followed by stimulation of $Na_V1.3$ responses using veratridine. Wells A1–I8 represent single replicates of various venoms and venom peptides. A large number of compounds elicited addition artefacts (yellow; *e.g.* wells A6, A18, A19, C20, C22, D2, E16, F19). Inhibition of $Na_V1.3$-mediated responses were apparent with several compounds (green; *e.g.* wells A1, A11, B9, B12, F8, G23) compared to control responses (blue, wells I9–P24).

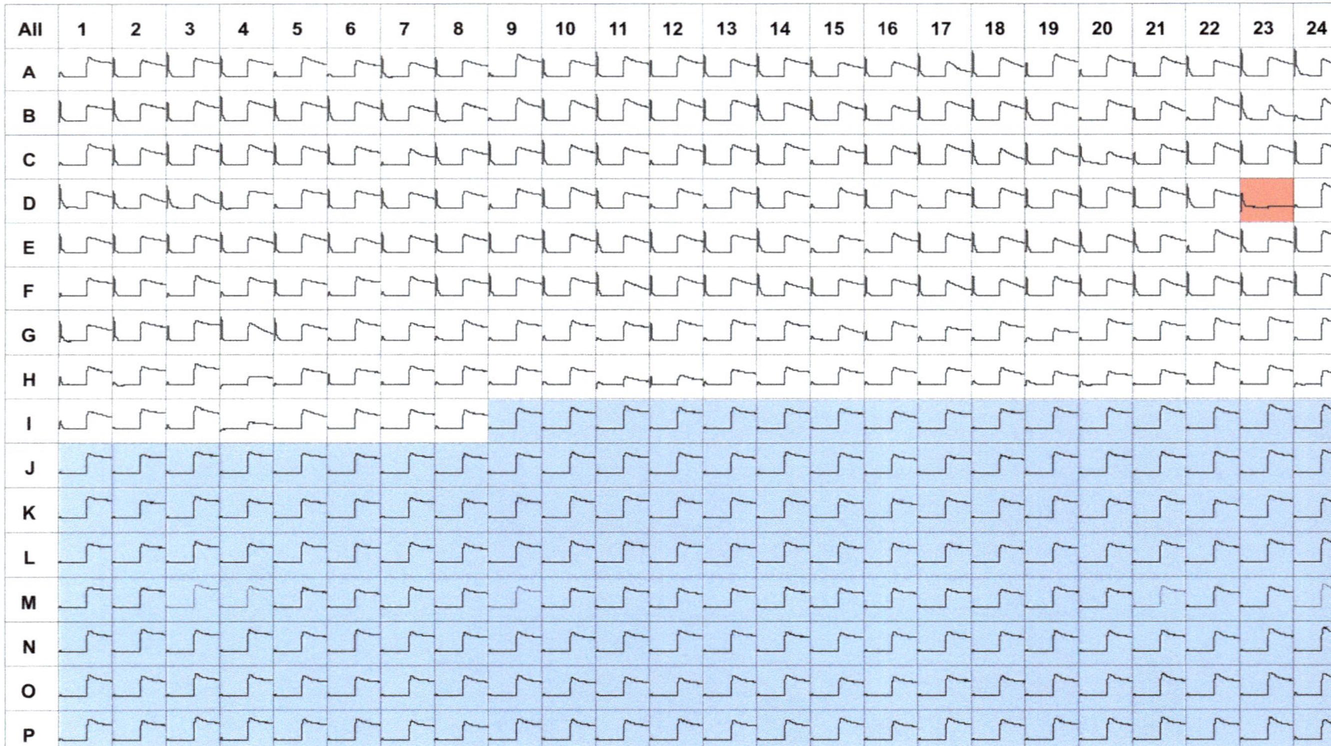

Figure 4.5 Example of a high-throughput FLIPR-based Ca^{2+} assay. The effect of venoms and venom peptides on TRPM3-mediated Ca^{2+} responses was assessed using a two-addition protocol, with test compounds added first, followed by stimulation of TRPM3. Wells A1–I8 represent single replicates of various venoms and venom peptides. Note that a number of compounds elicited an addition response, but no inhibition of TRPM3 responses was observed compared to control responses (blue, wells I9–P24). A false positive response (red, well D23) was detected with M-ctenitoxin-Cs1a, a linear cationic peptide with membranolytic activity.

The additional handling steps arising from lyophilisation and subsequent reconstitution of venom fractions in suitable assay buffers eliminate some of the inherent advantages of plate-based assays, in particular the ability to screen very large numbers of venom components. Nonetheless, fluorescence-based assays provide some of the highest throughput approaches to venoms-based drug discovery, and they can be adapted to a wide range of pharmacological targets. In combination with transcriptomic approaches, which eliminate the need for time-consuming and resource-intensive sequencing analysis, high-throughput screening could vastly accelerate the discovery of venom components with activity at defined pharmacological targets.

4.6.3.3 *Immunoassays*

ELISAs are end-point assays based on antibody-mediated detection of biochemical signalling molecules, including peptides, proteins, neurotransmitters and hormones, in multiwell plate format. In contrast to fluorescence-based assays, which provide real-time kinetic information and typically assess changes in fluorescence that occur over seconds or minutes, immunoassays quantify a single concentration, require many washing steps and typically take many hours to complete.

Immunoassays have, for example, been used to: quantify the antiviral activity of scorpion venom peptides against the replication-competent HIV-1 virus;[138] assess $\alpha 2\beta 1$ integrin inhibition by the snake venom protein EMS16;[139] and assess the effect of honey bee venom on interferon production.[140] Where venom components provoke an immune reaction in people or experimental animals, serum may be used in immunoassays to detect the immunogenic molecule in fractionated venoms, particularly snake venoms.[141] However, such approaches may not be suitable to smaller, peptidic venom components, such as conopeptides or conotoxins, which are often not immunogenic. Nonetheless, antibody-based assays have also been developed to quantify specific conotoxins, for example MVIIA,[142] as an alternative to HPLC-mediated quantification.

In summary, robust, accurate, sensitive and specific assays that can readily and reliably identify bioactives are essential for successful screening of venoms. Ideally, such assays should be high throughput, amenable to miniaturisation, and able to detect biological activity in crude or fractionated venoms. While no single assay can capture all of these characteristics, a range of assay systems has been adapted to isolate and characterise venom-derived peptides and proteins to aid development of novel drugs from the arsenal of exquisitely selective and potent bioactive molecules in animal venoms.

Acknowledgements

We thank Louise Pobjoy for careful reading of the manuscript and helpful suggestions to improve clarity.

References

1. J. M. McIntosh, B. M. Olivera, L. J. Cruz and W. R. Gray, *J. Biol. Chem.*, 1984, **259**, 14343–14346.
2. S. H. Ferreira, D. C. Bartelt and L. J. Greene, *Biochemistry*, 1970, **9**, 2583–2593.
3. R. C. Rodriguez de la Vega, E. F. Schwartz and L. D. Possani, *Toxicon*, 2010, **56**, 1155–1161.
4. O. Buczek, G. Bulaj and B. M. Olivera, *Cell. Mol. Life Sci.*, 2005, **62**, 3067–3079.
5. J. P. Bingham, E. Mitsunaga and Z. L. Bergeron, *Chem. Biol. Interact.*, 2010, **183**, 1–18.
6. T. S. Han, R. W. Teichert, B. M. Olivera and G. Bulaj, *Curr. Pharm. Des.*, 2008, **14**, 2462–2479.
7. S. Yang, Y. Xiao, D. Kang, J. Liu, Y. Li, E. A. B. Undheim, J. K. Klint, R. M., R. Lai and G. F. King, *Proc. Natl. Acad. Sci. U.S.A.*, 2013, **110**, 17534–17539.
8. I. Vetter, C. A. Mozar, T. Durek, J. S. Wingerd, P. F. Alewood, M. J. Christie and R. J. Lewis, *Biochem. Pharmacol.*, 2012, **83**, 1562–1571.
9. C. Clark, B. M. Olivera and L. J. Cruz, *Toxicon*, 1981, **19**, 691–699.
10. E. W. McCleskey, A. P. Fox, D. H. Feldman, L. J. Cruz, B. M. Olivera, R. W. Tsien and D. Yoshikami, *Proc. Natl. Acad. Sci. U.S.A.*, 1987, **84**, 4327–4331.
11. B. M. Olivera and L. J. Cruz, *Toxicon*, 2001, **39**, 7–14.
12. B. M. Olivera, J. M. McIntosh, L. J. Cruz, F. A. Luque and W. R. Gray, *Biochemistry*, 1984, **23**, 5087–5090.
13. E. Bulbring, *Br. J. Pharmacol. Chemother.*, 1946, **1**, 38–61.
14. B. L. Ginsborg and J. Warriner, *Br. J. Pharmacol. Chemother.*, 1960, **15**, 410–411.
15. A. L. Harvey, A. Barfaraz, E. Thomson, A. Faiz, S. Preston and J. B. Harris, *Toxicon*, 1994, **32**, 257–265.
16. A. Christopoulos, A. M. Parsons, M. J. Lew and E. E. El-Fakahany, *Eur. J. Pharmacol.*, 1999, **382**, 217–227.
17. M. J. Lew and J. A. Angus, *Trends Pharmacol. Sci.*, 1995, **16**, 328–337.
18. W. C. Hodgson and J. C. Wickramaratna, *Clin. Exp. Pharmacol. Physiol.*, 2002, **29**, 807–814.
19. M. A. Kamenskaya and S. Thesleff, *Acta Physiol. Scand.*, 1974, **90**, 716–724.
20. M. Rigoni, P. Pizzo, G. Schiavo, A. E. Weston, G. Zatti, P. Caccin, O. Rossetto, T. Pozzan and C. Montecucco, *J. Biol. Chem.*, 2007, **282**, 11238–11245.
21. C. C. Chang, J. D. Lee, D. Eaker and J. Fohlman, *Toxicon*, 1977, **15**, 571–576.
22. C. C. Chang, *Proc. Natl. Sci. Counc., Repub. China, Part B: Basic Sci.*, 1985, **9**, 126–142.
23. W. C. Hodgson, C. A. Dal Belo and E. G. Rowan, *Neuropharmacology*, 2007, **52**, 1229–1236.

24. A. L. Harvey, *Toxicon,* 2001, **39**, 15–26.
25. J. C. Barrett and A. L. Harvey, *Br. J. Pharmacol.,* 1979, **67**, 199–205.
26. A. L. Harvey and E. Karlsson, *Naunyn Schmiedebergs Arch. Pharmacol.,* 1980, **312**, 1–6.
27. R. Joseph, S. Pahari, W. C. Hodgson and R. M. Kini, *Curr. Drug. Targets. Cardiovasc. Haematol. Disord.,* 2004, **4**, 437–459.
28. S. H. Ferreira and M. Rocha e Silva, *Experientia,* 1965, **21**, 347–349.
29. H. Kato, T. Suzuki, K. Okada, T. Kimura and S. Sakakibara, *Experientia,* 1973, **29**, 574–575.
30. C. W. Chi, S. Z. Wang, L. G. Xu, M. Y. Wang, S. S. Lo and W. D. Huang, *Peptides,* 1985, **6**(3), 339–342.
31. L. A. Ferreira, E. W. Alves and O. B. Henriques, *Toxicon,* 1993, **31**, 941–947.
32. A. R. Meki, A. Y. Nassar and H. Rochat, *Peptides,* 1995, **16**, 1359–1365.
33. D. P. Westfall, R. E. Stitzel and J. N. Rowe, *Eur. J. Pharmacol.,* 1978, **50**, 27–38.
34. G. Burnstock and A. Verkhratsky, *Trends Pharmacol. Sci.,* 2010, **31**, 131–139.
35. P. Sneddon and G. Burnstock, *Eur. J. Pharmacol.,* 1984, **100**, 85–90.
36. R. J. Lewis and M. L. Garcia, *Nat. Rev. Drug Discov.,* 2003, **2**, 790–802.
37. I. A. Sharpe, J. Gehrmann, M. L. Loughnan, L. Thomas, D. A. Adams, A. Atkins, E. Palant, D. J. Craik, D. J. Adams, P. F. Alewood and R. J. Lewis, *Nat. Neurosci.,* 2001, **4**, 902–907.
38. A. Maiga, G. Mourier, L. Quinton, C. Rouget, C. Gales, C. Denis, P. Lluel, J. M. Senard, S. Palea, D. Servent and N. Gilles, *Toxicon,* 2012, **59**, 487–496.
39. H. Schweitz, P. Vigne, D. Moinier, C. Frelin and M. Lazdunski, *J. Biol. Chem.,* 1992, **267**, 13928–13932.
40. B. G. Fry, J. C. Wickramaratana, S. Lemme, A. Beuve, D. Garbers, W. C. Hodgson and P. Alewood, *Biochem. Biophys. Res. Commun.,* 2005, **327**, 1011–1015.
41. G. M. de Plater, R. L. Martin and P. J. Milburn, *Comp. Biochem. Physiol. C. Pharmacol. Toxicol. Endocrinol.,* 1998, **120**, 99–110.
42. B. G. Fry, N. Vidal, J. A. Norman, F. J. Vonk, H. Scheib, S. F. Ramjan, S. Kuruppu, K. Fung, S. B. Hedges, M. K. Richardson, W. C. Hodgson, V. Ignjatovic, R. Summerhayes and E. Kochva, *Nature,* 2006, **439**, 584–588.
43. S. Vink, A. H. Jin, K. J. Poth, G. A. Head and P. F. Alewood, *Toxicon,* 2012, **59**, 434–445.
44. Y. X. Wang, S. Bezprozvannaya, S. S. Bowersox, L. Nadasdi, G. Miljanich, G. Mezo, D. Silva, K. Tarczy-Hornoch and R. R. Luther, *Naunyn Schmiedebergs Arch. Pharmacol.,* 1998, **357**, 159–168.
45. S. Vink and P. F. Alewood, *Brit. J. Pharmacol.,* 2012, **167**, 970–989.
46. M. Amininasab, M. M. Elmi, N. Endlich, K. Endlich, N. Parekh, H. Naderi-Manesh, J. Schaller, H. Mostafavi, M. Sattler, M. N. Sarbolouki and C. Muhle-Goll, *FEBS Lett.,* 2004, **557**, 104–108.

47. A. C. Camargo, D. Ianzer, J. R. Guerreiro and S. M. Serrano, *Toxicon,* 2012, **59**, 516–523.
48. D. Ianzer, R. A. Santos, G. M. Etelvino, C. H. Xavier, J. de Almeida Santos, E. P. Mendes, L. T. Machado, B. C. Prezoto, V. Dive and A. C. de Camargo, *J. Pharmacol. Exp. Ther.,* 2007, **322**, 795–805.
49. J. E. Church and W. C. Hodgson, *Toxicon,* 2002, **40**, 1083–1093.
50. S. Ramasamy, G. K. Isbister, J. E. Seymour and W. C. Hodgson, *Toxicon,* 2004, **43**, 685–690.
51. S. Ramasamy, G. K. Isbister, J. E. Seymour and W. C. Hodgson, *Toxicol. Lett.,* 2005, **155**, 135–141.
52. J. Tibballs, *Toxicon,* 2006, **48**, 830–859.
53. K. L. Winter, G. K. Isbister, J. J. Schneider, N. Konstantakopoulos, J. E. Seymour and W. C. Hodgson, *Toxicol. Lett.,* 2008, **179**, 118–123.
54. K. S. Cole, *Arch. Sci. Physiol.,* 1949, **3**, 253–258.
55. G. Marmont, *J. Cell. Physiol.,* 1949, **34**, 351–382.
56. A. L. Hodgkin and A. F. Huxley, *J. Physiol.,* 1952, **117**, 500–544.
57. P. Rosenberg, *Toxicon,* 1965, **3**, 125–131.
58. P. Rosenberg and T. R. Podleski, *J. Pharmacol. Exp. Ther.,* 1962, **137**, 249–262.
59. M. H. Evans, *Nature,* 1963, **200**, 368.
60. K. R. Adam, H. Schmidt, R. Stampfli and C. Weiss, *Br. J. Pharmacol. Chemother.,* 1966, **26**, 666–677.
61. E. Koppenhofer and H. Schmidt, *Pflugers Archiv: Eur. J. Physiol.,* 1968, **303**, 133–149.
62. I. Vetter, F. Touska, A. Hess, R. Hinsbey, S. Sattler, A. Lampert, M. Sergejeva, A. Sharov, L. S. Collins, M. Eberhardt, M. Engel, P. J. Cabot, J. N. Wood, V. Vlachova, P. W. Reeh, R. J. Lewis and K. Zimmermann, *EMBO J.,* 2012, **31**, 3795–3808.
63. G. Ling and R. W. Gerard, *J. Cell. Physiol.,* 1949, **34**, 383–396.
64. R. D. Purves, *Microelectrode Methods for Intracellular Recording and Ionophoresis*, Academic Press, 1981.
65. R. Brette and A. Destexhe, in *Handbook of Neural Activity Measurement* eds. R. Brette and A. Destexhe, Cambridge University Press 2010.
66. M. C. Inserra, S. N. Kompella, I. Vetter, A. Brust, N. L. Daly, H. Cuny, D. J. Craik, P. F. Alewood, D. J. Adams and R. J. Lewis, *Biochem. Pharmacol.,* 2013, **86**, 791–799.
67. J. Ekberg, A. Jayamanne, C. W. Vaughan, S. Aslan, L. Thomas, J. Mould, R. Drinkwater, M. D. Baker, B. Abrahamsen, J. N. Wood, D. J. Adams, M. J. Christie and R. J. Lewis, *Proc. Natl. Acad. Sci. U.S.A.,* 2006, **103**, 17030–17035.
68. D. J. Orts, Y. Moran, C. T. Cologna, S. Peigneur, B. Madio, D. Praher, L. Quinton, E. De Pauw, J. E. Bicudo, J. Tytgat and J. C. de Freitas, *FEBS J.,* 2013, **280**, 4839–4852.
69. N. J. Saez, M. Mobli, M. Bieri, I. R. Chassagnon, A. K. Malde, R. Gamsjaeger, A. E. Mark, P. R. Gooley, L. D. Rash and G. F. King, *Mol. Pharmacol.,* 2011, **80**, 796–808.

70. G. Berecki, L. Motin, A. Haythornthwaite, S. Vink, P. Bansal, R. Drinkwater, C. I. Wang, M. Moretta, R. J. Lewis, P. F. Alewood, M. J. Christie and D. J. Adams, *Mol. Pharmacol.*, 2010, **77**, 139–148.
71. C. J. Bohlen, A. Priel, S. Zhou, D. King, J. Siemens and D. Julius, *Cell,* 2010, **141**, 834–845.
72. E. Engel, V. Barcilon and R. S. Eisenberg, *Biophys. J.*, 1972, **12**, 384–403.
73. J. J. B. Jack, D. Noble and R. W. Tsien. *Electrical Flow in Excitable Cells*, Clarendon Press, Oxford, 1975, pp. 1–502.
74. A. S. Finkel and S. J. Redman, in Voltage and Patch Clamping with Microelectrodes eds. T. Smith, H. Lecar, S. Redman and P. W. Gage, *American Physiological Society Bethesda*, 1985, pp. 95–121.
75. J. C. Bornstein, M. Costa, J. B. Furness and G. M. Lees, *J. Physiol.*, 1984, **351**, 313–325.
76. J. R. Keast, E. M. McLachlan and R. L. Meckler, *J. Neurophysiol.*, 1993, **69**, 384–394.
77. A. L. Hodgkin, A. F. Huxley and B. Katz, *J. Physiol.*, 1952, **116**, 424–448.
78. M. Field, P. L. Smith and J. E. Bolton, *J. Membr. Biol.*, 1980, **55**, 157–163.
79. W. A. Wilson and M. M. Goldner, *J. Neurobiol.*, 1975, **6**, 411–422.
80. A. S. Finkel and S. Redman, *J. Neurosci. Methods,* 1984, **11**, 101–127.
81. E. Neher and B. Sakmann, *Nature,* 1976, **260**, 799–802.
82. F. J. Sigworth and E. Neher, *Nature,* 1980, **287**, 447–449.
83. B. Sakmann and E. Neher, *Single-Channel Recording*, Plenum Press, New York, 1995.
84. O. P. Hamill, A. Marty, E. Neher, B. Sakmann and F. J. Sigworth, *Pflugers Archiv.: Eur. J. Physiol.*, 1981, **391**, 85–100.
85. P. H. Barry and J. W. Lynch, *J. Membr. Biol.*, 1991, **121**, 101–117.
86. E. Neher, A. Marty and E. Fenwick, *Progr. Brain Res.*, 1983, **58**, 39–48.
87. M. Pusch and E. Neher, *Pflugers Archiv.: Eur. J. Physiol.*, 1988, **411**, 204–211.
88. R. Horn and A. Marty, *J. Gen. Physiol.*, 1988, **92**, 145–159.
89. G. B. Yi, D. McClendon, D. Desaiah, J. Goddard, A. Lister, J. Moffitt, R. K. Meer, R. deShazo, K. S. Lee and R. W. Rockhold, *Int. J. Toxicol.*, 2003, **22**, 81–86.
90. R. J. Lewis, S. Dutertre, I. Vetter and M. J. Christie, *Pharmacol. Rev.*, 2012, **64**, 259–298.
91. G. F. King, P. Escoubas and G. M. Nicholson, *Channels (Austin),* 2008, **2**, 100–116.
92. C. Gotti, A. E. Ogando, W. Hanke, R. Schlue, M. Moretti and F. Clementi, *Proc. Natl. Acad. Sci. U.S.A.*, 1991, **88**, 3258–3262.
93. D. M. Michaelson and M. A. Raftery, *Proc. Natl. Acad. Sci. U.S.A.*, 1974, **71**, 4768–4772.
94. C. F. Stevens, *Cold Spring Harb. Symp. Quant. Biol.*, 1976, **40**, 169–173.
95. R. MacKinnon and C. Miller, *Science,* 1989, **245**, 1382–1385.
96. J. Barhanin, D. Pauron, A. Lombet, R. I. Norman, H. P. Vijverberg, J. R. Giglio and M. Lazdunski, *EMBO J.*, 1983, **2**, 915–920.

97. J. Siemens, S. Zhou, R. Piskorowski, T. Nikai, E. A. Lumpkin, A. I. Basbaum, D. King and D. Julius, *Nature,* 2006, **444**, 208–212.
98. J. J. Smith, I. Vetter, R. J. Lewis, S. Peigneur, J. Tytgat, A. Lam, E. M. Gallant, N. A. Beard, P. F. Alewood and A. F. Dulhunty, *Proc. Natl. Acad. Sci. U.S.A.,* 2013, **110**, 8906–8911.
99. A. A. Grishin, C. I. Wang, M. Muttenthaler, P. F. Alewood, R. J. Lewis and D. J. Adams, *J. Biol. Chem.,* 2010, **285**, 22254–22263.
100. W. A. Schmalhofer, J. Calhoun, R. Burrows, T. Bailey, M. G. Kohler, A. B. Weinglass, G. J. Kaczorowski, M. L. Garcia, M. Koltzenburg and B. T. Priest, *Mol. Pharmacol.,* 2008, **74**, 1476–1484.
101. J. J. Cox, F. Reimann, A. K. Nicholas, G. Thornton, E. Roberts, K. Springell, G. Karbani, H. Jafri, J. Mannan, Y. Raashid, L. Al-Gazali, H. Hamamy, E. M. Valente, S. Gorman, R. Williams, D. P. McHale, J. N. Wood, F. M. Gribble and C. G. Woods, *Nature,* 2006, **444**, 894–898.
102. R. J. Lewis, K. J. Nielsen, D. J. Craik, M. L. Loughnan, D. A. Adams, I. A. Sharpe, T. Luchian, D. J. Adams, T. Bond, L. Thomas, A. Jones, J. L. Matheson, R. Drinkwater, P. R. Andrews and P. F. Alewood, *J. Biol. Chem.,* 2000, **275**, 35335–35344.
103. M. Garcia-Calvo, R. J. Leonard, J. Novick, S. P. Stevens, W. Schmalhofer, G. J. Kaczorowski and M. L. Garcia, *J. Biol. Chem.,* 1993, **268**, 18866–18874.
104. J. Dunlop, M. Bowlby, R. Peri, G. Tawa, J. LaRocque, V. Soloveva and J. Morin, *Comb. Chem. High Throughput Screen.,* 2008, **11**, 514–522.
105. J. Dunlop, M. Bowlby, R. Peri, D. Vasilyev and R. Arias, *Nat. Rev. Drug Discov.,* 2008, 7, 358–368.
106. P. B. Bennett and H. R. Guthrie, *Trends Biotechnol.,* 2003, **21**, 563–569.
107. M. Hamelin, X. Meng, M. Cuddy, K. Korsun, J. Ondeyka, P. B. Simpson, D. F. Cully and B. T. Priest, *Assay Drug Dev. Technol.,* 2005, **3**, 59–64.
108. R. L. Papke and C. Smith-Maxwell, *Comb. Chem. High Throughput Screen.,* 2009, **12**, 38–50.
109. N. Castle, D. Printzenhoff, S. Zellmer, B. Antonio, A. Wickenden and C. Silvia, *Comb. Chem. High Throughput Screen.,* 2009, **12**, 107–122.
110. A. Finkel, A. Wittel, N. Yang, S. Handran, J. Hughes and J. Costantin, *J. Biomol. Screen.,* 2006, **11**, 488–496.
111. T. J. Dale, C. Townsend, E. C. Hollands and D. J. Trezise, *Mol. Biosyst.,* 2007, **3**, 714–722.
112. B. Fermini, in *Topics in Medicinal Chemistry*, Springer, 2008, vol. 3, pp. 1–25.
113. J. R. Penniman, D. C. Kim, J. J. Salata and J. P. Imredy, J. Pharmacol. Toxicol. Methods, 62, 107–118.
114. J. Xu, A. Guia, D. Rothwarf, M. Huang, K. Sithiphong, J. Ouang, G. Tao, X. Wang and L. Wu, *Assay Drug Dev. Technol.,* 2003, **1**, 675–684.
115. B. Balasubramanian, J. P. Imredy, D. Kim, J. Penniman, A. Lagrutta and J. J. Salata, *J. Pharmacol. Toxicol. Methods,* 2009, **59**, 62–72.
116. A. E. Dubin, N. Nasser, J. Rohrbacher, A. N. Hermans, R. Marrannes, C. Grantham, K. Van Rossem, M. Cik, S. R. Chaplan, D. Gallacher, J. Xu, A. Guia, N. G. Byrne and C. Mathes, *J. Biomol. Screen.,* 2005, **10**, 168–181.

117. J. Dunlop, R. Roncarati, B. Jow, H. Bothmann, T. Lock, D. Kowal, M. Bowlby and G. C. Terstappen, *Biochem. Pharmacol.,* 2007, **74**, 1172–1181.
118. Z. L. Mo, T. Faxel, Y. S. Yang, R. Gallavan, D. Messing and A. Bahinski, *J. Pharmacol. Toxicol. Methods,* 2009, **60**, 39–44.
119. E. S. Trepakova, M. G. Malik, J. P. Imredy, J. R. Penniman, S. J. Dech and J. J. Salata, *Assay Drug Dev. Technol.,* 2007, **5**, 617–627.
120. C. Farre and N. Fertig, *Expert Opin. Drug Discovery,* 2012, 7, 515–524.
121. N. Lenkey, R. Karoly, P. Lukacs, E. S. Vizi, M. Sunesen, L. Fodor and A. Mike, *PLoS One*, 5, e15568.
122. C. Mathes, S. Friis, M. Finley and Y. Liu, *Comb. Chem. High Throughput Screen.,* 2009, **12**, 78–95.
123. C. Mathes, *Expert Opin. Ther. Targets,* 2006, **10**, 319–327.
124. K. A. Jones, N. Garbati, H. Zhang and C. H. Large, *Methods Mol. Biol.,* 2009, **565**, 209–223.
125. S. Friis, C. Mathes, M. Sunesen, M. R. Bowlby and J. Dunlop, *J. Neurosci. Methods,* 2009, **177**, 142–148.
126. M. P. Korsgaard, D. Strobaek and P. Christophersen, *Comb. Chem. High Throughput Screen.,* 2009, **12**, 51–63.
127. J. Kutchinsky, S. Friis, M. Asmild, R. Taboryski, S. Pedersen, R. K. Vestergaard, R. B. Jacobsen, K. Krzywkowski, R. L. Schroder, T. Ljungstrom, N. Helix, C. B. Sorensen, M. Bech and N. J. Willumsen, *Assay Drug Dev. Technol.,* 2003, **1**, 685–693.
128. M. Fejtl, U. Czubayko, A. Hummer, T. Krauter and A. Lepple-Wienhues, *Methods Mol. Biol.,* 2007, **403**, 71–85.
129. J. H. Zhang, T. D. Chung and K. R. Oldenburg, *J. Biomol. Screen,* 1999, **4**, 67–73.
130. J. J. Clare, S. N. Tate, M. Nobbs and M. A. Romanos, *Drug Discovery Today,* 2000, **5**, 506–520.
131. J. Xu, X. Wang, B. Ensign, M. Li, L. Wu, A. Guia and J. Xu, *Drug Discovery Today,* 2001, **6**, 1278–1287.
132. A. Bhattacharya, Q. Wang, N. Wu and A. D. Wickenden, *FASEB J.,* 2009, **23**(Meeting Abstract Supplement), 998.31.
133. J. C. Meunier, R. Sealock, R. Olsen and J. P. Changeux, *Eur. J. Biochem.,* 1974, **45**, 371–394.
134. A. Brust, K. Sunagar, E. A. B. Undheim, I. Vetter, D. C. Yang, N. R. Casewell, T. Ruder, T. N. W. Jackson, I. Koludarov, P. F. Alewood, W. C. Hodgson, R. J. Lewis, G. F. King, A. Antunes, I. Hendrikx and B. G. Fry, *Mol. Cell. Proteomics,* 2012, **12**, 651–663.
135. F. Marcon, M. Leblanc, I. Vetter, R. J. Lewis, P. Escoubas and G. M. Nicholson, *Biochem. Pharmacol.,* 2012, **84**, 851–863.
136. I. Vetter, Z. Dekan, O. Knapp, D. J. Adams, P. F. Alewood and R. J. Lewis, *Biochem. Pharmacol.,* 2012, **84**, 540–548.
137. I. Vetter, *Adv. Exp. Med. Biol.,* 2012, **740**, 45–82.
138. Y. Chen, L. Cao, M. Zhong, Y. Zhang, C. Han, Q. Li, J. Yang, D. Zhou, W. Shi, B. He, F. Liu, J. Yu, Y. Sun, Y. Cao, Y. Li, W. Li, D. Guo, Z. Cao and H. Yan, *PLoS One,* 2012, 7, e34947.

139. C. Marcinkiewicz, R. R. Lobb, M. M. Marcinkiewicz, J. L. Daniel, J. B. Smith, C. Dangelmaier, P. H. Weinreb, D. A. Beacham and S. Niewiarowski, *Biochemistry,* 2000, **39**, 9859–9867.
140. M. Hamedani, A. Mirshafiey, H. Vatanpour, M. Khorramizadeh, F. Saadat, A. Berahmeh and F. Hadji-Ghasemi, *Iran J. Allergy Asthma Immunol.,* 2005, **4**, 9–14.
141. Z. E. Selvanayagam, S. G. Gnanavendhan, K. A. Ganesh, D. Rajagopal and P. V. Rao, *Toxicon,* 1999, **37**, 757–770.
142. F. Wang, G. Chen, Y. Zhu, Z. Guo, S. Zheng and J. Zhan, *J. Biochem. Biophys. Methods,* 2006, **67**, 49–56.

CHAPTER 5

Reptile Venoms as a Platform for Drug Development

RYAN J. R. McCLEARY[a], TSE S. KANG[b], AND R. MANJUNATHA KINI[a,c,d]*

[a]National University of Singapore, Department of Biological Sciences, 14 Science Drive 4, Singapore 117543, Singapore; [b]National University of Singapore, Department of Pharmacy, 18 Science Drive 4, Singapore 117543, Singapore; [c]Virginia Commonwealth University, Department of Biochemistry - Medical College of Virginia, Richmond, Virginia 23298-0614, USA; [d]University of South Australia, School of Pharmacy and Medical Sciences, Adelaide, South Australia 5001, Australia
*E-mail: dbskinim@nus.edu.sg

5.1 Introduction

Snakes elicit both fascination and disgust in humans, which is mainly due to their unusual serpentine movement, spine-tingling hissing or rattling, and deadly venoms. Since early Greek civilization, the rod of Asclepius and the caduceus—variations of a snake wrapped around a staff—have been used as symbols of medicine or medical professions. Although their origins are unrelated to the pharmaceuticals that have been—and will be—produced from reptile venoms, these symbols are *apropos* in that they highlight a growing connection between reptiles and human medicine.

Reptile venoms are sources for many different families of toxic proteins and peptides, some of which have been developed into research tools, diagnostics and/or pharmaceutical therapies. The venom from one snake

RSC Drug Discovery Series No. 42
Venoms to Drugs: Venom as a Source for the Development of Human Therapeutics
Edited by Glenn F. King

Published by the Royal Society of Chemistry, www.rsc.org

may contain a large variety of differently acting protein toxins, which, at times, exert synergistic effects on target prey. The high specificity of individual toxins makes them ideally suited for examination as, or development into, human therapeutics. However, only a small number of snake venom toxins have been thoroughly characterized and evaluated for their therapeutic potential. Thus, we have hardly scratched the surface and are a long way from realizing the tremendous potential inherent to these toxins. Furthermore, the venoms of many snake (>65%) and most lizard (>95%) species have not been even rudimentarily characterized.

In the past, most research focused on venoms from reptile species that had significant impacts on human health. Anti-venom research, for example, has understandably focused on those species that cause the largest numbers of human deaths. Further in-depth research has isolated and characterized specific toxins and/or enzymes from such potentially harmful reptiles (mainly elapid and viperid snakes, as well as helodermatid lizards). Thus, a large number of "harmless" snakes (including species from the decidedly polyphyletic taxa Colubridae) and most non-helodermatid lizards have been largely ignored because their venom components are less harmful, or because these species do not readily come into conflict with humans due to their behavior or remote location.

Due to advances in technology and our ability to characterize small amounts of protein toxins, there has been a recent barrage of newly described reptile toxins. A number of concurrent discoveries have helped to clarify the vastness of the potential pool of reptile toxins. First, some recently described toxins belong to entirely new toxin families,[1] indicating a wider diversity of types of toxins than previously known. Second, analyses of venoms and venom gland transcriptomes of "non-venomous" snakes belonging to the polyphyletic Colubridae family (currently split into four families[2]) have shown the presence of unusual toxins.[1,3–5] Third, venomous secretions from non-colubroid snakes[5] and lizards[6,7] have been discovered, indicating an even wider array of species with potentially interesting toxins. Overall, the incredible number of toxins and toxin families (Table 5.1) within species in conjunction with the large number of toxin-secreting taxa evolving independently over millions of years equates to a vast library of chemically and pharmacologically diverse toxins. This cornucopia of toxins, in turn, provides a diverse array of potential pharmaceutical leads.

The majority of pharmaceuticals developed from reptile venoms have been used to treat diseases or conditions relating to blood circulation (Table 5.2), particularly conditions such as high blood pressure and blood coagulopathies. However, many of the toxins currently being studied may offer treatment for a variety of other physiological systems and conditions, including anti-virals and analgesics.

Other venomous taxa, such as marine cone snails, spiders and scorpions, also have vast and diverse arrays of interesting toxins and toxin families that are useful for prey capture (see Chapters 6–8). However, these venoms have

Table 5.1 A listing of some of the major toxin families currently known from reptile venoms.

Toxin Family	*Abbreviation*
Acetylcholinesterases	AChE
AVIT proteins	AVIT
Bradykinin potentiating peptides	BPP
Celestoxin	–
Cholecystoxins	–
Cobra venom factors	CVF
Crotamine (β-defensin)	–
Cystatins	–
Cysteine-rich secretory proteins (helveprins)	CRISP
Epididymal secretory proteins	–
Exendins	–
Factor V-like proteins	FV
Goannatyrotoxins	–
Helofensins	–
Hyaluronidases	–
Inflamins	–
L-amino acid oxidases	LAAO
Natriuretic peptides	NP
Nerve growth factors	NGF
Phosphodiesterases	PDE
Phospholipases A_2	PLA_2
Phospholipases B	PLB
Renin aspartate proteases	–
Ribonucleases	–
Sarafotoxins	SFTx
Serine proteases	SP
Factor Xa-like	FXa
Kallikrein-related	–
Serine protease inhibitors (kunitz)	–
Snake C-type lectin-like proteins	Snaclec
Snake venom metalloproteinases	SVMP
Three-finger toxins	3FTx
Vascular endothelial growth factors	VEGF
Veficolins	–
Vespryns	–
Waglerin	–
Waprins	–

evolved, in most cases, to target invertebrates and lower vertebrates. In contrast, the majority of snake species have evolved as predators of vertebrates, including mammals. Thus, snake venoms may contain a myriad of toxins that have evolved to specifically affect the physiological systems of vertebrates. In turn, this makes snake and other reptile venoms ideal for development as platforms for pharmaceuticals directed toward one mammal in particular: *Homo sapiens*.

Table 5.2 Past, present and future therapeutic treatments developed from reptile venoms

Name	*Source*	*Species*	*Toxin family*	*Target*	*Effect*	*Indication*	*Notes, other names*
Developed as therapeutic							
Captopril, enalapril, lisinopril, perindopril, ramipril, *etc.* (ACE inhibitors)	Synthetic	*Bothrops* sp., including *B. jararaca* (snake)	BPP	Bradykinin/angiotensin converting enzyme	Reduction in blood pressure	Hypertension	Other drugs available based on same platform
Tirofiban	Synthetic	*Echis carinatus* (snake)	Disintegrin (SVMP)	Glycoprotein Iia/IIIb	Antiplatelet, anticoagulation	Unstable angina/myocardial infarction	Aggrastat
Eptifibatide	Synthetic	*Sistrurus miliarius barbouri* (snake)	Disintegrin (SVMP)	Glycoprotein Iia/IIIb	Antiplatelet, anticoagulation	Unstable angina/myocardial infarction	Integrillin
Batroxobin[a]	Purified from venom	*Bothrops atrox*	SVSP	Fibrinogen	Anticoagulation	Thrombotic disorders	Hemocoagulase, Baquting[b], Reptilase
Batroxobin *moojeni*[a]	Purified from venom	*Bothrops moojeni*	SVSP	Fibrinogen	Anticoagulation	Thrombotic disorders	Defibrase
Exenatide	Synthetic	*Heloderma horridum* (lizard)	Incretin	Glucagon-like peptide-1 (agonist)	Increase insulin secretion, decrease glucagon release, decrease gastric emptying, *etc.*	Type II diabetes mellitus	Byetta, Bydureon
In development							
CD-NP	Synthetic	*Dendroaspis angusticeps* (snake)/*Homo sapiens* chimera	NP	Natriuretic peptide receptor	Natriuresis/diuresis; preservation of renal blood flow and filtration rate	Decompensated heart failure, myocardial infarction	Cenderitide

Hemocoagulase agkistrodon	Purified from venom	*Deinagkistrodon acutus* (snake)	SVSP	Fibrinogen	Blood coagulation	Capillary hemorrhage	Acutin
Crotalase	Purified from venom	*Crotalus adamanteus*	SVSP	Fibrinogen	Blood coagulation	Bleeding	
Textilinin-1	Recombinant, expressed in *E. coli*	*Pseudonaja textilis* (snake)	SVSP (Kunitz)	Plasmin	Blood coagulation	Bleeding	Q8008
Factor Xa-like protein	Purified from venom	*Pseudonaja textilis*	SVSP (FXa-like)	Prothrombinase complex, thrombin	Blood coagulation	Bleeding	Haempatch, Q8009
Factor Va-like protein	Purified from venom	*Pseudonaja textilis*	SVSP (FVa-like)	Prothrombinase complex, thrombin	Blood coagulation	Non-compresible hemorrhage	CoVase, V0801
ρ-Da1a	Purified from venom	*Dendroaspis angusticeps*	3FTx	α_{1A} β-adrenergic receptor (antagonist)	Urethral relaxation	Prostate hypertrophy	
ρ-Da1b	Purified from venom	*Dendroaspis angusticeps*	3FTx	α2 β-adrenergic receptors	Hypotension, gastrointestinal tract motility	Hypertension, impaired gastrointestinal tract motility	
Mambalgins	Recombinant	*Dendroaspis polylepis polylepis*	3FTx	Neuronal acid-sensing ion channels	Analgesic	Pain	
Prohanin	Synthetic	*Ophiophagus hannah* (snake)	3FTx	Unknown, thought to be nNOS-dependent	Analgesic, activation of enkephalin system?	Pain	Derived from hannalgesin; THA903
Nyloxin	Purified from venom	"cobra"[c]	3FTx	Nicotinc acetylcholine receptors (antagonist)	Analgesic, anti-inflammatory	Pain, inflammation	Derived from cobratoxin; Cobroxin

(continued)

Table 5.2 *(continued)*

Name	*Source*	*Species*	*Toxin family*	*Target*	*Effect*	*Indication*	*Notes, other names*
RPI-MN	Purified from venom, chemically modified	*Naja atra*[c]	3FTx	Nicotinc acetylcholine receptors (antagonist)	Undetermined	HIV, other viral infections	Derived from α-cobrotoxin; Pepteron
RPI-78M	Purified from venom, chemically modified	*Naja kaouthia*[c]	3FTx	Nicotinc acetylcholine receptors (antagonist)	Undetermined	Neurodegenerative diseases	Derived from α-cobratoxin; Receptin
Oxynor	Purified from venom	*Oxyuranus scutellatus*	PLA_2	Unknown	Wound healing	Wounds	β isoforms of taipoxin
Vicrostatin	Recombinant, expressed in *E. coli*	*Echis carinatus/ Agkistrodon contortrix contortrix* chimera	SVMP	Integrins α5β1, αvβ3, and αvβ5	Reduced tumor growth and microvascular density	Cancer	Combines portions of echistatin and contortrostatin
Development currently halted or discontinued							
Ancrod	Purified from venom	*Calloselasma rhodostoma* (snake)	SVSP	Fibrinogen	Defibrinogenation	Ischaemic stroke	Viprinex, Arvin, Arwin
Alfimeprase	Synthetic	*Agkistrodon contortrix contortrix*	SVMP	Fibrinogen	Defibrinogenation	Bleeding	Synthetic form of fibrolase

See text for references. Abbreviations: 3FTx = three-finger toxin; ACE = angiotensin-converting enzyme; BPP = bradykinin-potentiating protein; NP = natriuretic peptide; SVSP = snake venom serine protease; SVMP = snake venom metalloproteinase.

[a]Batroxobins are also used for diagnostic purposes.

[b]Baquting is batroxobin with purified factor Xa from the same venom.

[c]Species determination is difficult due to use of common names and/or older taxonomic terminology.

5.2 Anti-Hypertensive Agents

One of the most common reactions to snake bites is the immediate fear of envenomation and death. While such fear may be associated with a variety of psychological consequences such as nausea, vomiting and syncope[8] it is interesting to note that venomous snakes have evolved to possess specific venom proteins that dramatically lower the blood pressure of their victims.[9,10] These proteins include the bradykinin-potentiating peptides (BPP), natriuretic peptides (NP), and blockers of L-type (Ca_V1) voltage-gated calcium channels.

5.2.1 Bradykinin-Potentiating Peptides

Bradykinin is a proteolytic product of plasma kininogen with potent vasodilatory effects. This important molecule was identified through the study of envenomation by the South American pit viper *Bothrops jararaca*. It was shown that bradykinin exhibits strong hypotensive effects, and the "bradykinin-potentiating factors" from *B. jararaca* venom were found to enhance such effects.[11] This initial discovery led to the realization of the complex kallikrein–kinin system, as well as the development of captopril, one of the most successful examples of a venom-derived therapeutic.[12]

The development of captopril dates back to 1965, when the "bradykinin-potentiating factors" were first identified in the venom of *B. jararaca*.[13] These factors were subsequently identified to be peptides capable of potentiating the hypotensive activity of endogenous bradykinin molecules as well as inhibiting a zinc metalloproteinase known as angiotensin-converting enzyme (ACE). ACE converts the endogenous angiotensin I peptide into angiotensin II, thereby achieving a reduction in systemic blood pressure.[14,15] The history on the chemistry and development of the initial bradykinin-potentiating peptides into one of the earliest venom-derived therapeutics has been extensively reviewed.[12,16–20] Essentially, a pentapeptide from *B. jararaca* venom was identified as the lead molecule for developing an anti-hypertensive drug.[21] The minimal stable pharmacophoric element of the lead peptide was subsequently determined to be the Phe–Ala–Pro tripeptide motif (Figure 5.1F), which led to the D-2-methyl succinyl-L-proline analogue of the Ala–Pro dipeptide (Figure 5.1A).[21] This designed synthetic analogue was termed captopril, and it is capable of potently reducing blood pressure in humans. Captopril and several subsequent analogues such as enalapril (Figure 5.1B), lisinopril (Figure 5.1C), perindopril (Figure 5.1D), and ramipril (Figure 5.1E) were sequentially introduced into the market, playing distinct treatment roles in hypertension[22,23] (Figure 5.1). The discovery and development of captopril from snake venom provided the impetus to search for novel toxins for developing a broad range of novel therapeutics.[24]

The identification of bradykinin, and subsequently BPP, demonstrated that valuable pharmacologically active leads can be isolated from the plethora of peptides and proteins present in the venom of a single snake species. Although the first BPP was isolated half a century ago, novel analogs

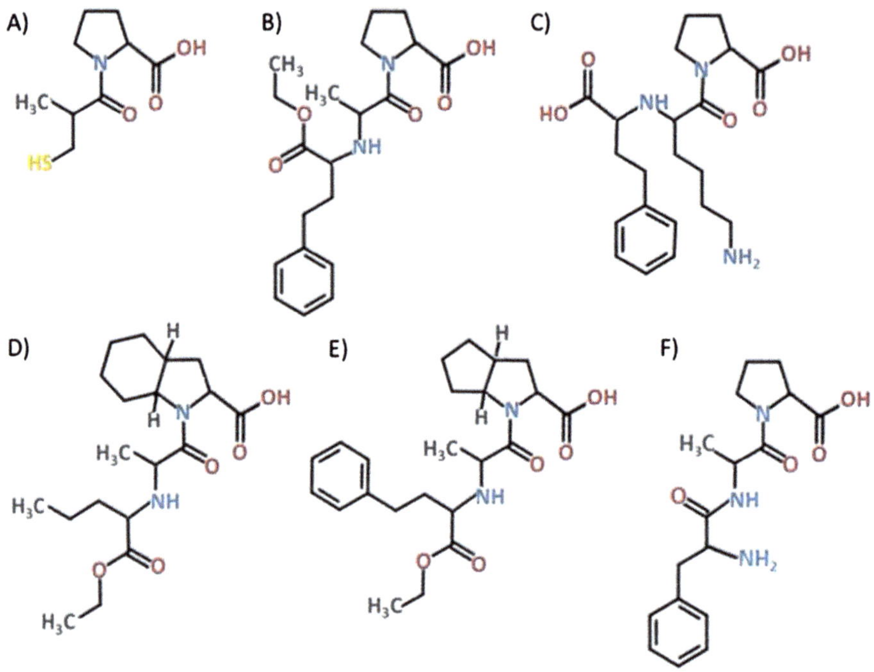

Figure 5.1 ACE inhibiting drugs developed based on the structure of a toxin isolated from snake venom. (A) captopril; (B) enalapril; (C) lisinopril; (D) perindopril; (E) ramipril; (F) Phe–Ala–Pro (FAP), the minimal stable pharmacophore of the bradykinin-potentiating pentapeptide isolated from the venom of *Bothrops jararaca*.[21]

are being identified from the same venom. For example, a recently identified decameric BPP analog was shown to interact with a novel functional target in the kidney, argininosuccinate synthetase, which is important for regulating blood pressure on a novel pathway.[25]

5.2.2 Natriuretic Peptides

NPs are cardio-protective hormones that are released from myocardial cells in response to stretching of the myocardium.[26] They play an important role in modulating blood pressure.[27] Since identification of the first NP isolated from rat atrial tissue,[27] NPs have been isolated from various tissues. The mammalian NPs include atrial natriuretic peptide (ANP),[28] B-type natriuretic peptide (BNP) isolated from porcine brain homogenate,[29] C-type natriuretic peptide (CNP) from porcine brain,[30] and urodilatin isolated from human urine.[31] Since NPs exhibit diverse effects on the cardiovascular system by not just enhancing the sodium and fluid excretion (natriuresis and diuresis) but also *via* vasorelaxation and inhibition of the renin-angiotensin system, this group of peptides has stimulated much research.

Structurally, all known NPs have a highly conserved 17-residue ring-like structure formed by a critical disulfide bridge between two intrachain cysteine residues.[32] Among the 17 residues found in this loop, over half are highly conserved across the various classes of NPs identified from various species. ANPs and BNPs have N- and C-terminal extensions beyond the conserved loop, while CNPs have only an N-terminal extension. These extensions are well conserved between members of each class.[33] The C-terminal extensions, in particular, influence receptor selectivity and hence the biological activity of the NPs, as sequential deletion of C-terminal residues reduces biological activity.[34]

Two variants of human ANP, anaritide and carperitide, as well as the recombinant form of human BNP, nesiritide, have been examined extensively in clinical trials.[35–39] While all three products have been registered in a limited number of countries for the treatment of acute decompensated heart failure,[40] there have been concerns about the efficacy and the potential adverse effects of these therapeutics on renal function and increased mortality associated with the use of NPs.[41–43] It was therefore recommended that further clinical studies be performed to examine the implications of NP usage on patient survival.[40]

In addition to the mammalian NPs, NPs from reptile venoms are also of immense research interest, particularly because not only do they have greater stability compared to mammalian homologs, but some of them also have greater potency.[40] The first reptilian NP was isolated from the venom of the Eastern green mamba (*Dendroaspis angusticeps*), and was termed *Dendroaspis* natriuretic peptide (DNP).[44] Subsequently, reptilian NPs were purified from the venoms of *Pseudocerastes persicus*,[45] *Oxyuranus microlepidotus*,[46] *Crotalus durissus cascavella* (=*C. d. terrificus*),[47] *Micrurus corallinus*,[48] *Bungarus flaviceps*,[49] and *Agkistrodon halys blomhoffii* (=*Gloydius blomhoffii blomhoffii*).[50] Recently, NPs have also been identified in the venom glands of the Mexican beaded lizard (*Heloderma horridum*).[7,51]

Venom-derived NPs were found to exhibit *in vivo* functions that are similar to mammalian NPs. In view of their hypotensive potency and resistance to renal endopeptidase degradation,[52] venom NPs such as DNP are interesting leads for the development of novel therapeutic agents. As such, chimeric peptides aimed at optimizing the functional and physical attributes of venom NPs have been designed. These include vasonatrin, which comprises a CNP molecule with an attached C-terminal tail of ANP,[53] as well as CD-NP, a fusion of CNP with the 15-residue C-terminal tail of DNP.[54] CD-NP, in particular, is currently being investigated extensively for treatment of acute decompensated heart failure as well as acute myocardial infarction. Unlike the parent NP molecules, CD-NP affects natriuresis and diuresis, while having a renal protective effect by preserving renal blood flow and filtration rate.[40,55] Through careful protein engineering, the venom-derived DNP has contributed to the design of a chimeric peptide that bypasses the adverse effects of both its progenitor molecules.

5.2.3 L-Type Ca^{2+} Channel Blocker

In addition to BPPs and NPs, another class of venom-derived protein that can influence blood pressure is peptides that specifically target L-type voltage-gated calcium (Ca_V1) channels without affecting other Ca_V channel subtypes. Ca_V channels are critical for mediating a multitude of physiological processes such as muscle contraction and neurotransmitter release.

Ca_V1 channel blockers that were isolated and identified from snake venoms include calciseptine and FS2 toxin isolated from the venom of the black mamba (*Dendroaspis polylepis polylepis*),[56,57] $C_{10}S_2C_2$ isolated from the venom of the eastern green mamba (*Dendroaspis angusticeps*),[58] and S4C8 isolated from the venom of the eastern Jameson's mamba (*Dendroaspis jamesoni kaimosae*).[59] Similar to classical Ca_V1 channel blockers such as dihydropyridine and nifedipine, calciseptine exhibits tissue-specific inhibition of Ca_V1 channels, with specific and complete inhibition of calcium channels in the cardiovascular system, partial inhibition of neuronal calcium channels, and no activity against skeletal muscle tissue.[56] In fact, calciseptine and FS2 toxin were suggested to be even more potent in inducing sustained hypotension than nifedipine.[60]

These Ca_V1 channel blockers belong to the family of three-finger toxins (3FTxs).[61] This family of non-enzymatic proteins is characterized by the three finger-like β-stranded loops held together by a central core interlaced by four conserved intramolecular disulfide bridges. While 3FTxs as a whole may present with diverse functions, ranging from neurotoxicity to ion channel blocking activity or even cytolysis, it is well established that the finger-like loops serve as conformational scaffolds for presenting bioactive motifs for interaction with their target receptors.[62] Using "proline bracket theory" (which theorizes that proline residues are frequently observed on the flanks of bioactive motifs involved in protein–protein recognition and interaction[63]), Kini *et al.* predicted that the Ca_V1 channel-binding motif of calciseptine would be located between Pro42 and Pro47.[64] A synthetic 8-mer peptide, L-calchin, was designed based on this prediction, and the peptide exhibited a similar dose-dependent and voltage-independent inhibition of Ca_V1 channels.[64] Although venom-derived Ca_V1 channel blockers were eventually not pursued as therapeutic leads due to similarity in mechanisms of action to established chemical drugs, the route of discovery and design of L-calchin demonstrated that it is feasible to develop venom toxins into potential therapeutic agents, as further illustrated by the many other examples in this chapter.

5.3 Antiplatelet Agents

Thrombocytes, or platelets, are found circulating in the bloodstream. Although these small discoid bodies are not true cells and lack a cell nucleus, they contain many receptors and molecules that are important in executing their principal task of initiating platelet aggregation, which complements blood coagulation and prevents excessive blood loss from vascular injuries.

The response of platelets to vascular damage is divided into three phases. The first phase involves the adhesion of the resting platelet to multiple ligands and factors of the subendothelial layer—including collagen, von Willebrand factor (vWF) and fibronectin—through multiple receptors such as glycoprotein Ib-IX and integrin $\alpha_5\beta_1$ present on platelet surfaces. This initial interaction between the exposed subendothelial matrices and the platelet receptors triggers the platelet to release a panel of soluble agonists, including collagen, thrombin, thromboxane A_2, and ADP, which together initiate the active recruitment of more platelets to begin the platelet aggregation process. Platelet aggregation then occurs *via* the interaction between the platelet integrin receptor $\alpha_{IIB}\beta_3$ and fibrinogen to form the platelet thrombus needed to plug the gaping vessel wall. This initial platelet plug also triggers the subsequent activation of the blood coagulation cascade.[65] As platelets are involved in the critical role of hemostasis, it is understandable that any deviation from this well-regulated pathway can result in either excessive bleeding or thrombogenic vascular diseases.

Both inducers and inhibitors of platelet aggregation have been identified from the venoms of reptiles. In particular, snake venoms have been shown to possess a wide array of toxin molecules that can affect platelet aggregation. Some of these have been extensively reviewed.[66] Venom-derived platelet aggregation inducers include toxins with enzymatic activities such as serine proteinases and PLA_2s as well as non-enzymatic toxins such as lectins and other factors that interact with plasma factors to initiate platelet activation. There is an even more impressive array of venom-derived platelet aggregation inhibitors comprising numerous toxins purified from multiple species. These inhibitors can be categorized into the following six classes: (1) non-enzymatic toxin Class I—antagonists of glycoprotein IIb/IIIa complex ($\alpha_{IIB}\beta_3$ integrin); (2) non-enzymatic toxin Class II—antagonists of vWF receptor glycoprotein Ib; (3) non-enzymatic toxin Class III—antagonists of the collagen-platelet interaction; (4) non-enzymatic toxin Class IV—antagonists of thrombin–thrombin receptor; (5) non-enzymatic toxin Class V; and (6) enzymatic toxin inhibitors.[66] The most well-studied group is the non-enzymatic toxin Class I, named disintegrins, which antagonize the interaction between fibrinogen and its platelet integrin receptor $\alpha_{IIB}\beta_3$. These low molecular weight proteins were first discovered from the venoms of vipers in the late 1980s and were shown to be potent inhibitors of platelet aggregation.[67] Since then, over 50 different disintegrins have been purified from snake venoms,[68] and they are classified into three functional groups according to their integrin selectivity and their bioactive tripeptide motifs.[69] They have an Arg–Gly–Asp (RGD) tripeptide motif on a loop structure and interfere in the final common step of platelet aggregation in which the integrin $\alpha_{IIB}\beta_3$ receptor binds to the RGD motif present on target proteins such as fibrinogen, fibronectin, vitronectin and vWF.[70] It was subsequently shown that the neighboring residues and the consequent "microenvironment" around the RGD motif do influence the binding affinity.[71,72] Disintegrins isolated from snake venoms were

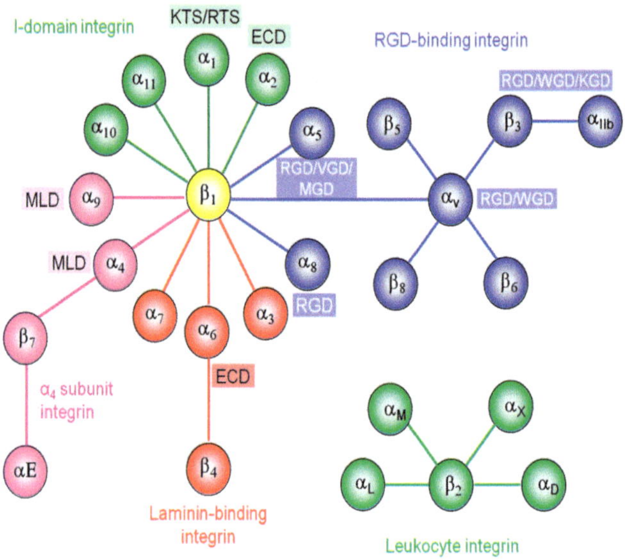

Figure 5.2 Schematic of the integrin family and corresponding tripeptide motif inhibitors. Snake venom disintegrins are potent antagonists of integrins, which are heterodimers with an α- and a β-subunit. The RGD tripeptide represents the most common bioactive motif on the integrin-binding loop of the disintegrin molecule, and it is specific towards $\alpha_{IIb}\beta_3$, $\alpha_V\beta_3$, and $\alpha_5\beta_1$ integrins. Substitution of the crucial tripeptide motif within the disintegrin molecule's binding loop results in a change of binding specificity to the remaining integrin heterodimeric receptors. The permutation of the 18 α-subunits and 8 β-subunits known to date readily generates at least 24 possible heterodimeric combinations. Adapted from Calvete *et al.*[75] and Rüegg *et al.*[73]

among the first described RGD-containing exogenous inhibitors of platelet aggregation.

In some disintegrins, although not so common, the RGD tripeptide is replaced by several distinct tripeptide motifs, including KGD, MLD, WGD, VGD, ECD, MGD, RTS, or KTS.[73–76] Depending on the active tripeptide motif, the various disintegrins can interact with specific subtypes of integrin receptors, thereby inhibiting the corresponding biological activity (Figure 5.2). Although most disintegrins in snake venoms are monomers, some are found as homo- or heterodimers.[77–79] Except for two disintegrin molecules that bear the KGD tripeptide motif, all disintegrin molecules that target the $\alpha_{IIB}\beta_3$ integrin bear the RGD tripeptide motif.[66] Inspired by the potency of disintegrins, a series of inhibitors of platelet aggregation were designed, eventually giving rise to two $\alpha_{IIB}\beta_3$ inhibitors approved for use as antiplatelet drugs in humans.[80]

Although the RGD motif is critical for binding and inhibiting the activity of the integrin $\alpha_{IIB}\beta_3$ receptor, the generic RGD motif can potentially interact with other integrin receptors (Figure 5.2), leading to undesirable side effects.[81] Therefore, various disintegrins were screened for specificity in integrin binding. The first antiplatelet drug derived from a snake venom disintegrin was Tirofiban (Aggrastat®). Tirofiban is a non-peptidic molecule that was designed based on the structure of the RGD tripeptide motif present in the parent disintegrin molecule, echistatin, purified from the venom of *Echis carinatus*.[82] Through the synthesis and optimization of the RGD-bearing lead peptide, a peptidomimetic was developed and optimized, and this ultimately led to the development of this selective inhibitor of platelet aggregation.[83,84]

The second disintegrin-inspired antiplatelet drug was eptifibatide, which was designed based on the structure of barbourin from the venom of the southeastern pigmy rattlesnake, *Sistrurus miliarius barbouri*.[81,85] Unlike the vast majority of disintegrins that interact with the $\alpha_{IIB}\beta_3$ receptors, barbourin possesses a KGD tripeptide motif that exhibits better specificity towards the $\alpha_{IIB}\beta_3$ receptor, albeit at slightly lower activity compared to the RGD motif.[85] Through extensive modification and optimization, the KGD lead peptide from barbourin was eventually designed into a cyclic heptapeptide analog, eptifibatide (Integrillin®), which not only demonstrated high specificity and potency towards the $\alpha_{IIB}\beta_3$ receptor, but also showed resistance to proteolytic degradation.[86,87]

5.4 Disintegrins and Cancer Metastasis

Noting the broad range of possible permutations in the formation of integrin heterodimers (Figure 5.2) as well as the wide distribution of integrin receptors in various tissues in the body, it is not surprising that integrin receptors are involved in a wide variety of signaling pathways as well as cell–cell or cell–matrix interactions. These integrin-mediated physiological pathways include, but are not limited to, antiplatelet effects,[67] cytoskeletal organization of cells,[88] neutrophil migration and apoptosis,[89,90] angiogenesis,[91] and tumor cell metastasis.[92–94]

One of the most important, and consequently most extensively studied, activities of snake venom disintegrins is their potent inhibitory effect on angiogenesis. The first snake-venom disintegrin shown to have potent anti-angiogenic properties was triflavin, derived from the venom of *Trimeresurus flavoviridis* (currently *Protobothrops flavoviridis*). The RGD-bearing triflavin was shown to be highly potent in inhibiting endothelial cell migration, as well as in inhibiting TNF-α induced angiogenesis in chicken chorioallantoic membrane assays.[95] Subsequently, several other RGD- or non-RGD bearing heterodimeric or homodimeric disintegrins were similarly found to have anti-angiogenic, anti-metastatic, and/or anti-tumor effects in both *in vitro* and *in vivo* models.[96] A particularly well-studied toxin from the family of snake-venom disintegrins is contortrostatin, isolated from the venom of the North American copperhead (*Agkistrodon contortrix contortrix*), which provides an exemplary model of disintegrin-mediated inhibition of angiogenesis. Contortrostatin was first characterized by Zhou *et al.*, who demonstrated inhibition of angiogenesis and tumor metastasis in several mouse models of human cancers as well as inhibition of adhesion and migration of human umbilical vein endothelial cells.[78,91–93] Subsequently, a chimeric recombinant version of contortrostatin was generated by creating a fusion protein of the C-terminal tail of another viperid disintegrin, echistatin, to the original contortrostatin sequence.[94] Not only was this chimeric disintegrin easily produced by recombinant expression, it also was shown to be capable of inhibiting multiple integrin targets, thus achieving potent anti-angiogenic and pro-apoptotic effects.

As illustrated in Figure 5.2, disintegrins with differing bioactive tripeptide motifs have significantly different affinity towards the various subtypes of integrin receptors. In selected cases, promiscuity of the disintegrin to various integrin subtypes was also observed. It would be interesting to determine if this group of molecules could generate a novel class of anti-tumor agents *via* the integrin or integrin-related pathways.

Over the past two decades, a substantial number of disintegrin molecules with various structural attributes that exhibit potent anti-angiogenic effects have been identified and characterized. The reader is referred to the list of disintegrin molecules that has been concisely reviewed by Selistre-de-Araujo *et al.*[96]

5.5 Pro- and Anti-Coagulant Factors from Snake Venom

Proper hemostasis is essential for vertebrates, and several snake venom toxins have evolved to disrupt hemostasis in their prey. Cessation of blood flow to essential organs (especially the brain, heart, and lungs) can rapidly impair the prey's ability to function, thereby reducing its capacity to flee from the predatory reptile. The blood coagulation cascade is a highly complex biological system that allows for repair of compromised circulation through a number of factors. When working properly, the coagulation cascade causes blood to clot at the site of injury, thereby reducing overall loss of blood. Later, after repair of the injury, it also dissolves the clot in order to restore proper blood flow.

The coagulation cascade culminates in the activation of prothrombin to thrombin, which plays a key role in blood coagulation and platelet aggregation. Thrombin is a serine protease that is responsible for cleaving fibrinogen into fibrin and activating factor XIII to factor XIIIa. The fibrin polymerizes to form a mesh that is cross-linked by the action of factor XIIIa, thereby creating a blood clot. These fibrin clots are essential for wound closure and eventual healing.

Fibrin clots are naturally degraded by plasmin, which breaks fibrin molecules into smaller pieces that can then be metabolized by circulating proteases or cleared through the liver and kidneys. However, clots that do not degrade in a normal and timely fashion can become dislodged from one site to travel through the circulatory system. If this occurs, it is possible for clots to lodge in blood vessels and cause their obstruction (ischemia), thereby leading to serious complications, including pulmonary embolism, myocardial infarction and stroke. In some cases, blood clots can form without a causative injury due to some form of hypercoagulopathy, such as in deep vein thrombosis.

5.5.1 Snake Venom Thrombin-Like Enzymes (TLEs)

Thrombin-like enzymes (TLEs) are single-chain serine proteases (for example, see ref. 97) that are widely distributed within several pit viper genera (*e.g. Agkistrodon*, *Bothrops*, *Lachesis* and *Trimeresurus*) as well as some true viper genera (*Bitis* and *Cerastes*) and one colubrid species, *Dispholidus typus* (for an inventory and reviews, see ref. 98–100). The isoform from *Cerastes cerastes* is a rare exception that is reported to consist of two identical disulfide-linked chains.[101] TLEs preferentially release either fibrinopeptide A or B but rarely both with equal efficiency, unlike thrombin.[99,102] They act on blood plasma, usually forming friable and translucent clots presumably due to lack of cross-linking of fibrin by factor XIIIa. TLEs are not inhibited by classical serine protease inhibitors, either endogenous thrombin inhibitor antithrombin III or exogenous thrombin inhibitor hirudin.[99,102,103] TLEs have been used to develop defibrinogenating therapeutic drugs to reduce clot formation under various circumstances. One of these drugs, ancrod, is isolated and purified from the venom of the Malayan pitviper *Calloselasma rhodostoma*, and it has been marketed under the names Viprinex, Arwin, and Arvin. Ancrod has the potential to cause defibrinogenation, and it has been examined for use in various blood-clot-related conditions. An examination of ancrod administration to healthy individuals indicated that it degrades fibrinogen, thereby releasing fibrinopeptide A. This leads to the formation of the desAA-fibrin monomer but also significant amounts of desA-profibrin (at low concentrations of ancrod), which is converted to desAA-fibrin at high concentrations of ancrod.[104] It has been considered for treatment of deep vein thrombosis[105] as well as retinal vein thrombosis,[106] although clinical studies did not show a significant improvement in treated patients. Positive results were noted in a study examining ancrod's effects on circulatory insufficiency, with patients exhibiting improvements in measurements of blood flow.[107] This same study noted that the intravenous route for ancrod therapy makes it undesirable to use in the long term, but states that it could be useful for periods of time when increased blood flow is necessary, such as during skin healing after limb amputation.[107]

For the most part, ancrod has been examined as a treatment for ischemic stroke, with equivocal results over the course of multiple clinical trials. Initially, there were indications that ancrod treatment in stroke sufferers led to greater mobility after 3 months,[108,109] the positive results of which led to further, larger-scale studies. In the first follow-up study, those patients receiving ancrod had reduced levels of plasminogen, but there was no difference between them and those receiving placebo in terms of neurological stroke scale score.[110] For patients with fibrinogen levels below the median of the treatment group (130 mg dl^{-1}), the stroke scale score was significantly better than placebo. Overall, major bleeding complications were not seen in this study. Although results were encouraging, they were not conclusive, and the study size was relatively small ($n = 64$ treatment; $n = 68$ placebo). Other clinical studies have indicated that ancrod use within 3 h of stroke has a slight benefit over placebo,[111] but not with a larger sample size and with initiation time extended to 6 h.[112,113] A review of all clinical studies of ancrod use indicated that stroke recurrence, dependency, and death were all reduced with use, but that more data are needed regarding these trends.[114] A major problem in determining overall trends with ancrod use is that many of the studies conducted thus far have confounding factors that preclude definitive conclusions (different study end points, dosage, regimens, *etc.*). It is possible that ancrod may find use for only certain indications, rather than as a blanket treatment for all stroke sufferers,[114] and it has been hypothesized that secondary blockage of microvasculature of the brain may account for some of the negative effects of ancrod.[115]

The amino acid sequence of the mature ancrod protein consists of 234 residues with five confirmed glycosylation sites.[116] The full sequence derived from cDNA includes a total of 258 amino acid residues with an 18-residue signal peptide and a six-residue propeptide.[97] This arrangement is similar to that of another snake-venom TLE, batroxobin (described below), with 58% identity in the mature proteins, six disulfide bridges, and conserved putative functional catalytic residues (His43, Asp88, and Ser182). An interesting difference between the two is the presence of an extra (13th) cysteine residue (Cys180) in ancrod near Ser182, which indicates the potential for dimerization.

Batroxobin was isolated from the venoms of South American pit vipers of the genus *Bothrops*. There are currently two forms of batroxobin utilized as therapeutic agents, one from *B. atrox* (termed batroxobin, Hemocoagulase®) and one from *B. moojeni* (termed batroxobin moojeni or Defibrase®), and a third from *B. marajoensis* has also been isolated.[117] All have similar modes of action, although they are of different size, with the *atrox* form at approximately 43 kDa in size, the *moojeni* form at around 36 kDa, and the *marajoensis* at about 41.5 kDa. The two major forms (*atrox* and *moojeni*) both release fibrinopeptide A through cleavage of the Aα-chain of fibrinogen between the Arg16 and Gly17 bond. These snake-venom TLEs belong to class A.[118–120] Although it initially causes creation of thrombi, batroxobin ultimately leads to dissolution of these clots and prevention of further clots by both degrading available fibrinogen and causing tissue plasminogen activator (t-PA) release from blood vessel endothelial cells. Freshly released t-PA is responsible for converting plasminogen into plasmin, which degrades the fibrin clots.

Batroxobin is currently used for treatment of many thrombotic disorders, and is being examined for many others. It is not approved by the United States Food and Drug Administration (FDA), so it is currently only available outside the USA.[121] Batroxobin from both *B. atrox* (Hemocoagulase®) and *B. moojeni* (Defibrase®) are utilized for various diagnostic purposes (Pentapharm, Switzerland; www.pentapharm.com), and the *atrox* form is also combined with factor Xa and produced as Baquting (Nuokang Biopharma, China).[121] Among the diverse uses for which batroxobin has been examined are acute ischemic stroke,[114] pulmonary hemorrhage in premature infants,[122,123] sudden sensorineural hearing loss,[124,125] deep vein thrombosis,[126] autoimmune encephalomyelitis,[127] and peri-operative bleeding,[121,128] among others.[129,130] Unlike some other pharmaceutical leads arising from reptile venoms, batroxobin is purified from the crude venom of the snake and not in an optimized or synthetic form. The important catalytic residues of batroxobin are His41, Asp86, and Ser178,[131] all of which originate from different exons,[132] and it has 12 cysteine residues leading to six disulfide bonds.[131]

A third snake-venom TLE, hemocoagulase agkistrodon, was isolated from the Chinese moccasin snake *Deinagkistrodon acutus* (formerly *Agkistrodon acutus*). It was also called acutin and is currently being examined for its coagulant function for use in decreasing clotting times for surgical incisions.[133] The snake-venom TLE crotalase (from *Crotalus adamanteus*) has been examined for its coagulant function, and its amino acid sequence has been reported.[134] However, it has not yet been examined for human therapeutic effects. Other snake-venom TLEs have been discovered but have not yet undergone therapeutic-type testing (*e.g.*, shedaonase,[135] agacutase,[136] Cdc SI and Cdc SII[137]).

Defibrinogenating enzymes thus far developed from snake venom have shown some positive results, but it is generally difficult to provide conclusive interpretations of their medical use. Batroxobin has not undergone the rigorous trials mandated by the FDA, so its potential has not benefited from stringent scientific testing. Ancrod, on the other hand, has undergone such tests, but with equivocal results—it may be that it has therapeutic value for patients within strict criteria. However, these are only a few of over 100 isolated proteins from this family,[118,119] so the potential for toxins of therapeutic value remains high.

5.5.2 Factor Xa-Like Proteins

Physiologically, for conversion of prothrombin to thrombin to occur, factor Xa and factor Va must combine as the prothrombinase complex in the presence of Ca^{2+} ions and phospholipids. Thrombin can then be utilized in the remainder of the coagulation pathway and lead to a clot. Some snake venoms have factor Xa-like proteins that mimic the effects of factor Xa, thereby leading to clot formation.[138–142] The venom of the Australian brown snake, *Pseudonaja textilis*, was shown to have a substance with factor Xa-like activity, which was later found to be composed of both factor Xa-like and factor Va-like subunits.[143,144] The factor Xa-like substance was later purified from the complex, and it is currently being developed as a treatment to stop bleeding at surgical sites under the name Haempatch (or Q8009).[145]

Q8009 has been sequenced by cloning of the cDNA transcript as well as by N-terminal amino acid sequencing of its two different heavy and light chain components.[146] It is reported to have 48% sequence similarity to human factor Xa and cleave prothrombin at the same site as human factor Xa, but it is active in clotting citrated blood without the need for Ca^{2+}, phospholipid, or factor Va. Its use has been shown to reduce blood loss in many different physiological models,[147,148] and its overall clotting effect is 50 to 70 times greater than that of thrombin.[145] A recombinant form of Q8009, expressed in mammalian cells, has shown similar activity, but it has not yet been produced in amounts large enough for commercial use.[145]

5.5.3 Factor Va-Like Proteins

The factor Va-like component of the venom of *P. textilis* is also being developed as a potential pharmaceutical under the name V0801 or CoVase.[145] In the mammalian system, factor Va is an essential part of the prothrombinase complex, and its level of availability may limit the formation of the complex and thus later blood clotting. Unlike human factor V, V0801 is in the active form, and it appears to be resistant to activated Protein C cleavage.[149,150] This molecule may, therefore, have blood clotting ability, and it is being examined in cases where systemic clotting factors are more useful than site-specific ones, such as in non-compressible haemorrhage.[145]

5.5.4 Fibrin(Ogen)Olytic Metalloproteinases (FMPs)

Besides snake-venom TLEs, snake venom metalloproteinases (SVMPs) have been examined for their potential therapeutic use in reducing blood clot formation. The main example of this is the FMP fibrolase, which was isolated from the venom of the North American copperhead snake, *Agkistrodon contortrix contortrix*. A synthetic construct of fibrolase, alfimeprase, has been examined for its human therapeutic value. It contains 201 amino acid residues with the first two N-terminal residues of fibrolase removed and a subsequent substitution of arginine to serine. This modification gives it greater stability and a prolonged shelf life. Alfimeprase targets the Aα-chain of fibrinogen with much greater affinity than the Bβ-chain, and it has no affinity at all for the γ-chain. It is inhibited by α_2-macroglobulin, similar to other serum proteases. Thus, outside the area of therapuetic interest (where a thrombus may be), alfimeprase is rapidly trapped/inhibited by the body's normal defenses, and so does not cause secondary side effects.[151]

Alfimeprase was examined as a treatment for blockage of both peripheral arteries and central venous access devices (catheters).[152] It gave promising results for both conditions in Phase I and II clinical trials,[153,154] but failed to reach the targets for Phase II trials.[155,156] However, some still think that it may be useful as a treatment, and that further study selecting different end points is still warranted.[152,156]

5.5.5 Anti-Fibrinolytics

Besides their ability to dissolve clots, many snake venom toxins have the capacity to maintain the presence of clots by acting on plasmin, the main factor responsible for dissolution of blood clots. One specific plasmin inhibitor from the Australian brown snake, *P. textilis*, is currently under development for its potential to reduce blood loss.[145] Textilinin-1 is a serine protease inhibitor that was initially found to have specific inhibitory activities towards plasmin and trypsin, but with minimal effects towards other serine proteases.[157] It is a Kunitz-type serine protease that is 6688 Da in mass, and it is one of two isoforms of textilinin found in the venom. Both of these forms were found to reduce blood loss using a mouse tail model, and four additional forms of textilinin were isolated from a cDNA library of the venom gland.[158] However, of these isoforms, only textilinin-1 and -2 were found to inhibit plasmin and result in reduction of blood loss.[158]

Textilinin-1 is being developed under the moniker Q8008 by Venomics Pty Ltd (a subsidiary of QRxPharma) as an alternative to aprotinin, a bovine plasmin inhibitor that was once a leading anti-fibrinolytic drug.[145] The non-specific effects of aprotinin, and its tendency to accumulate in the kidneys, are thought to be the reasons for increased risk of heart attack, stroke, and kidney failure with its use. Because of these safety concerns, the marketing of aprotinin (under the name of Trasylol and produced by Bayer) as a pharmaceutical was suspended.[159] Although recent recommendations have been made to lift the suspension of aprotinin,[160] the more specific actions of textilinin-1 make it attractive for drug development.[161,162] Recombinant Q8008 is expressed in *Escherichia coli*, and its structure, determined by X-ray crystallography, shows a similar overall fold to aprotinin[163,164] (Figure 5.3).

5.5.6 Direct Thrombin Inhibitors

The drug Exanta, produced by AstraZeneca, has been previously reported to be a direct thrombin inhibitor based on the structure of a toxin from cobra venom.[22,121] Exanta works directly on thrombin by binding to its active site,[165] inactivating both the bound and free forms[166] (thereby reducing blood coagulation). It appears, however, that the idea that Exanta was derived from a snake venom protein is an error reported in a non-peer-reviewed article,[167] because the pro-drug (ximelegatran) and active forms (melagatran) were based on structures related to human fibrinopeptide A (D. Gustafsson, pers. comm.). Recently, a PLA_2 toxin isolated from the venom of the cobra *Naja haje* has been shown as the first known direct thrombin inhibitor from snake venom, which indicates further potential for therapeutic development.[168]

5.6 Anti-Diabetic Agents

Diabetes mellitus type 2 is an important disease affecting an increasing number of people in developed and developing countries as high-fat and high-sugar foods become more prevalent in the diet. It accounts for about

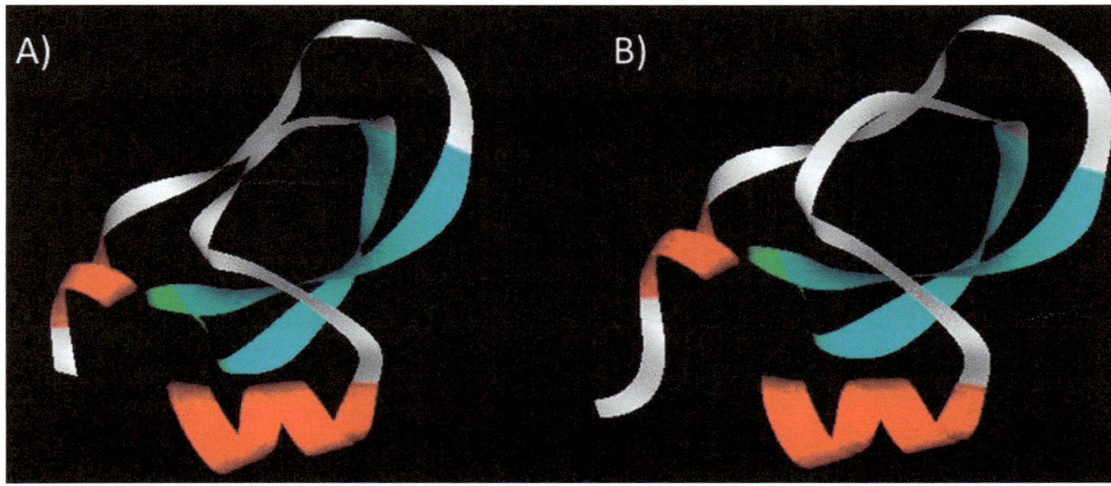

Figure 5.3 Structural comparison of (A) aprotinin and (B) textilinin-1. Aprotinin (3LDJ) is from *Bos taurus* pancreas[164] and textilinin-1 is from *Pseudonaja textilis* venom.[163] Colors indicate secondary structure, with α-helices in red, β strands in blue, T turns in green, and non-regular secondary structure in white.

90% of diabetes occurrences, and it is more common in obese individuals with a genetic disposition towards it. The disease arises due to frequent and sustained hyperglycemia, which causes insulin resistance and reduced insulin production. With insulin resistance, there is a reduction of the capability for cells to utilize glucose for normal functions, and such impairment results in long-term problems such as blindness, kidney failure, and other conditions that can eventually lead to death.

Incretins are gastrointestinal hormones that function to keep insulin at appropriate levels through various mechanisms. They can increase the activities of insulin-secreting beta cells in the kidneys, reduce gastric emptying (which limits the entrance of glucose into the bloodstream), and inhibit release of glucagon from renal alpha cells. In general, incretins help to maintain proper levels of insulin, thereby reducing hyperglycemic conditions. Two hormones, glucagon-like peptide-1 (GLP-1) and gastric inhibitory peptide (GIP), appear to function as incretins, but their efficacy is limited due to degradation by the enzyme dipeptidyl peptidase.

The Gila monster (*Heloderma suspectum*) is a venomous lizard found in the southwestern United States and northern Mexico. Within its venom is a toxin known as exendin-4 that has structural similarity with GLP-1, but it is degraded much more slowly than GLP-1 in the presence of dipeptidyl peptidase.[169] A similar toxin, exendin-3, is found in the venom of the Mexican beaded lizard (*Heloderma horridum*), a closely related species, and it has similar stability.[169] A truncated, synthetic form of exendin-4, known as exenatide and marketed as Byetta or Bydureon, is currently used to treat Type 2 diabetes mellitus in patients that do not respond to exercise and diet changes in combination with other drugs.[170] It has been on the market since 2005, and very few serious side effects have been reported thus far. Although there have been a few reported cases of pancreatitis in exenatide users, no definitive causation has been proven; rather, the major side effect appears to be gastric issues such as nausea.[171]

Exenatide comprises 39 amino acid residues (with exendin-3 and -4 differing only in the second and third N-terminal residues), and it has a C-terminal serine amide residue.[172] Exendin-4, which exenatide was based upon, shares 53% sequence identity with GLP-1 and 45% with glucagon (Figure 5.4)[172] and it has a mass of 4186.6 Da. There are five ways in which exenatide is thought to help regulate glucose, including increasing blood insulin concentration (especially effective when taken before a meal), reducing gastric emptying, reducing release of glucagon from the pancreas, decreasing hunger, and reducing the fat content of the liver.[173] Although originally exenatide use required twice daily injections of the drug, an extended release formulation (Bydureon) has been approved that requires only once-weekly injections. It was released in the USA in January 2012, and it initially appears to be more efficacious with fewer side effects than Byetta.[174]

Exenatide	------------------------HGEGTFTSDLSKQMEEEAVRLFIEWLKNGGPSSGAPPPS-NH_2	
Exendin-4	MPVESGLSSEDSASSESFASKIKRHGEGTFTSDLSKQMEEEAVRLFIEWLKNGGPSSGAPPPSG	
Exendin-3	MPVESGLSSEDSASSESFASKIKRHSDGTFTSDLSKQMEEEAVRLFIEWLKNGGPSSGAPPPSG	93.3% (96.9%)
GLP-1	------------------------HAEGTFTSDVSSYLEGQAAKEFIAWLVKGR---------	53.3%
Glucagon	------------------------HSQGTFTSDYSKYLDSRRAQDFVQWLMNT----------	44.8%

Figure 5.4 Alignment of the amino acid sequence of exendin-4 with similar proteins. Exenatide (top line) is the reduced, synthetic, pharmaceutical form of exendin-4, which contains an N-terminal serine amide residue. Percentages indicate similarity with exendin-4 over the shaded 30 or 29 (in the case of glucagon) functional amino acid residue segment. The percentage in parentheses indicates similarity with the full-length sequence of exendin-4. GLP-1 = glucagon-like peptide-1.

5.7 G-protein Coupled Receptor (GPCR) Antagonists/Agonists

It has been estimated that up to 2% of the human genome codes for upwards of 800 G-protein coupled receptors (GPCRs), and that between 30%[175] and 50%[176] of current drugs on the market act in some way through GPCRs. Because of this high prevalence, agonists and antagonists of GPCRs are exciting potential lead molecules for pharmaceutical development, although their targets and effects are highly variable. Structurally, GPCRs have seven transmembrane helices with six loops (three internal, three external).[175] They have wide-ranging physiological functions, although many are still considered orphan receptors with no currently known role.[177] Snake venoms have been shown to contain a few different types of protein that affect GPCRs: (a) sarafotoxins of *Atractaspis* spp., which act on endothelin receptors;[178] (b) muscarinic toxins of *Dendroaspis* spp., which act on muscarinic acetylcholine receptors;[179] (c) β-cardiotoxin of *Ophiophagus hannah*, which acts on β-adrenergic receptors (ARs);[180] (d) MTα from *Dendroaspis polylepis*, which acts specifically on the α_{2B} AR;[181] and (e) two novel toxins from *Dendroaspis angusticeps* that act on the α_{1A} or α_2 ARs,[182,183] termed ρ-Da1a and ρ-Da1b, respectively. MTα was shown to target the α_{2B} AR with high affinity, and did not affect the α_{1A}, α_{1B}, α_{2A}, or α_{2C} ARs, indicating a high level of specificity.[181]

The α_{1A} AR antagonist ρ-Da1a is currently being examined as a potential treatment for prostrate hypertrophy, as blocking the α_{1A} AR allows for relaxation of the urethra, thereby allowing urine flow.[184] The currently available drug treatment, tamsulosin, also affects the α_{1B} AR, but has the side effect of hypotension, so an agent that more specifically targets the α_{1A} AR may result in a better pharmaceutical treatment.[177] Along similar lines of research, ρ-Da1b affects α2 ARs and therefore is being examined for use in reducing hypotension and increasing gastrointestinal tract motility following surgery.[177] These two recent examples, and the prevalence of GPCRs with as yet unknown function, makes potential GPCR agonists and antagonists intriguing pharmaceutical leads.

5.8 Analgesics

Although the potential antinociceptic effects of snake venom were noted some time ago with the description of the 3FTx hannalgesin from the venom of the king cobra (*O. hannah*),[185] interest in developing analgesics from snake venom has greatly expanded recently. Prohanin, a peptide designed based on a 3FTx from *O. hannah* venom[186] (P.T.H. Wong and R.M. Kini, unpublished observations) and licensed to Theralpha (THA903) is being pursued as a potential analgesic drug. Another 3FTx, cobratoxin (CTX), is also being examined for its analgesic effects,[187,188] although the current analgesic forms of these, Cobroxin and Nyloxin, are marketed and developed as homeopathic remedies so are not subject to the rigorous standards required for FDA approval.

A recently discovered group of 3FTxs from the black mamba (*Dendroaspis polylepis polylepis*) and the green mamba (*Dendroaspis angusticeps*), termed mambalgins,[189] are being developed for their extremely strong, non-addictive analgesic effects. These toxins, as well as a toxin from the Texas coral snake (*Micrurus tener tener*) known as MitTx, are of interest because they belong to a family of toxins that specifically affect acid-sensing ion channels (ASICs), which are key mediators of inflammatory pain.[190] MitTx, however, induces hyperalgesia in animals.[191] Because these toxins affect ASICs, they hold the potential for development as analgesics targeting many different localized types of pain.[192] Although king cobras, mambas, and coral snakes all belong to the family Elapidae, and although the drug leads have analgesic effects, THA903 and the mambalgins work through entirely different mechanisms.

Besides elapid 3FTxs, work is also currently being done on the analgesic effects of toxins from snakes of the viperid genus *Crotalus*. Specifically, it was noted that venom from the neotropical rattlesnake *Crotalus durissus terrificus* had an analgesic effect in mice, and that the responsible component was of low molecular weight (approximately 3 kDa).[193] Later it was determined that this component is an agonist of peripheral κ- and δ-opioid receptors and that the nitric oxide-cyclic GMP pathway is at least partially responsible for its effects,[194,195] as is peripheral ATP-sensitive K^+ channel activation.[196] The peptide responsible for this activity has been determined to be a peptide 14 amino acid residues in length, and it has been termed crotalphine.[197] Among the factors that make this an interesting drug lead are the lack of tolerance over long treatments periods and the lack of abstinence symptoms after cessation of treatment in mice.

5.9 Other Uses—Limitless Potential

The potential therapeutic uses for toxins from reptile venoms have been realized in a few very successful instances, but can perhaps be exhibited even better through the amount and types of research now underway. Currently, snake toxins are being examined as prophylactics or treatments against a myriad of viruses (including dengue and yellow fever,[198] herpes,[199] and HIV[200]), bacteria (*Staphylococcus aureus*, *Leishmania amazonensis*, and *L. chagasi*) and flagellated parasites (*Trypanosoma cruzi*).[201]

CTX (mentioned above) is an example of the potential versatility of using reptile toxins as pharmaceuticals, as modified forms of it are currently being developed not only for analgesic purposes, but also for use in preventing viruses from entering cells. This is meant to prevent neurological damage such as that associated with encephalitis or AIDS, and the toxin is being pursued as a treatment for HIV infection (as RPI-MN, www.nutrapharma.com). Further, another form of CTX is being examined as a treatment for other neurodegenerative diseases such as myasthenia gravis, multiple sclerosis, muscular dystrophy, adrenomyeloneuropathy, and amyotrophic lateral sclerosis (as RPI-78M, www.nutrapharma.com). These products are currently undergoing clinical testing, so only the future will tell if they deliver on their potential.

Within a different toxin family, the PLA_2s, there is a toxin from the Australian coastal taipan (*Oxyuranus scutullatus*). The PLA_2 of this species, termed taipoxin, is composed of three subunits (α, β, γ), with the α unit being the major toxic portion that can have its action increased by the γ unit. The β subunit of taipoxin, however, is not lethal, and two different isoforms, β-1 and β-2, have been isolated and combined to create Oxynor. β-Taipoxin is mitogenically active and basically acts as an epidermal growth factor,[202] so Oxynor is being developed as a therapeutic for enhancing wound healing.[121]

5.10 Conclusions

Many snakes and some lizards are predators specifically adapted to incapacitate vertebrate prey through the use of venom from their oral cavities. The complex synergistic action of venom secreted by the venom gland and injected into prey by fangs or teeth has enabled such reptiles to kill and capture prey that would otherwise be unattainable. Although most venoms are complex cocktails of toxic components, many of these toxins are specific in their targeting and activities. The evolution of venom through various taxonomic lineages with different prey species has resulted overall in a plethora of components with different physiological effects. While some of these have been co-opted for use in human medicine, it is only the tip of the proverbial iceberg. Discoveries of major impact and the development of toxins into marketable products have the potential to save millions of lives, alleviate suffering, and produce huge economic benefits.

Recent advances in techniques for screening venoms for specific bioactivities (see Chapter 4) have greatly increased the prospects for targeted discovery of pharmaceutical leads.[203] With collaborations between field researchers undertaking ecological studies,[204] evolutionary biologists understanding phylogenetic relationships[2] and protein structure–function relationships,[205,206] and clinical scientists identifying needs and examining drug leads, the potential for discovering and developing new drugs from reptile venoms is very promising. The fairly recent discovery of toxic secretions from lizard species other than *Heloderma* species,[7] the discovery of new toxin families in snake venoms,[1,207] and the presence of toxins in snakes typically considered harmless all indicate that the evolutionary library of toxins is much greater than previously thought.

Acknowledgements

We would like to thank the following agencies for funding throughout the years: Biomedical Research Council (Singapore), Ministry of Education (Singapore), and the National University of Singapore Academic Grants.

References

1. G. OmPraba, A. Chapeaurouge, R. Doley, K. R. Devi, P. Padmanaban, C. Venkatraman, D. Velmurugan, Q. Lin and R. M. Kini, *J. Proteome Res.,* 2010, **9**, 1882–1893.
2. R. A. Pyron, F. T. Burbrink, G. R. Colli, A. N. de Oca, L. J. Vitt, C. A. Kuczynski and J. J. Wiens, *Mol. Phylogenet. Evol.,* 2011, **58**, 329–342.
3. A. T. Ching, A. F. Paes Leme, A. Zelanis, M. M. Rocha, F. Furtado Mde, D. A. Silva, M. R. Trugilho, S. L. da Rocha, J. Perales, P. L. Ho, S. M. Serrano and I. L. Junqueira-de-Azevedo, *J. Proteome Res.,* 2012, **11**, 1152–1162.
4. A. T. Ching, M. M. Rocha, A. F. Paes Leme, D. C. Pimenta, D. F. M. de Fatima, S. M. Serrano, P. L. Ho and I. L. Junqueira-de-Azevedo, *FEBS Lett.,* 2006, **580**, 4417–4422.
5. B. G. Fry, E. A. Undheim, S. A. Ali, J. Debono, H. Scheib, T. Ruder, T. N. Jackson, D. Morgenstern, L. Cadwallader, D. Whitehead, R. Nabuurs, L. van der Weerd, N. Vidal, K. Roelants, I. Hendrikx, S. P. Gonzalez, A. Jones, G. F. King, A. Antunes and K. Sunagar, *Mol. Cell. Proteomics,* 2013, **12**, 1881–1899.
6. B. G. Fry, N. Vidal, J. A. Norman, F. J. Vonk, H. Scheib, S. F. R. Ramjan, S. Kuruppu, K. Fung, S. B. Hedges, M. K. Richardson, W. C. Hodgson, V. Ignjatovic, R. Summerhayes and E. Kochva, *Nature,* 2006, **439**, 584–588.
7. B. G. Fry, K. Winter, J. A. Norman, K. Roelants, R. J. Nabuurs, M. J. van Osch, W. M. Teeuwisse, L. van der Weerd, J. E. McNaughtan, H. F. Kwok, H. Scheib, L. Greisman, E. Kochva, L. J. Miller, F. Gao, J. Karas, D. Scanlon, F. Lin, S. Kuruppu, C. Shaw, L. Wong and W. C. Hodgson, *Mol. Cell. Proteomics,* 2010, **9**, 2369–2390.
8. B. S. Gold, R. C. Dart and R. A. Barish, *N. Engl. J. Med.,* 2002, **347**, 347–356.
9. N. Marsh and B. Whaler, *Int. J. Biochem.,* 1978, **9**, 217–220.
10. L. C. de Mesquita, H. S. Selistre and J. R. Giglio, *Am. J. Trop. Med. Hyg.,* 1991, **44**, 345–353.

11. E. S. M. Rocha, W. T. Beraldo and G. Rosenfeld, *Am. J. Physiol.*, 1949, **156**, 261–273.
12. D. W. Cushman, M. A. Ondetti, E. M. Gordon, S. Natarajan, D. S. Karanewsky, J. Krapcho and E. W. Petrillo, Jr., *J. Cardiovasc. Pharmacol.*, 1987, **10**(Suppl. 7), S17–S30.
13. S. H. Ferreira, *Br. J. Pharmacol. Chemother.*, 1965, **24**, 163–169.
14. S. H. Ferreira, D. C. Bartelt and L. J. Greene, *Biochemistry*, 1970, **9**, 2583–2593.
15. M. A. Hayashi and A. C. Camargo, *Toxicon*, 2005, **45**, 1163–1170.
16. D. W. Cushman, H. S. Cheung, E. F. Sabo and M. A. Ondetti, *Am. J. Cardiol.*, 1982, **49**, 1390–1394.
17. D. W. Cushman and M. A. Ondetti, *Hypertension*, 1991, **17**, 589–592.
18. D. W. Cushman and M. A. Ondetti, *Nat. Med.*, 1999, **5**, 1110–1113.
19. C. Y. Koh and R. M. Kini, *Toxicon*, 2012, **59**, 497–506.
20. E. W. Petrillo, Jr., J. R. Powell, D. W. Cushman and M. A. Ondetti, *Clin. Exp. Hypertens. A*, 1987, **9**, 235–241.
21. D. W. Cushman, J. Pluscec, N. J. Williams, E. R. Weaver, E. F. Sabo, O. Kocy, H. S. Cheung and M. A. Ondetti, *Experientia*, 1973, **29**, 1032–1035.
22. J. W. Fox and S. M. Serrano, *Curr. Pharm. Design*, 2007, **13**, 2927–2934.
23. R. E. Smith and M. Ashiya, *Nat. Rev. Drug Discov.*, 2007, **6**, 597–598.
24. A. C. Camargo, D. Ianzer, J. R. Guerreiro and S. M. Serrano, *Toxicon*, 2012, **59**, 516–523.
25. J. R. Guerreiro, C. Lameu, E. F. Oliveira, C. F. Klitzke, R. L. Melo, E. Linares, O. Augusto, J. W. Fox, I. Lebrun, S. M. Serrano and A. C. Camargo, *J. Biol. Chem.*, 2009, **284**, 20022–20033.
26. P. Kinnunen, O. Vuolteenaho and H. Ruskoaho, *Endocrinology*, 1993, **132**, 1961–1970.
27. A. J. de Bold, H. B. Borenstein, A. T. Veress and H. Sonnenberg, *Life Sci.*, 1981, **28**, 89–94.
28. S. A. Atlas, H. D. Kleinert, M. J. Camargo, A. Januszewicz, J. E. Sealey, J. H. Laragh, J. W. Schilling, J. A. Lewicki, L. K. Johnson and T. Maack, *Nature*, 1984, **309**, 717–719.
29. T. Sudoh, K. Kangawa, N. Minamino and H. Matsuo, *Nature*, 1988, **332**, 78–81.
30. T. Sudoh, N. Minamino, K. Kangawa and H. Matsuo, *Biochem. Biophys. Res. Commun.*, 1990, **168**, 863–870.
31. H. Saxenhofer, A. Raselli, P. Weidmann, W. G. Forssmann, A. Bub, P. Ferrari and S. G. Shaw, *Am. J. Physiol.*, 1990, **259**, F832–F838.
32. K. S. Misono, H. Fukumi, R. T. Grammer and T. Inagami, *Biochem. Biophys. Res. Commun.*, 1984, **119**, 524–529.
33. R. Joseph, S. Pahari, W. C. Hodgson and R. M. Kini, *Curr. Drug Targets. Cardiovasc. Haematol. Disord.*, 2004, **4**, 437–459.
34. P. W. Schiller, F. Bellini, G. Dionne, L. A. Maziak, R. Garcia, A. DeLean and M. Cantin, *Biochem. Biophys. Res. Commun.*, 1986, **138**, 880–886.
35. R. J. Cody, S. A. Atlas, J. H. Laragh, S. H. Kubo, A. B. Covit, K. S. Ryman, A. Shaknovich, K. Pondolfino, M. Clark, M. J. F. Camargo, R. M. Scarborough and J. A. Lewicki, *J. Clin. Invest.*, 1986, **78**, 1362–1374.

36. M. A. Fifer, C. R. Molina, A. C. Quiroz, T. D. Giles, H. C. Herrmann, I. R. De Scheerder, D. L. Clement, S. Kubo, R. J. Cody, J. N. Cohn and M. B. Fowler, *Am. J. Cardiol.*, 1990, **65**, 211–216.
37. N. Hata, Y. Seino, T. Tsutamoto, S. Hiramitsu, N. Kaneko, T. Yoshikawa, H. Yokoyama, K. Tanaka, K. Mizuno, J. Nejima and M. Kinoshita, *Circ. J.*, 2008, **72**, 1787–1793.
38. Y. Saito, K. Nakao, K. Nishimura, A. Sugawara, K. Okumura, K. Obata, R. Sonoda, T. Ban, H. Yasue and H. Imura, *Circulation*, 1987, **76**, 115–124.
39. M. Yoshimura, H. Yasue, E. Morita, N. Sakaino, M. Jougasaki, M. Kurose, M. Mukoyama, Y. Saito, K. Nakao and H. Imura, *Circulation*, 1991, **84**, 1581–1588.
40. S. Vink, A. H. Jin, K. J. Poth, G. A. Head and P. F. Alewood, *Toxicon*, 2012, **59**, 434–445.
41. J. D. Sackner-Bernstein, M. Kowalski, M. Fox and K. Aaronson, *J. Am. Med. Assoc.*, 2005, **293**, 1900–1905.
42. J. D. Sackner-Bernstein, H. A. Skopicki and K. D. Aaronson, *Circulation*, 2005, **111**, 1487–1491.
43. M. Suwa, Y. Seino, Y. Nomachi, S. Matsuki and K. Funahashi, *Circ. J.*, 2005, **69**, 283–290.
44. H. Schweitz, P. Vigne, D. Moinier, C. Frelin and M. Lazdunski, *J. Biol. Chem.*, 1992, **267**, 13928–13932.
45. M. Amininasab, M. M. Elmi, N. Endlich, K. Endlich, N. Parekh, H. Naderi-Manesh, J. Schaller, H. Mostafavi, M. Sattler, M. N. Sarbolouki and C. Muhle-Goll, *FEBS Lett.*, 2004, **557**, 104–108.
46. B. G. Fry, J. C. Wickramaratana, S. Lemme, A. Beuve, D. Garbers, W. C. Hodgson and P. Alewood, *Biochem. Biophys. Res. Commun.*, 2005, **327**, 1011–1015.
47. J. S. Evangelista, A. M. Martins, N. R. Nascimento, C. M. Sousa, R. S. Alves, D. O. Toyama, M. H. Toyama, J. J. Evangelista, D. B. Menezes, M. C. Fonteles, M. E. Moraes and H. S. Monteiro, *Toxicon*, 2008, **52**, 737–744.
48. P. L. Ho, M. B. Soares, T. Maack, I. Gimenez, G. Puorto, M. F. Furtado and I. Raw, *Eur. J. Biochem.*, 1997, **250**, 144–149.
49. A. S. Siang, R. Doley, F. J. Vonk and R. M. Kini, *BMC Mol. Biol.*, 2010, **11**, 24.
50. S. Higuchi, N. Murayama, K. Saguchi, H. Ohi, Y. Fujita, A. C. Camargo, T. Ogawa, M. Deshimaru and M. Ohno, *Immunopharmacol.*, 1999, **44**, 129–135.
51. C. Ma, M. Yang, M. Zhou, Y. Wu, L. Wang, T. Chen, A. Ding and C. Shaw, *Peptides*, 2011, **32**, 1166–1171.
52. H. H. Chen, J. G. Lainchbury and J. C. Burnett, Jr, *J. Am. Coll. Cardiol.*, 2002, **40**, 1186–1191.
53. C. M. Wei, C. H. Kim, V. M. Miller and J. C. Burnett, Jr., *J. Clin. Invest.*, 1993, **92**, 2048–2052.
54. O. Lisy, B. K. Huntley, D. J. McCormick, P. A. Kurlansky and J. C. Burnett, Jr., *J. Am. Coll. Cardiol.*, 2008, **52**, 60–68.

55. C. Y. Lee, H. H. Chen, O. Lisy, S. Swan, C. Cannon, H. D. Lieu and J. C. Burnett, Jr., *J. Clin. Pharmacol.*, 2009, **49**, 668–673.
56. J. R. de Weille, H. Schweitz, P. Maes, A. Tartar and M. Lazdunski, *Proc. Natl. Acad. Sci. U.S.A.*, 1991, **88**, 2437–2440.
57. D. J. Strydom, *Eur. J. Biochem.*, 1977, **76**, 99–106.
58. F. J. Joubert and N. Taljaard, *Biochim. Biophys. Acta*, 1980, **623**, 449–456.
59. F. J. Joubert and N. Taljaard, *Int. J. Biochem.*, 1980, **12**, 567–574.
60. T. X. Watanabe, Y. Itahara, H. Kuroda, Y. N. Chen, T. Kimura and S. Sakakibara, *Jpn. J. Pharmacol.*, 1995, **68**, 305–313.
61. B. G. Fry, W. Wuster, R. M. Kini, V. Brusic, A. Khan, D. Venkataraman and A. P. Rooney, *J. Mol. Evol.*, 2003, **57**, 110–129.
62. R. M. Kini, *Clin. Exp. Pharmacol. Physiol.*, 2002, **29**, 815–822.
63. R. M. Kini, *Toxicon*, 1998, **36**, 1659–1670.
64. R. M. Kini, R. A. Caldwell, Q. Y. Wu, C. M. Baumgarten, J. J. Feher and H. J. Evans, *Biochemistry*, 1998, **37**, 9058–9063.
65. M. H. Buch, B. D. Prendergast and R. F. Storey, *Ther. Adv. Cardiovasc. Dis.*, 2010, **4**, 249–275.
66. R. M. Kini, *Curr. Drug Targets Cardiovasc. Haematol. Disord.*, 2004, **4**, 301–325.
67. C. H. Ouyang, Y. H. Ma, H. C. Jih and C. M. Teng, *Biochim. Biophys. Acta*, 1985, **841**, 1–7.
68. K. J. Clemetson, Q. Lu and J. M. Clemetson, *Curr. Pharm. Des.*, 2007, **13**, 2887–2892.
69. C. Marcinkiewicz, *Curr. Pharm. Des.*, 2005, **11**, 815–827.
70. D. R. Phillips, I. F. Charo, L. V. Parise and L. A. Fitzgerald, *Blood*, 1988, **71**, 831–843.
71. D. H. Farrell, P. Thiagarajan, D. W. Chung and E. W. Davie, *Proc. Natl. Acad. Sci. U.S.A.*, 1992, **89**, 10729–10732.
72. D. H. Farrell and P. Thiagarajan, *J. Biol. Chem.*, 1994, **269**, 226–231.
73. C. Ruegg, O. Dormond and A. Mariotti, *Biochim. Biophys. Acta*, 2004, **1654**, 51–67.
74. M. A. McLane, E. E. Sanchez, A. Wong, C. Paquette-Straub and J. C. Perez, *Curr. Drug Targets Cardiovasc. Haematol. Disord.*, 2004, **4**, 327–355.
75. J. J. Calvete, C. Marcinkiewicz, D. Monleon, V. Esteve, B. Celda, P. Juarez and L. Sanz, *Toxicon*, 2005, **45**, 1063–1074.
76. L. Sanz, R. Q. Chen, A. Perez, R. Hilario, P. Juarez, C. Marcinkiewicz, D. Monleon, B. Celda, Y. L. Xiong, E. Perez-Paya and J. J. Calvete, *J. Biol. Chem.*, 2005, **280**, 40714–40722.
77. N. Moiseeva, R. Bau, S. D. Swenson, F. S. Markland, Jr., J. Y. Choe, Z. J. Liu and M. Allaire, *Acta Crystallogr. D*, 2008, **64**, 466–470.
78. S. Swenson, S. Ramu and F. S. Markland, *Curr. Pharm. Design*, 2007, **13**, 2860–2871.
79. J. J. Calvete, M. P. Moreno-Murciano, R. D. Theakston, D. G. Kisiel and C. Marcinkiewicz, *Biochem. J.*, 2003, **372**, 725–734.
80. G. Proimos, *J. Thromb. Thrombolysis*, 2001, **11**, 99–110.

81. R. M. Scarborough, J. W. Rose, M. A. Naughton, D. R. Phillips, L. Nannizzi, A. Arfsten, A. M. Campbell and I. F. Charo, *J. Biol. Chem.*, 1993, **268**, 1058–1065.
82. V. Saudek, R. A. Atkinson and J. T. Pelton, *Biochemistry,* 1991, **30**, 7369–7372.
83. K. Peerlinck, I. De Lepeleire, M. Goldberg, D. Farrell, J. Barrett, E. Hand, D. Panebianco, H. Deckmyn, J. Vermylen and J. Arnout, *Circulation,* 1993, **88**, 1512–1517.
84. J. S. Barrett, G. Murphy, K. Peerlinck, I. De Lepeleire, R. J. Gould, D. Panebianco, E. Hand, H. Deckmyn, J. Vermylen and J. Arnout, *Clin. Pharmacol. Ther.*, 1994, **56**, 377–388.
85. R. M. Scarborough, J. W. Rose, M. A. Hsu, D. R. Phillips, V. A. Fried, A. M. Campbell, L. Nannizzi and I. F. Charo, *J. Biol. Chem.*, 1991, **266**, 9359–9362.
86. R. M. Scarborough, M. A. Naughton, W. Teng, J. W. Rose, D. R. Phillips, L. Nannizzi, A. Arfsten, A. M. Campbell and I. F. Charo, *J. Biol. Chem.*, 1993, **268**, 1066–1073.
87. R. M. Scarborough, *Am. Heart J.*, 1999, **138**, 1093–1104.
88. A. L. Coelho, M. S. de Freitas, A. L. Oliveira-Carvalho, V. Moura-Neto, R. B. Zingali and C. Barja-Fidalgo, *Exp. Cell Res.*, 1999, **251**, 379–387.
89. M. A. Williams and J. S. Solomkin, *J. Leukocyte Biol.*, 1999, **65**, 725–736.
90. A. L. Coelho, M. S. De Freitas, A. Mariano-Oliveira, D. C. Rapozo, L. F. Pinto, S. Niewiarowski, R. B. Zingali, C. Marcinkiewicz and C. Barja-Fidalgo, *Exp. Cell Res.*, 2004, **292**, 371–384.
91. Q. Zhou, R. P. Sherwin, C. Parrish, V. Richters, S. G. Groshen, D. Tsao-Wei and F. S. Markland, *Breast Cancer Res. Treat.*, 2000, **61**, 249–260.
92. F. S. Markland, K. Shieh, Q. Zhou, V. Golubkov, R. P. Sherwin, V. Richters and R. Sposto, *Haemostasis,* 2001, **31**, 183–191.
93. S. Swenson, F. Costa, W. Ernst, G. Fujii and F. S. Markland, *Pathophysiol. Haemost. Thromb.*, 2005, **34**, 169–176.
94. R. O. Minea, C. M. Helchowski, S. J. Zidovetzki, F. K. Costa, S. D. Swenson and F. S. Markland, Jr., *PloS One,* 2010, **5**, e10929.
95. J. R. Sheu, M. H. Yen, Y. C. Kan, W. C. Hung, P. T. Chang and H. N. Luk, *Biochim. Biophys. Acta,* 1997, **1336**, 445–454.
96. H. S. Selistre-de-Araujo, C. L. Pontes, C. F. Montenegro and A. C. Martin, *Toxins,* 2010, **2**, 2606–2621.
97. L. C. Au, S. B. Lin, J. S. Chou, G. W. Teh, K. J. Chang and C. M. Shih, *Biochem. J.*, 1993, **294**, 387–390.
98. H. Pirkle and K. Stocker, *Thromb. Haemost.*, 1991, **65**, 444–450.
99. W. R. Bell, Jr, *Drugs,* 1997, **54**(Suppl. 3), 18–30.
100. H. T. Pirkle, in *Enzymes from Snake Venom*, ed. G.S. Bailey, Alaken Inc., Fort Collins, CO, 1998, pp. 39–69.
101. F. Laraba-Djebari, M. F. Martin-Eauclaire and P. Marchot, *Toxicon,* 1992, **30**, 1399–1410.
102. D. L. Aronson, *Thromb. Haemost.*, 1976, **36**, 9–13.
103. R. A. Hutton and D. A. Warrell, *Blood Rev.*, 1993, 7, 176–189.

104. C. E. Dempfle, S. Argiriou, K. Kucher, H. Muller-Peltzer, K. Rubsamen and D. L. Heene, *Blood,* 2000, **96**, 2793–2802.
105. A. Gent, G. I. Ingram, C. L. Arocha-Pinango, P. Fenton and R. J. Buckley, *Thromb. Res.,* 1979, **14**, 61–66.
106. J. A. Davies, M. V. Merrick, A. A. Sharp and J. M. Holt, *Br. J. Haematol.,* 1972, **22**, 640.
107. J. A. Dormandy and H. L. Reid, *Angiology,* 1978, **29**, 80–88.
108. C. P. Olinger, T. G. Brott, W. G. Barsan, J. R. Hedges, P. Glas-Greenwalt, V. E. Pollak, J. Spilker and R. Eberle, *Ann. Emerg. Med.,* 1988, **17**, 1208–1209.
109. V. Hossmann, W. D. Heiss, H. Bewermeyer and G. Wiedemann, *Arch. Neurol.,* 1983, **40**, 803–808.
110. Ancrod Study Stroke Investigators, *Stroke*, 1994, **25**, 1755–1759.
111. D. G. Sherman, R. P. Atkinson, T. Chippendale, K. A. Levin, K. Ng, N. Futrell, C. Y. Hsu and D. E. Levy, *J. Am. Med. Assoc.,* 2000, **283**, 2395–2403.
112. M. G. Hennerici, R. Kay, J. Bogousslavsky, G. L. Lenzi, M. Verstraete and J. M. Orgogozo, *Lancet,* 2006, **368**, 1871–1878.
113. D. E. Levy, G. J. del Zoppo, B. M. Demaerschalk, A. M. Demchuk, H. C. Diener, G. Howard, M. Kaste, A. M. Pancioli, E. B. Ringelstein, C. Spatareanu and W. W. Wasiewski, *Stroke,* 2009, **40**, 3796–3803.
114. Z. Hao, M. Liu, C. Counsell, J. M. Wardlaw, S. Lin and X. Zhao, *Cochrane Database Systematic Rev.,* 2012, **3**, CD000091.
115. S. Liu, V. J. Marder, D. E. Levy, S. J. Wang, F. Yang, A. Paganini-Hill and M. J. Fisher, *Stroke,* 2011, **42**, 3277–3280.
116. W. Burkhart, G. F. Smith, J. L. Su, I. Parikh and H. LeVine, 3rd, *FEBS Lett.,* 1992, **297**, 297–301.
117. K. Stocker and G. H. Barlow, *Methods Enzymol.,* 1976, **45**, 214–223.
118. H. C. Castro, R. B. Zingali, M. G. Albuquerque, M. Pujol-Luz and C. R. Rodrigues, *Cell. Mol. Life Sci.,* 2004, **61**, 843–856.
119. N. Marsh and V. Williams, *Toxicon,* 2005, **45**, 1171–1181.
120. F. S. Markland, *Toxicon,* 1998, **36**, 1749–1800.
121. G. F. King, *Expert Opin. Biol. Ther.,* 2011, **11**, 1469–1484.
122. A. Lodha, M. Kamaluddeen, A. Akierman and H. Amin, *Indian J. Pediatrics,* 2011, **78**, 838–844.
123. Y. Shi, J. Zhao, S. Tang, F. Pan, L. Liu, Z. Tian and H. Li, *Indian J. Pediatrics,* 2008, **45**, 199–202.
124. T. Kubo, T. Matsunaga, H. Asai, K. Kawamoto, J. Kusakari, Y. Nomura, M. Oda, N. Yanagita, H. Niwa, T. Uemura and S. Kimune, *Arch. Otolaryngol.,* 1988, **114**, 649–652.
125. A. E. Conlin and L. S. Parnes, *Arch. Otolaryngol.,* 2007, **133**, 582–586.
126. Z. Lei, L. Shi Hong, L. Li, Y. G. Tao, W. Yong Ling, H. Senga, Y. Renchi and H. Zhong Chao, *Clin. Appl. Thromb.-Hemost.,* 2011, **17**, 75–79.
127. Y. Yang, S. J. Tian, L. Wu, D. H. Huang and W. P. Wu, *Cell. Mol. Neurobiol.,* 2011, **31**, 437–448.
128. C. Xu, A. Wu and Y. Yue, *Arch. Orthop. Trauma Surg.,* 2012, **132**, 25–31.

129. J. Wang, Y. Q. Zhu, M. H. Li, J. G. Zhao, H. Q. Tan, J. B. Wang, F. Liu and Y. S. Cheng, *J. Vasc. Interv. Radiol.*, 2011, **22**, 987–994.
130. J. Wang, Y. Q. Zhu, F. Liu, M. H. Li, J. G. Zhao, H. Q. Tan, J. B. Wang, Y. S. Cheng and P. L. Zhang, *Ann. Vasc. Surg.*, 2010, **24**, 876–884.
131. N. Itoh, N. Tanaka, S. Mihashi and I. Yamashina, *J. Biol. Chem.*, 1987, **262**, 3132–3135.
132. N. Itoh, N. Tanaka, I. Funakoshi, T. Kawasaki, S. Mihashi and I. Yamashina, *J. Biol. Chem.*, 1988, **263**, 7628–7631.
133. J. M. Wei, M. W. Zhu, Z. T. Zhang, Z. G. Jia, X. D. He, Y. L. Wan, S. Wang, D. R. Xiu, Y. Tang, J. Li, J. Y. Xu and Q. S. Heng, *Chinese Med. J.*, 2010, **123**, 589–593.
134. A. H. Henschen-Edman, I. Theodor, B. F. Edwards and H. Pirkle, *Thromb. Haemost.*, 1999, **81**, 81–86.
135. H. M. Jiao, L. X. Yang, B. Lu, Y. Q. Wu and Y. C. Zhou, *Acta Biochim. Biophys. Sin.*, 2005, **37**, 835–842.
136. S. S. Tang, X. H. Wang, J. H. Zhang, B. S. Tang, L. Qian, P. Y. Li and L. W. Luo, *Eur. J. Pharm. Sci.*, 2013, **49**, 90–98.
137. A. C. Patino, J. A. Pereanez, J. M. Gutierrez and A. Rucavado, *Toxicon*, 2013, **63**, 32–43.
138. P. P. Masci, A. N. Whitaker and J. de Jersey, *Biochem. Int.*, 1988, **17**, 825–835.
139. V. S. Rao, J. S. Joseph and R. M. Kini, *Biochem. J.*, 2003, **369**, 635–642.
140. G. Tans, J. W. Govers-Riemslag, J. L. van Rijn and J. Rosing, *J. Biol. Chem.*, 1985, **260**, 9366–9372.
141. H. Speijer, J. W. Govers-Riemslag, R. F. Zwaal and J. Rosing, *J. Biol. Chem.*, 1986, **261**, 13258–13267.
142. H. Speijer, J. W. Govers-Riemslag, R. F. Zwaal and J. Rosing, *Thromb. Haemost.*, 1987, **57**, 349–355.
143. V. S. Rao and R. M. Kini, *Thromb. Haemost.*, 2002, **88**, 611–619.
144. V. S. Rao, S. Swarup and R. M. Kini, *Thromb. Haemost.*, 2004, **92**, 509–521.
145. S. T. Earl, P. P. Masci, J. de Jersey, M. F. Lavin and J. Dixon, *Toxicon*, 2012, **59**, 456–463.
146. I. Filippovich, N. Sorokina, L. St Pierre, S. Flight, J. de Jersey, N. Perry, P. P. Masci and M. F. Lavin, *Br. J. Haematol.*, 2005, **131**, 237–246.
147. R. L. Warner, S. D. McClintock, A. G. Barron and F. A. de la Iglesia, *Exp. Mol. Pathol.*, 2007, **83**, 241–248.
148. R. L. Warner, S. D. McClintock, A. G. Barron and F. A. de la Iglesia, *Exp. Mol. Pathol.*, 2009, **87**, 204–211.
149. M. H. Bos, M. Boltz, L. St Pierre, P. P. Masci, J. de Jersey, M. F. Lavin and R. M. Camire, *Blood*, 2009, **114**, 686–692.
150. V. S. Rao, S. Swarup and R. M. Kini, *Blood*, 2003, **102**, 1347–1354.
151. C. F. Toombs, *Haemostasis*, 2001, **31**, 141–147.
152. J. L. Baskin, U. Reiss, J. A. Wilimas, M. L. Metzger, R. C. Ribeiro, C. H. Pui and S. C. Howard, *Haematologica*, 2012, **97**, 641–650.
153. K. Ouriel, J. Cynamon, F. A. Weaver, H. Dardik, D. Akers, J. Blebea, L. Gruneiro, C. F. Toombs, F. Wang-Clow, M. Mohler, L. Pena, C. Y. Wan and S. R. Deitcher, *J. Vasc. Interv. Radiol.*, 2005, **16**, 1075–1083.

154. M. A. Moise and V. S. Kashyap, *Expert Opin. Biol. Ther.*, 2008, **8**, 683–689.
155. F. S. Markland and S. Swenson, *Toxins*, 2010, **2**, 793–808.
156. S. M. Han, F. A. Weaver, A. J. Comerota, B. A. Perler and M. Joing, *J. Vasc. Surg.*, 2010, **51**, 600–609.
157. N. Willmott, P. Gaffney, P. Masci and A. Whitaker, *Fibrinolysis*, 1995, **9**, 1–8.
158. I. Filippovich, N. Sorokina, P. P. Masci, J. de Jersey, A. N. Whitaker, D. J. Winzor, P. J. Gaffney and M. F. Lavin, *Br. J. Haematol.*, 2002, **119**, 376–384.
159. D. T. Mangano, I. C. Tudor and C. Dietzel, *N. Engl. J. Med.*, 2006, **354**, 353–365.
160. N. Howell, E. Senanayake, N. Freemantle and D. Pagano, *J. Thoracic Cardiovasc. Surg*, 2012, **145**, 234–240.
161. S. Flight, L. Johnson, M. Trabi, P. Gaffney, M. Lavin, J. de Jersey and P. Masci, *Pathophysiol. Haemost. Thromb.*, 2005, **34**, 188–193.
162. S. M. Flight, L. A. Johnson, Q. S. Du, R. L. Warner, M. Trabi, P. J. Gaffney, M. F. Lavin, J. de Jersey and P. P. Masci, *Br. J. Haematol.*, 2009, **145**, 207–211.
163. E. K. Millers, M. Trabi, P. P. Masci, M. F. Lavin, J. de Jersey and L. W. Guddat, *FEBS J.*, 2009, **276**, 3163–3175.
164. I. S. Yang, T. G. Kim, B. S. Park, K. J. Cho, J. H. Lee, Y. Park and K. H. Kim, *Biochem. Biophys. Res. Commun.*, 2010, **397**, 429–435.
165. D. Gustafsson and M. Elg, *Thromb. Res.*, 2003, **109**(Suppl. 1), S9–S15.
166. J. Hauptmann and J. Sturzebecher, *Thromb. Res.*, 1999, **93**, 203–241.
167. P. J. Skerrett and T. H. Lee, *Harvard Heart Lett.*, 2005, **15**, 3.
168. A. V. Osipov, S. Y. Filkin, Y. V. Makarova, V. I. Tsetlin and Y. N. Utkin, *Toxicon*, 2010, **55**, 186–194.
169. A. Thum, K. Hupe-Sodmann, R. Goke, K. Voigt, B. Goke and G. P. McGregor, *Exp. Clin. Endocrinol. Diab.*, 2002, **110**, 113–118.
170. B. L. Furman, *Toxicon*, 2012, **59**, 464–471.
171. C. Ellero, J. Han, S. Bhavsar, B. B. Cirincione, M. B. Deyoung, A. L. Gray, I. Yushmanova and P. W. Anderson, *Diabet. Med.*, 2010, **27**, 1168–1173.
172. J. Eng, W. A. Kleinman, L. Singh, G. Singh and J. P. Raufman, *J. Biol. Chem.*, 1992, **267**, 7402–7405.
173. M. E. Tushuizen, M. C. Bunck, P. J. Pouwels, J. H. van Waesberghe, M. Diamant and R. J. Heine, *Liver Int.*, 2006, **26**, 1015–1017.
174. J. R. Taylor, E. Dietrich and J. G. Powell, *Clin. Ther.*, 2013, **35**, A3–A17.
175. E. Jacoby, R. Bouhelal, M. Gerspacher and K. Seuwen, *ChemMedChem*, 2006, **1**, 761–782.
176. L. Chen, L. Jin and N. Zhou, *Expert Opin. Drug Discovery*, 2012, **7**, 791–806.
177. A. Maiga, G. Mourier, L. Quinton, C. Rouget, C. Gales, C. Denis, P. Lluel, J. M. Senard, S. Palea, D. Servent and N. Gilles, *Toxicon*, 2012, **59**, 487–496.
178. E. Kochva, A. Bdolah and Z. Wollberg, *Toxicon*, 1993, **31**, 541–568.

179. D. Servent and C. Fruchart-Gaillard, *J. Neurochem.,* 2009, **109**, 1193–1202.
180. N. Rajagopalan, Y. F. Pung, Y. Z. Zhu, P. T. Wong, P. P. Kumar and R. M. Kini, *FASEB J.,* 2007, **21**, 3685–3695.
181. K. Koivula, S. Rondinelli and J. Nasman, *Toxicon,* 2010, **56**, 440–447.
182. L. Quinton, E. Girard, A. Maiga, M. Rekik, P. Lluel, G. Masuyer, M. Larregola, C. Marquer, J. Ciolek, T. Magnin, R. Wagner, J. Molgo, R. Thai, C. Fruchart-Gaillard, G. Mourier, J. Chamot-Rooke, A. Menez, S. Palea, D. Servent and N. Gilles, *Br. J. Pharmacol.,* 2010, **159**, 316–325.
183. C. Rouget, L. Quinton, A. Maiga, C. Gales, G. Masuyer, C. Malosse, J. Chamot-Rooke, R. Thai, G. Mourier, E. De Pauw, N. Gilles and D. Servent, *Br. J. Pharmacol.,* 2010, **161**, 1361–1374.
184. T. L. Bullock and G. L. Andriole, Jr., *Expert Opin. Emerg. Drugs,* 2006, **11**, 111–123.
185. X. C. Pu, P. T. Wong and P. Gopalakrishnakone, *Toxicon,* 1995, **33**, 1425–1431.
186. P. Gopalakrishnakone, X. C. Pu, P. T. H. Wong, M. C. Gwee and R. M. Kini, U.S. Patent 6,613,745, Singapore Patent 129211.
187. B. C. Cheng, X. P. Zhou, Q. Zhu, S. Gong, Z. H. Qin, P. F. Reid, L. N. Raymond, Q. Z. Yin and X. H. Jiang, *Toxicon,* 2009, **54**, 224–232.
188. G. N. Shi, Y. L. Liu, H. M. Lin, S. L. Yang, Y. L. Feng, P. F. Reid and Z. H. Qin, *Acta Pharmacol. Sin.,* 2011, **32**, 1233–1238.
189. S. Diochot, A. Baron, M. Salinas, D. Douguet, S. Scarzello, A. S. Dabert-Gay, D. Debayle, V. Friend, A. Alloui, M. Lazdunski and E. Lingueglia, *Nature,* 2012, **490**, 552–555.
190. H. J. Kweon and B. C. Suh, *BMB Reports,* 2013, **46**, 295–304.
191. C. J. Bohlen, A. T. Chesler, R. Sharif-Naeini, K. F. Medzihradszky, S. Zhou, D. King, E. E. Sanchez, A. L. Burlingame, A. I. Basbaum and D. Julius, *Nature,* 2011, **479**, 410–414.
192. A. Baron, S. Diochot, M. Salinas, E. Deval, J. Noël and E. Lingueglia, *Toxicon*, 2013, **75**, 187–204.
193. R. Giorgi, M. M. Bernardi and Y. Cury, *Toxicon,* 1993, **31**, 1257–1265.
194. G. Picolo and Y. Cury, *Life Sci.,* 2004, **75**, 559–573.
195. V. P. Gutierrez, V. O. Zambelli, G. Picolo, M. Chacur, S. C. Sampaio, P. Brigatte, K. Konno and Y. Cury, *Behav. Pharmacol.,* 2012, **23**, 14–24.
196. G. Picolo, A. C. Cassola and Y. Cury, *Eur. J. Pharmacol.,* 2003, **469**, 57–64.
197. K. Konno, G. Picolo, V. P. Gutierrez, P. Brigatte, V. O. Zambelli, A. C. Camargo and Y. Cury, *Peptides,* 2008, **29**, 1293–1304.
198. V. D. Muller, R. R. Russo, A. C. Cintra, M. A. Sartim, M. Alves-Paiva Rde, L. T. Figueiredo, S. V. Sampaio and V. H. Aquino, *Toxicon,* 2012, **59**, 507–515.
199. S. Hubbard, S. Choudhary, E. Maus, D. Shukla, S. Swenson, F. S. Markland, Jr. and V. Tiwari, *Antiviral Ther.,* 2012, **17**, 1319–1326.
200. Y. J. Zhang, J. H. Wang, W. H. Lee, Q. Wang, H. Liu, Y. T. Zheng and Y. Zhang, *Biochem. Biophys. Res. Commun.,* 2003, **309**, 598–604.

201. R. R. de Menezes, A. F. Torres, T. S. da Silva, D. F. de Sousa, D. B. Lima, D. B. Norjosa, N. A. Nogueira, M. F. Oliveira, M. R. de Oliveira, H. S. Monteiro and A. M. Martins, *Nat. Prod. Commun.,* 2012, **7**, 71–74.
202. B. V. Lipps, *Toxicon,* 2000, **38**, 1845–1854.
203. F. Heus, F. Vonk, R. A. Otvos, B. Bruyneel, A. B. Smit, H. Lingeman, M. Richardson, W. M. Niessen and J. Kool, *Toxicon,* 2012, **61**, 112–124.
204. F. J. Vonk, K. Jackson, R. Doley, F. Madaras, P. J. Mirtschin and N. Vidal, *BioEssays,* 2011, **33**, 269–279.
205. R. M. Kini and R. Doley, *Toxicon,* 2010, **56**, 855–867.
206. R. M. Kini, *Toxicon,* 2005, **45**, 1147–1161.
207. B. G. Fry, H. Scheib, I. L. M. Junqueira de Azevedo, D. A. Silva and N. R. Casewell, *Toxicon,* 2012, **59**, 696–708.

CHAPTER 6

The Molecular Diversity of Conoidean Venom Peptides and their Targets: From Basic Research to Therapeutic Applications

RUSSELL W. TEICHERT[a], BALDOMERO M. OLIVERA[a], J. MICHAEL McINTOSH[a], GRZEGORZ BULAJ[b], AND MARTIN P. HORVATH*[a]

[a]University of Utah, Department of Biology, Salt Lake City, Utah 84112, USA; [b]University of Utah, Department of Medicinal Chemistry, Salt Lake City, Utah 84112, USA
*E-mail: martin.horvath@utah.edu

6.1 Introduction

Venomous marine snails (superfamily Conoidea) have evolved a sophisticated chemical strategy to interact with other animals in their environment. Insofar as evidence is available, all venomous marine snails appear to be predatory. Because they move slowly and, unlike many other venomous predators, are devoid of mechanical weaponry for prey capture, they are particularly dependent upon the efficacy of their venom for capturing prey. This requirement for extremely effective venom provides a rationale for why the venom components have been able to achieve remarkable molecular

RSC Drug Discovery Series No. 42
Venoms to Drugs: Venom as a Source for the Development of Human Therapeutics
Edited by Glenn F. King

Published by the Royal Society of Chemistry, www.rsc.org

specificity over the course of evolution.[1,2] The ability of conoidean venom components to differentiate among closely related molecular targets makes them both important research tools and attractive lead compounds for therapeutic drugs. This chapter presents an overview of the venom components of cone snails and other venomous marine molluscs, with an emphasis on those features particularly germane to drug development.

The potential for developing individual conoidean venom components as therapeutic drugs and drug leads is already well validated. One such venom component, ω-conotoxin MVIIA, has become an approved commercial drug for the treatment of neuropathic pain (see Chapter 7 for more details). The commercial product, known as Prialt® (ziconotide), is chemically identical to the native peptide extracted from the venom of *Conus magus*. Several cone snail venom peptides have advanced from preclinical development to human clinical trials (see Table 6.1), while many others are being evaluated preclinically. In addition to their direct use as therapeutics, components of cone snail venoms may be used for target validation. Many of the peptides from *Conus* venoms are highly selective for a single molecular target. Consequently, they can be used to assess whether a particular molecular target is appropriate for initiating a drug-development program. As illustrated in Table 6.1, several *Conus* peptides have been used to identify and validate novel molecular targets for analgesia in animal pain models. ω-Conotoxin MVIIA again provides a powerful example. Although useful as an intrathecal drug for morphine-resistant patients, Prialt is a peptide and therefore not orally available, which limits its potential as a therapeutic. An incredibly valuable insight provided by ω-conotoxin MVIIA is identification of the molecular target responsible for analgesia. ω-Conotoxin MVIIA specifically inhibits the N-type voltage-gated calcium (Ca_V) channel ($Ca_V2.2$) with high selectivity, and for this reason, this particular Ca_V channel is now a major target for drug development programs to obtain orally available small molecule inhibitors (see Chapter 7 for more details). These examples demonstrate that basic scientific research into the mechanisms of conotoxins (and their unprecedented ability to discriminate among closely related targets) can lead to drugs and also to identification of drug targets.

6.1.1 Biodiversity and Phylogeny of Venomous Molluscs

A consideration of conoidean biodiversity is needed to appreciate the vast wealth of potential drugs yet to be discovered within the natural resource represented by conoidean venom components. Evolution of new species is driven by biotic interactions that also drive the differentiation of toxins. These pressures result in a huge and complex set of evolved conoidean venom toxins, making it highly probable that many other drugs will be discovered by continued study of venomous molluscs.

There are probably well over 10 000 species of venomous marine snail. All venomous snails are traditionally assigned to the superfamily Conoidea (in the older literature, the suborder Toxoglossa). The most familiar of

Table 6.1 Pipeline of *Conus* peptides that have reached preclinical and clinical development status

Conopeptide (indication) [Molecular Target]	**Preclinical**	**Phase I**	**Phase II**	**Phase III**	**Comments**
ω-MVIIA (Prialt)[101§] **(neuropathic pain)** [Cav2.2]					**Elan, approved in the U.S. and the E.U.**
ω-CVID (AM336)[102] **(cancer pain)** [Cav2.2]					**Amrad, Inc.**
χ-MrIA (XEN2174)[102] **(neuropathic pain)** [Norepinephrine Transporter]					**Xenome, Inc.**
α-Vc1.1 (ACV1)[103*] **(neuropathic pain)** [α9α10 Nicotinic Receptor]					**Metabolic, Inc.**
Contulakin-G (CGX1160)[104] **(neuropathic pain)** [Neurotensin Receptor]					**Cognetix, Inc.**
Conantokin-G (CGX1007)[105] **(epilepsy)** [NMDA Receptor]					**Cognetix, Inc.**
κ-PVIIA (CGX1051)[106] **(cardioprotection)** [Kv1 Subfamily]					**Cognetix, Inc.**
α-RgIA[107*] **(nerve injury pain)** [α9α10 Nicotinic Receptor]					**Kineta, Inc.**

§Details on the molecular targets and results from preclinical and clinical studies are reviewed in the references provided (superscripted numbers).

*An alternative target for these peptides has been suggested to be the $GABA_B$ receptor. For a discussion of these alternative targets see the review by Vetter and Lewis.[100]

venomous molluscs are the cone snails (genus *Conus*, family Conidae), and these have been the subject of recent molecular phylogenetic analyses.[3,4] Although all cone snails belong to the family Conidae, they apparently do not, as was previously thought, belong to the same genus within that family. The majority of cone snail species (~600) were assigned to the established genus *Conus*, while a substantial minority of cone snail species (~100) were assigned to one of three other genera: *Conasprella*, *Californiconus* or *Profundiconus*. Practically all of the biomedically relevant work has been done with venoms from cone snail species. While they currently dominate the biomedical literature, new exploration of other families belonging to Conoidea makes it clear that cone snails comprise only a minor fraction (less than 5%) of the biodiversity of venomous molluscs.

It is widely accepted that a large number of conoidean species have not yet been named, and indeed, it is anticipated that many small, deep-water forms are yet to be discovered.[5] Traditionally, the superfamily Conoidea was divided into three groups, each assigned family rank, as follows: the cone snails (family Conidae), the auger snails (family Terebridae) and the turrid snails (the traditional family Turridae). It has become clear that the greatest species diversity is in the last group, the turrids. Professor Gisela Concepcion and co-workers at the University of the Philippines have pioneered the exploration of the turrids. Species richness and diversity in the turrids may, in fact, reflect the existence of multiple distinct families. Recently obtained molecular phylogenetic (see Figure 6.1) and anatomical data have revealed that the traditional family Turridae is a polyphyletic assemblage.[6–8] Thus, the traditional family Turridae has become progressively more narrowly defined, and additional family groups have been proposed, such as Raphitomidae, Mangelidae, Clathurellidae, Borsonidae, Drillidae, Crassispiridae and Clavatulidae.[5,6] Each of the major new "turrid" family lineages has hundreds, if not thousands, of species.

Although the "turrids" (broadly defined) are the most biodiverse conoideans, they are the least well understood. Their remarkable adaptive radiation has produced mostly small or very small species, many in deep-water habitats. Most turrid species were previously inaccessible for investigation. While cone snails and auger snails are prominent features of the shallow-water tropical marine biota, at greater depths (>200 m), only the turrids have a significant biodiversity among the venomous molluscs. In addition, several turrid lineages have adapted better to cooler marine environments; cone snails and auger snails are not found in significant numbers outside the tropics. Because many *Conus* species are abundant in the tropical marine environments where they occur, it was far easier to access significant amounts of cone snail venoms than the venom of any turrid. However, recent advances in transcriptomics and proteomics make it increasingly facile to access peptide toxin sequences even from the tiniest venomous molluscs. The recent characterization of a venom peptide from a small turrid species, *Crassispira cerithina*,[9] is only the first of what will undoubtedly be a continuing series of such studies.

All conoidean venoms previously analyzed have proven to be extremely complex mixtures of peptidic toxins and small proteins. A recent insight from transcriptome analyses is that there is no correlation between the size of a venomous mollusc and the complexity of its venom. It appears that even a turrid species less than 3 mm in length may have a venom that is just as complex as venom from a cone snail over 15 cm in length. It should be noted, however, that in some conoidean lineages, notably the Terebridae (auger snails) and the turrid family Raphitomidae, a significant number of species have secondarily lost their venom ducts and do not produce any venom at all.

6.1.2 Overview

Conoidean venoms are highly effective because of two properties: (1) venom components work together; and (2) venom components are highly specific. Early work on *Conus* venom showed that toxins work together in a synergistic way to overthrow the neurological system of prey, predators and competitors. Section 6.2 of this chapter reviews the concept of a cabal – a set of toxins working together to achieve a physiological end-point such as paralysis. Equally important for venom efficacy, toxins have evolved to achieve ultra-precise specificity and can in many cases differentiate one target among several closely related subtypes found within a family of signaling molecules. Section 6.3 reviews this concept of molecular target specificity, drawing upon particular examples such as *Conus* toxins that inhibit the nicotinic acetylcholine receptor (nAChR) and others that interfere with normal function of the voltage-gated sodium (Na_V) channel. The chapter continues with a review of the diversity of conoidean venom components in Section 6.4. The inescapable conclusions from these sections is that conoidean venoms, in total, probably contain innumerable molecules that act on the nervous system and are evolved to be highly selective, and that many of these would be helpful for the study of behavior and treatment of neurological disease. The final Section 6.5, looks to the future and attempts to explain how we are constructing a new highly parallelized platform to find more quickly the most valuable neuroactive ligands from the vast natural resource represented by conoidean peptides.

6.2 Physiology of Envenomation

6.2.1 Cabals and Constellations

After more than three decades of research into cone snail venoms, a rationale for why these venoms are such complex mixtures of peptides has gradually emerged. Cone snail venoms contain groups of peptides that are functionally coupled; all of the peptides in such a group work in concert to produce a specific physiological end-point. A group of venom peptides that act on a set of functionally coupled molecular targets are known as "cabals",[10,11] and the corresponding set of molecular targets of a given cabal is known as a "target constellation".[12] The word *cabal* refers to a secret group of plotters collaborating to overthrow current authority. The metaphor is apt in that toxins within a cabal work together to derail nervous system control, for instance control over muscle contraction. The biological activity of any given conoidean venom can be deconvoluted through identification of the cabals present. Furthermore, the mapping of cabals to their target constellations provides powerful inferential tools for deducing the likely mechanism of action for a given venom component.

For illustrative purposes, we describe some specific aspects of cabals and target constellations using fish-hunting cone snails as our biological focus. The nAChR at the neuromuscular junction provides an example of a specific

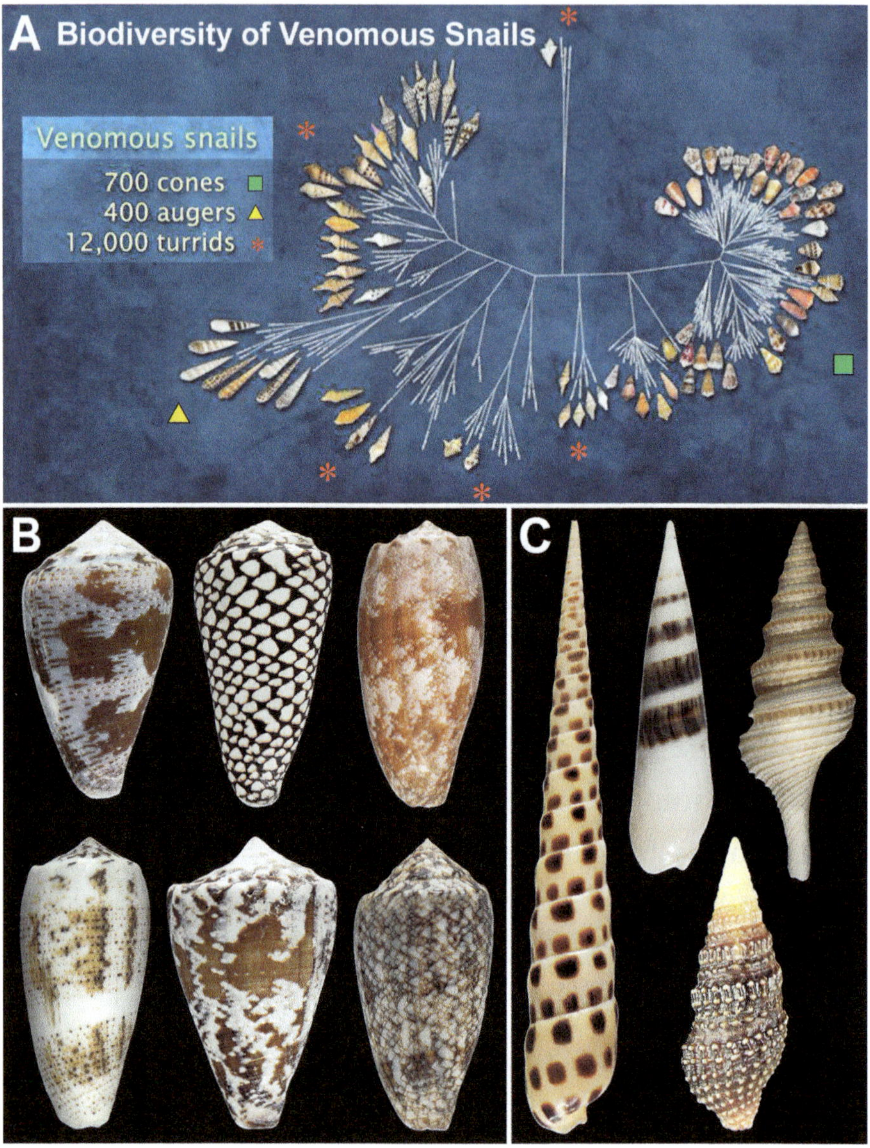

Figure 6.1 Biodiversity and pharmaceutical potential of venomous snails. (A) Phylogenetic tree of the superfamily Conoidea. A phylogenetic tree based on 12S ribosomal RNA sequences illustrates how the three classical groups of venomous marine snails assort using molecular markers. The cone snails (family Conidae, green square) comprising ~700 species, and the auger snails (family Terebridae, yellow triangle) comprising ~400 species, are monophyletic branches. This phylogenetic tree demonstrates that the classical family Turridae ("turrids", red asterisk) is not monophyletic. In general, each distinctive lineage of venomous snails has its own set of gene

molecular target in a constellation of ion channels and receptors inhibited by toxin cabals of fish-hunting cone snails. The venom of such snails contains what is known as the "motor cabal" that causes neuromuscular block in the fish prey, resulting in paralysis of the fish after venom has been injected. Typically, the motor cabal of a fish-hunting cone snail contains a peptide that blocks Na_V channels on the muscle membrane, as well as a peptide that blocks presynaptic Ca_V channels in the motor axon, thereby preventing release of acetylcholine into the synaptic cleft. Finally, there are invariably one or more conopeptide antagonists of the nAChR at the neuromuscular junction. Nicotinic antagonists that compete with the neurotransmitter acetylcholine are always present, and in certain venoms, additional peptides that are non-competitive inhibitors of the same receptor are also found. Each of the peptides of the motor cabal targets a different ligand-binding site with high affinity and selectivity. Collectively, the venom peptides act upon

---

superfamilies that are expressed in venom. Thus, identification of where in the phylogenetic tree a species belongs immediately suggests which gene superfamilies will be found in its venom. Of the relatively small proportion of turrid species analyzed and shown in this phylogenetic tree, five discrete branches can be defined (as illustrated by the asterisks). A new nomenclature has been suggested for the major turrid branches shown, which from the top center going clockwise are now designated as the following families: Raphitomidae, Borsonidae, Drillidae, Crassispiridae (or alternatively, Psuedomelatomidae), and Turridae (*sensu stricto*). When this phylogenetic tree was assembled, relatively few turrids were available for molecular analysis. Consequently, the number of species shown in each branch are not at all proportional to their true biodiversity, and several other major families formerly regarded as turrids are absent from the phylogenetic tree altogether. (B) *Conus* species that are the source of peptides with therapeutic potential. Top row, from left to right: *Conus purpurascens*, West Coast of Mexico. Peptide: κ-conotoxin PVIIA, in preclinical development for cardioprotection. *Conus marmoreus*, Philippines. Peptide: χ-conotoxin MrIA, in Phase II clinical trials for neuropathic pain. *Conus geographus*, Marinduque Island, Philippines. Peptides: Contulakin G, Phase I human clinical trials for neuropathic pain. Conantokin G, Phase I clinical trials for intractable epilepsy. Lower row, left to right: *Conus magus*, Bantayan Island, Philippines. Peptide: ω-conotoxin MVIIA, approved drug for intractable pain (Prialt; ziconotide). *Conus regius*, Florida, USA. Peptide: α-conopeptide RgIA, Preclinical development for neuropathic pain from nerve injury. *Conus victoriae*, Northern Australia. Peptide: Vc1.1, Phase I clinical trials for neuropathic pain from nerve injury. (C) Other conoideans with characterized venom components. The four species shown are among the few conoideans that are not cone snails and for which a systematic characterization of venom components has been initiated. Clockwise from far left: *Terebra subalata*, Marinduque Island, Philippines; *Hastula hectica*, Panglao Island, Philippines; *Gemmula speciosa*, Samar Island, Philippines; *Crassispira cerithina*, Olango Island, Philippines. Outside *Conus*, peptides have yet to be developed for therapeutic purposes.

a constellation of functionally linked receptors and ion channels to efficiently block neuromuscular transmission.

The importance of the cabal strategy – use of multiple toxins all dedicated to overthrow the same neuronal circuit – becomes clear if we consider a more individualistic approach. For instance, if a competitive antagonist of the neuromuscular nAChR were acting as a "lone ranger" instead of as part of a cabal, it would need to block nearly all of the nAChRs at many of the neuromuscular junctions in the fin musculature so as to prevent prey from escaping. Because there is a high safety margin of nAChRs in the muscle membrane, effective block of these neuromuscular synapses would require delivery of a high concentration of toxin throughout the entire fish—something that is simply not biochemically feasible for a comparatively small snail that must balance investment into toxin production with available resources.

The actual situation found in nature uses the nAChR inhibitor as part of a toxin cabal working together with other toxins that target a more extensive constellation of different signaling molecules driving coordinated muscle contraction. In contrast to the lone ranger nAChR inhibitor described above, a much lower concentration of the cabal inhibitor, which produces a partial block of the nAChRs, is highly effective when combined with blockers of pre-synaptic Ca_V channels and post-synaptic Na_V channels. Each of these peptides may individually produce only a partial block of the collection of targets present, but cumulatively the cabal effectively prevents muscle contraction. The cabal strategy thus allows the cone snail to achieve paralysis of prey more quickly and efficiently.

6.2.2 Diversity of Cabals

The motor cabal effectively prevents muscle contraction, necessary for capture of prey; however, even the multiple components of this cabal appear to be insufficient. The need for an additional cabal is apparent if one considers the physiology of toxin delivery. Components of the motor cabal must circulate through blood vessels to reach the targeted neuromuscular junctions. This takes some time, and acting by itself the motor cabal would probably make a fish-hunting cone snail—a relatively slow moving animal—vulnerable to attack by the snail's own predators as it searches for paralyzed prey. Furthermore, snails lack mechanical weaponry for prey capture, such as fangs and distensible jaws found in other venomous animals, relying instead on a tenuous tether (the radular tooth) to maintain a connection to envenomated prey. For these reasons, efficiency of venom action is absolutely critical, leading to unusually strong selective pressure for evolution of venoms with multiple toxin cabals.

The venoms of many fish-hunting cone snails contain a cabal that immobilizes the fish nearly instantly, prior to the onset of the effects of the motor cabal. This cabal, which has been called the "lightning strike cabal", includes peptides that block voltage-gated potassium (K_V) channels, others that activate Na_V channels, and additional peptides that delay inactivation of Na_V channels. The cumulative physiological effect is the equivalent of an electric shock.[11] All axons adjacent to the point of injection are massively depolarized and fire retrograde action potentials that rapidly immobilize the fish in a tetanic state. The quick acting lightning strike cabal provides crucial time for the components of the motor cabal to arrive at neuromuscular junctions, where they produce a longer lasting, irreversible paralysis.

Cabals comprising multiple toxins working together and the concerted action of two or more cabals provides part of the rationale for why there are so many components in conoidean venoms. Further consideration of the potential for off-target binding sites and mixed signals between cabals explains why venom components have evolved such a high degree of molecular specificity. A toxin evolved to block nAChRs at the neuromuscular junction of a particular prey would be ill suited if it also interacted with homologous acetylcholine receptors in the neurons of that prey since these off-target binding sites could then deplete the toxin as it circulates. Likewise, homologous Na_V channels are found in muscle cells and neurons. Components of the motor cabal and the lightning strike cabal would be maladapted if they acted on muscle and neuronal isoforms indiscriminately, since one cabal depends on inhibiting these channels in muscles and the other acts to keep these channels open in peripheral axons. The next section reviews our current understanding of molecular specificity for conoidean peptides.

6.3 Molecular Targets of Venom Peptide Families of Conoidea

6.3.1 A Conopeptide Family Targets a Corresponding Ion-Channel or Receptor Family

The *Conus* venom peptide literature routinely refers to conopeptide or conotoxin families designated by a Greek letter in front of the specific name of the peptide (*e.g.* ω-conotoxin MVIIA, α-conotoxin G1, κM-conotoxin RIIIJ, where ω-, α- and κM- are different family groups). A summary of known *Conus* peptide toxin families and their respective molecular targets is shown in Table 6.2. Peptides that belong to the same family have two common features—one related to structure and the other related to mechanism of action. The shared structural feature is a conserved disulfide framework necessary for maintaining stability, or in the case of the conantokin family, a conserved pattern of post-translationally modified glutamate residues important for helical structure, with marked variability in non-cysteine (or non-glutamate) residues. Peptides within a toxin family also share a common function defined by the molecular target and mechanism of action. For example, all ω-conotoxins are Ca_V channel antagonists that block the channel pore, and all α-conotoxins are competitive nAChR antagonists.

Table 6.2 *Conus* peptide toxin families grouped by molecular target

Conopeptide Family	*Example Conotoxin*	*References*
Nicotinic acetylcholine receptors		
α	α-GI	108
αA	αA-PIVA	109
αC	αC-PrXA	110
αS	αS-RVIIIA	110
αD	αD-VxXIIA	111
Ψ	Ψ-PIIIE	112
Voltage-gated sodium channels		
μ	μ-GIIIA	113
μO	μO-MrVIA	80
δ	δ-TVIA	114
ι	ι−RXIA	14
Voltage-gated potassium channels		
κ	κ-PVIIA	11
κA	κA(S)-PIVE	115
κM	κM-RIIIJ	116
κJ	κJ-PlXIVA	117
–	Conkunitzin-S	75
Voltage-gated calcium channels		
ω	ω-GVIA	118
ωX	ωX-RsXXIVA	119
G-protein coupled receptors		
σ	σ-GVIIIA	120
ρ	ρ-TIA	121
–	Contulakin-G	55
–	Conopressin-G	122
Voltage-gated pacemaker channels		
γ	γ-PnVIIIA	123
Glutamate receptors		
–	Conantokin-G	124
–	Con-ikot-ikot	78
Monoamine transporters		
χ	MrIA	121

By convention, the first family of peptides that was discovered to interact with a particular type of molecular target at a specific site (*e.g.* K_V channel pore blocker) was named with a single Greek letter to designate the peptide-family name (*e.g.* κ-conotoxins). Subsequently, when a structurally unrelated but functionally analogous peptide family is discovered, this is indicated by adding a gene superfamily designation to the Greek letter family designation. For example, the κ-conotoxins were the first conopeptides discovered to block K_V channels; the κM-conotoxins that were subsequently discovered to block K_V channels have a different structural scaffold from κ-conotoxins and are members of a different conopeptide superfamily, the M-superfamily of venom peptides (see also the section on molecular genetic framework in Section 6.4.1).

Strictly speaking, a peptide should not be assigned to a family until its physiological activity has been elucidated, since two peptides may have quite similar structures and belong to the same venom peptide gene superfamily but have different mechanisms of action. For example, the ω-conotoxins and κ-conotoxins both have three disulfide bonds, with similar disulfide cross-linking patterns, namely the inhibitor cystine knot (ICK) motif (see Chapter 2).[13] Nevertheless, these peptides are assigned to two separate families, based on their targeting selectivity for Ca_V channels (ω-conotoxins) *versus* K_V channels (κ-conotoxins).

Why are there so many diverse toxin families within Conoidea? After all, in other venomous groups, including snakes, spiders, or scorpions, a smaller number of structural scaffolds are used. For example, in the venoms of elapid snakes, the characteristic "three-finger" scaffold is dominant. Part of the answer relates to the lack of mechanical structures for prey capture and the cabal strategy, already described above, which relies on targeting multiple molecular targets simultaneously. Each additional molecular target included in this cabal-based strategy encourages innovation of a new structural scaffold. Also, for the more physiologically important molecular targets, multiple structural scaffolds emerged so as to be able to act at diverse binding sites on an ion channel or receptor target. For example, there are four structurally divergent families of conotoxins identified so far that each act on Na_V channels but at different binding sites and with different molecular consequences: μ-conotoxins are pore blockers, μO-conotoxins prevent activation upon binding the voltage sensor, and δ-conotoxins inhibit inactivation, whereas ι-conotoxins promote activation.[14]

The availability of diverse biological niches probably provided additional selective pressure driving the emergence of diverse toxin families and also the diversification of conotoxins within any particular conopeptide family. For each new niche, success required optimization of existing scaffolds, or the creation of new structural scaffolds to target the particular signaling molecules present in the prey, competitors and predators particular to that niche. In the case of the α-conotoxin family, this structural scaffold was well suited to allow adaptation in targeting nAChRs from diverse vertebrates and invertebrates, resulting in widespread distribution of α-conotoxins throughout the genus *Conus*. While the family is conserved, and all α-conotoxins compete with acetylcholine for binding to nAChRs, each *Conus* species evolved its own distinct set of α-conotoxins to act on the particular molecular isoforms of nAChRs present in niche-defined prey, predators and competitors. The result is a large family with numerous individual members that differ in their amino-acid sequences and binding selectivities.

Other cone snail venom peptide families differ from α-conotoxins considerably in their breadth of distribution. For example, the κ-conotoxin family, which targets K_V channels, may be restricted to a single clade of fish-hunting cone snails. Narrow distribution for this particular κ-conotoxin family reflects a different evolutionary outcome compared with that seen for the α-conotoxin family. Whereas most nAChR antagonists in cone snail venoms belong to the α-conotoxin family, making this family broadly distributed, antagonists of K_V channels belong to several different and structurally distinct families, each of which is more narrowly distributed. Which particular K_V channel-targeting conopeptide family is present varies from one cone snail clade to another, although all species within a particular subgeneric clade appear to use the same scaffold for homologous purposes.

As described above, the evolutionary success of Conoidea likely depended both on rapid adaptive changes within a preserved structural scaffold (within one conopeptide family) and creation of new structural scaffolds. The rich biodiversity of conoideans therefore reflects the rich diversity of the arsenal of neuroactive compounds. Molecular diversity is driving and being driven by the emergence of new species in new biological niches. The implication important for drug discovery is that a broad survey of the biodiversity of venomous animals makes it possible to identify diverse venom peptides in families that have evolved to interact with families of molecular targets, described previously.[15] The diversity of venom peptides distributed across the thousands of species of marine snails evolved to match the diversity of the receptor subtypes and ion channel subtypes present in the thousands of species of marine-snail prey, predators and competitors. Consequently, the more broadly the biodiversity of conoideans is sampled (especially the incredibly large biodiversity of turrids), the more likely will be the discovery of new ligands with novel subtype-selectivity.

6.3.2 Individual Peptides of Conopeptide Families Can Be Used to Distinguish Between Closely Related Ion-Channel or Receptor Subtypes

Because *Conus* peptides selectively target a specific subtype of an ion channel or receptor family, they have been used to identify the particular receptor and ion channel subtypes expressed in individual native neurons. Historically, conotoxins were important for differentiating between Ca_V channel subtypes. The N-type Ca_V channel ($Ca_V2.2$) was biochemically purified as the ω-conotoxin GVIA receptor.[16,17] Currently, many additional conopeptides are used by the neuroscience community to identify specific Na_V channel subtypes and nAChR subtypes expressed in neurons and other cells (Table 6.3).

In mammals, there are nine genes that encode α-subunits of Na_V channels. Different neuronal cell types express various combinations of these. The classical pharmacological tool for differentiating between Na_V channel subtypes is tetrodotoxin (or its pharmacological equivalent, saxitoxin). However, six Na_V channel subtypes are tetrodotoxin sensitive ($Na_V1.1$–1.4, 1.6, and 1.7), and three are tetrodotoxin insensitive ($Na_V1.5$, 1.8, 1.9), meaning that tetrodotoxin serves as a relatively low-resolution tool for discrimination among Na_V channel isoforms. For high-resolution discrimination, a set of ligands with enhanced molecular selectivity would be highly desirable. At present, μ-conotoxins are arguably the most promising pharmacological tools for differentiating between the tetrodotoxin-sensitive Na_V

Table 6.3 Examples of subtype-selective conopeptides

Conopeptide	Molecular Target(s)	Conopeptide	Molecular Target(s)
	Nicotinic Receptors		**Sodium Channels**
α-RgIA	α9α10 nAChR	μ-TIIIA	$Na_V1.1$
α-BuIA	nAChR containing β4	μ-PIIIA	$Na_V1.1$, $Na_V1.6$
α-ArIB [V11L; V16D]	α7 nAChR	μ-KIIIA, μ-SmIIIA	$Na_V1.1$, $Na_V1.6$, $Na_V1.7$
α-MII [H9A; L15A]	α6β2 nAChR	μO-MrVIB	$Na_V1.8$
α-BuIA [T5A; P6O]	α6β4 nAChR		
α-AuIB	α3β4 nAChR		**Potassium Channels**
α-MI	α1δ	κM-RIIIJ	$K_V1.2$
αA-OIVA	α1γ	Cce9a	$K_V1.6$
		PI14a	$K_V1.6$
	Calcium Channels	Conkunitsin-S1	$K_V1.7$
ω-GVIA	N-type calcium channel ($Ca_V2.2$)		
ω-MVIIC	P/Q-type calcium channel ($Ca_V2.1$)		**Glutamate Receptors (NMDA)**
		Conantokin-RI-B	NR2B
	Neurotensin Receptors	Conantokin-R	NR2B, NR2A
Contulakin-G	NTS1	Conantokin-RI-A	NR2A, NR2B, NR2D

channel subtypes expressed in neurons.[18–21] (But see also work describing spider toxins with Na_V subtype specificity.[22,23]) Table 6.3 provides examples of a few μ-conotoxins that have been used to identify the complement of Na_V channel subtypes expressed in sensory and sympathetic neurons. Increasingly, analogs of native μ-conotoxins have been synthesized for structure activity studies and to understand subtype selectivity.[24–27] The μ-conotoxins also have potential for dissecting specific combinations of Na_V channel α- and β-subunits that are co-expressed in neurons.[18,19] Thus, the μ-conotoxins comprise a key component of the pharmacological toolkit necessary for differentiating neuronal subclasses on the basis of the different Na_V channel expression in individual neurons.

The pharmacological diversity of the peptides targeted to nAChRs from *Conus* venoms (Table 6.3) is even more impressive than that discovered for Na_V channels. Enriched pharmacological diversity probably relates to the potential molecular complexity of nAChR subtypes, which is even greater than the complexity of Na_V channel subtypes. A functional nAChR comprises five subunits, and the pentamers may be homomeric or heteromeric combinations of subunits encoded by 16 different genes (Figure 6.2). At the neuromuscular synapse, the post-synaptic nAChR has two α1-subunits, and one subunit each of β1 (γ or ε) and δ-subunits in a defined arrangement that is known.[28] However, in other tissues the precise subunit composition of nAChR subtypes is largely unknown. Although certain heteromeric-nAChR subtypes that contain specific α- and β-subunits are known to be expressed *in vivo*, presumably many other nAChR pentameric isoforms are expressed *in vivo*, and the subunit composition and stoichiometry of many of these heteromeric nAChRs is presently unclear. It is evident that that there is a vast molecular complexity of nAChR subtypes to explore. The α-conotoxins are increasingly used to identify specific nAChR subtypes expressed *in vivo* and for exploring the functional roles of each.

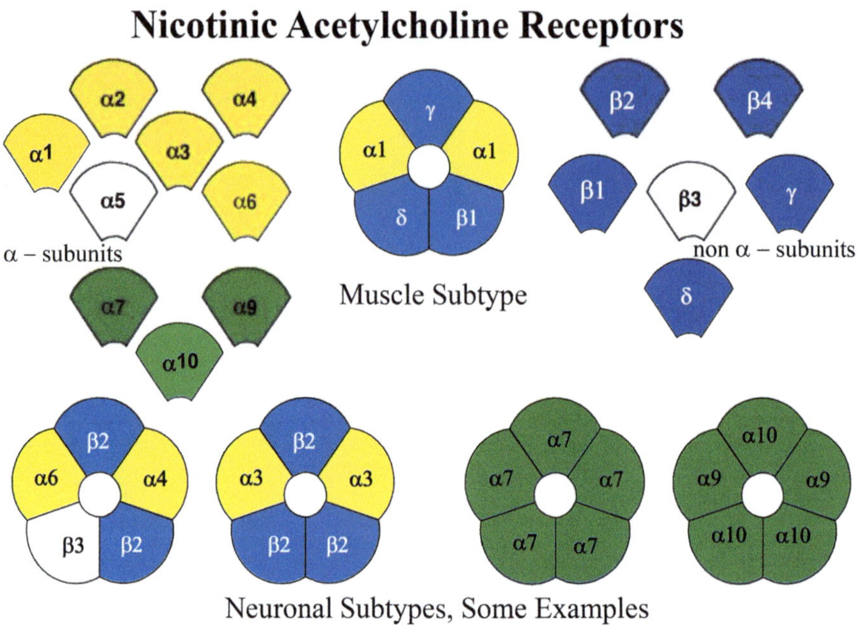

Figure 6.2 Molecular complexity of nAChRs. Shown diagrammatically are the subunits of nAChRs expressed in mammals. The functional receptor is a pentamer. At the mammalian neuromuscular junction, the subunits are assembled as illustrated with two copies of α1-, β1-, δ-, and γ-subunits (the γ-subunit is only found in fetal muscle and is replaced with an ε-subunit in mature muscle). The nAChR found in muscle was the first ion channel to have been biochemically characterized. Many other nAChR subtypes are expressed *in vivo*. For most subtypes, the stoichiometry of subunits has not been determined. Four potential subtypes believed to be present in the mammalian nervous system are shown. It is generally believed that the α7-, α9- and α10-subunits (green) can only assemble with each other to form functional receptors. Different from these all-α neuronal receptors, many other neuronal subtypes comprise a combination of two α-subunits (yellow) and three non-α-subunits (blue or white), as illustrated in the figure. Even with these restrictions, an enormous number of potential nAChR subtypes can be expressed. *Conus* peptides have played a leading role in defining these molecular isoforms. Notably, each of the four neuronal subtypes shown in the figure is blocked by different α-conotoxins.

The standard method used for analyzing function at the molecular level is genetic perturbations (*i.e.* gene knockouts) in model organisms, but this has serious shortcomings when dealing with heteromeric complexes such as nAChRs. Knocking out a specific nAChR subunit (for example, the gene encoding the α6-subunit) would be expected to produce an animal with a phenotype, but interpreting the phenotype is problematic because most nAChRs are heteromeric complexes. Therefore, the observed phenotype could be due to the elimination of any or all of the various nAChR subtypes expressed *in vivo* that contain an α6-subunit (*e.g.* α6β2, α6β4, α6α4β2, α6α4β4, α6α5β2, *etc.*). Thus, to interpret gene knockout experiments, a complementary pharmacological approach is required. In turn, ligands that can distinguish selectively between various heteromeric subtypes are urgently needed. As shown in Table 6.3, several α-conopeptides can differentiate between closely related nAChR subtypes. Some specific examples are discussed below.

6.3.3 Linking Biodiversity to Discovery: α-Conotoxins and nAChR Subtypes

The peptide toxin cabals found in the venoms of conoidean species should provide a rich harvest of subtype-selective ligands. The best indicator of this is provided by the peptide toxins already characterized that target the nAChR family. The most rapid progress has been made with this group of venom components because the relevant family of peptides, the α-conotoxins, are structurally simpler than most other venom peptides with only two disulfide cross-links, which makes them relatively straightforward to synthesize. Thus, structure/function studies are significantly less difficult than for longer peptides with three or more disulfide cross-links.

The preceding sections described many α-conotoxins that selectively target various nAChR subtypes. These discoveries were enabled, in part, through the analysis of a broad biodiversity of cone snails. In Figure 6.3, three different nAChR subtypes are illustrated, and for each of these, α-conotoxins targeting particular subunit interfaces have been identified. The α-conotoxins are homologs that share a conserved structure and mechanism of action but diverge in amino-acid sequences to target different isoforms of a particular family of signaling molecules. Other families of conotoxins reinforce this pattern, first described for the α-conotoxins, of structure/mechanism conservation with sequence variation to achieve isoform specificity (Table 6.3).

The rationale for the evolution of α-conotoxins that inhibit the nAChR at the vertebrate neuromuscular junction is clear: fish-hunting cone snails need to paralyze their prey. However, the evolutionary rationale for α-conotoxins that target neuronal nAChR subtypes that are not widely distributed is less obvious because *a priori* we would not expect such peptides to facilitate prey capture. Why do cone snail venoms contain a peptide that targets the α9α10 nAChR, which occurs only in restricted loci of the mammalian nervous system (*e.g.* the auditory circuitry) and a few non-excitable tissues? How can a peptide that selectively targets this receptor subtype be generated through natural selection?

A rationale is suggested by the observation that α-conotoxins that selectively inhibit the α9α10 nAChR were all found in the venoms of worm-hunting cone snail species, and these are particularly enriched in those species that prey on fireworms (also known as amphinomid polychaetes), such as *Conus regius* (Figure 6.1) and *C. imperialis*. Unlike mammals, many

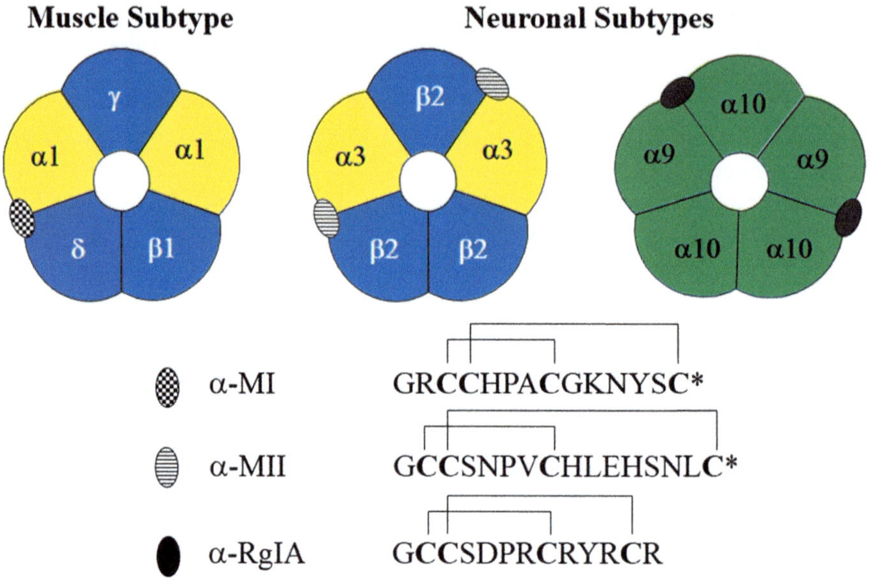

Figure 6.3 Three nAChR subtypes with selective α-conotoxins. The three pentameric nAChRs are the muscle subtype, found at neuromuscular junctions of mammals, a ganglionic subtype (α3β2) and a subtype found in auditory circuitry (α9α10). The primary structures of three different α conotoxins are shown that target these nAChR isoforms: α-conotoxin MI for the muscle subtype, α-conotoxin MII for the α3β2 neuronal subtype and α-conotoxin RgIA for the α9α10 subtype. Both MI and MII were originally identified from the venom of *Conus magus*. RgIA comes from the worm-hunting species *Conus regius*.

invertebrates express a homolog of the mammalian α9α10 nAChR at their neuromuscular junctions.[29] Thus, the α-conotoxins that inhibit this nAChR homolog would presumably provide a selective advantage because they produce neuromuscular paralysis in fireworms (though not in mammals). The α9α10 nAChR belongs to a special class of nAChR subtypes that only contain α-subunits and no β-subunits (these are illustrated in Figures 6.2 and 6.3 with green-colored α-subunits). In mammals, such nAChR subtypes contain α7-, α9- or α10-subunits (α1–α6-subunits always form heteromeric receptors in combination with β-subunits). However, invertebrates, such as *Caenorhabditis*, express numerous nAChR homologs of mammalian α7, α9 and α10, both at the neuromuscular junction and other loci in their nervous systems.[29] A key lesson from this example is that the structure of homologous ion channels and receptors may be conserved across species, even when patterns of expression and physiological roles are not conserved. Thus, a peptide toxin that selectively targets a particular receptor subtype to enable prey capture may also function as a therapeutic drug in humans, where the homologous receptor subtype might be expressed in a different tissue, with a different physiological role.

6.3.4 Translational Applications of α-Conotoxins

The following example demonstrates the type of translational application made possible through identification of novel venom peptides that target specific receptor subtypes and ion channel subtypes. It has been established that smokers are protected against Parkinson's disease, and the degree of protection appears to be a function of how much an individual smokes. Furthermore, if smokers quit, this protection is abolished.[30,31] Clearly, understanding the mechanistic basis of the protective effect of smoking has biomedical relevance. Evidence that the protective effects of smoking are due to the effects of nAChRs has been obtained.[32] The question is, Which are the relevant subtypes of nicotinic receptor?

This question has been explored with conopeptides. There are several different subtypes of nicotinic receptor that, when activated, result in dopamine release from the relevant circuitry in the striatum.[33] In order to identify these specific nAChR isoforms, it was necessary to use highly selective conopeptides that belong to the α-conotoxin family. Thus, the availability of native conopeptides and their derivatives that can discriminate between molecular isoforms of the large and complex nAChR family has been essential for exploring the functional role of specific isoforms under normal physiological conditions and in the etiology of Parkinson's disease.[34] The experimental application of subtype-selective α-conotoxins suggested that two closely related nAChR subtypes, α6β3β2 and α6α4β3β2, play critical and functionally different roles in the relevant dopaminergic circuitry.[34–37]

6.3.5 Molecular Determinants of Specificity

The α-conotoxins and their nAChR targets offer a unique view of the molecular interactions leading to subtype specificity and serve as a model for the evolutionary mechanisms by which other families of conoidean peptide toxins have achieved molecular specificity. Additionally, structural models of α-conotoxins in complex with the acetylcholine-binding protein (AChBP), a homolog of the nAChR, have yielded mechanistic insights.[38–42] Here, structures of AChBP complexed with α-conotoxins will be discussed in order to illustrate subtype-selectivity mechanisms (Figure 6.4 and Table 6.4).

The structures show the AChBP subunits arranged with five-fold symmetry. Peptide toxin is bound at the interface between subunits and occludes the agonist-binding (acetylcholine-binding) site (Figure 6.4A). At each subunit interface, one subunit contributes the principal (+) binding surface, and the other subunit contributes the complementary (−) binding surface (Figure 6.4B). Loop C, which is part of the principle (+) surface, is found in different dispositions depending on the occupancy of the agonist-binding pocket. For the purpose of this discussion, loop C refers to the beta-turn-beta element that includes two segments of extended beta-strand structure flanking a connecting loop, with the vicinal cysteine residues at the very tip of this loop being a conserved feature of α-subunits and AChBPs. Compared

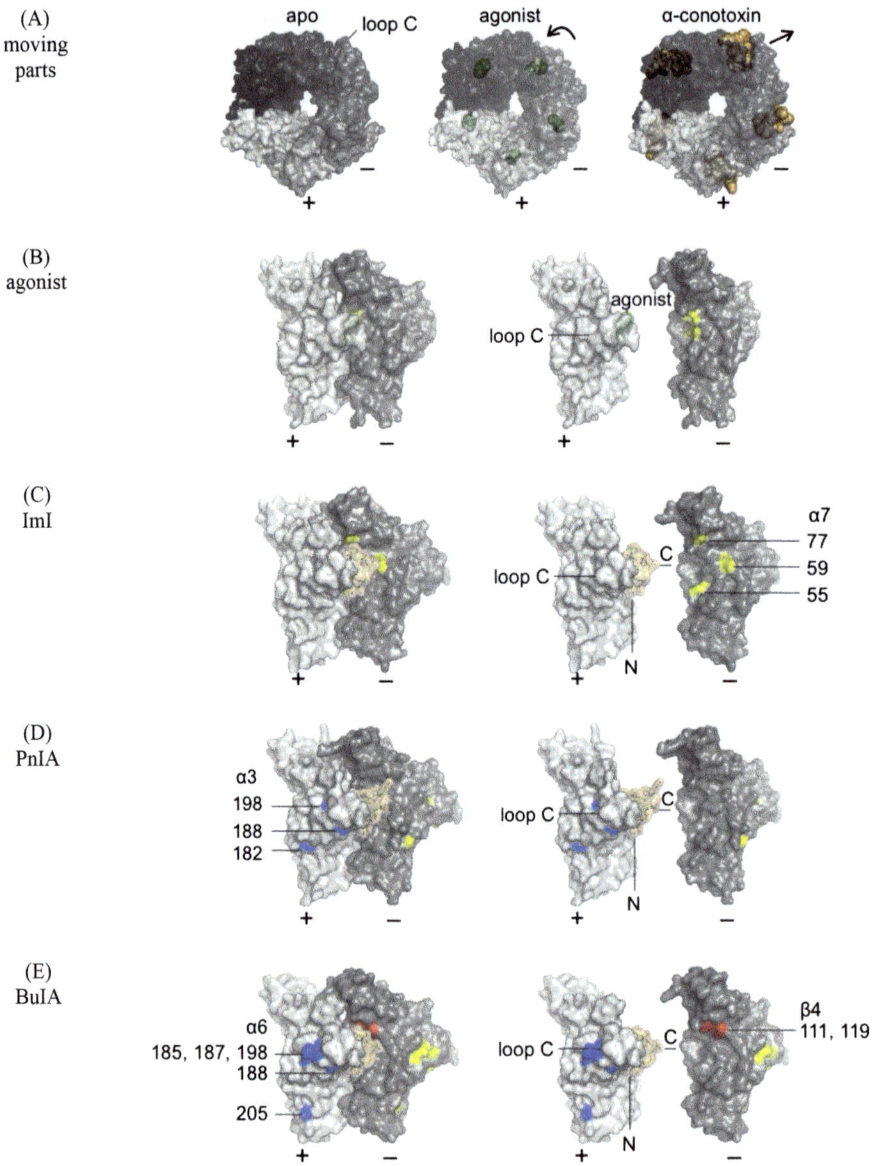

Figure 6.4 X-ray crystal structures of the AChBP in complex with α-conotoxins. Subunits of the AChBP (shades of grey) and α-conotoxin (gold) are depicted as water-accessible surfaces. (A) Pentameric assembly in different structural states. The AChBP assembly is viewed slightly askance from the five-fold axis so as to highlight movement of loop C that is coupled to binding agonist and α-conotoxin. In the context of a nAChR, the membrane would be below the ligand-binding units shown here. Loop C in each of the five subunits closes upon interaction with agonist. Occupancy by α-conotoxin PnIA[A10L, D14K] requires opening of loop C. Movement of loop C is accompanied by

with the apo state, loop C closes to make interactions with agonist in the epibatidine complex (Figure 6.4A). This same element is forced into a more open position to accommodate an α-conotoxin antagonist. Movement of loop C is necessary for agonist dissociation and for α-conotoxin binding, as there is otherwise no unobstructed route out of and into the binding pocket for these ligands. Movement of loop C (coupled to movement of other structural elements) has been inferred from comparison of several AChBP structures and likely explains channel gating in nAChRs.[43] As will be described below, movement of loop C also probably contributes to an allosteric mechanism for establishing subtype specificity characterizing several α-conotoxins.

---

global adjustments within the assembly that communicate agonist/antagonist binding events to the ion channel. (B) Two subunits in complex with agonist (epibatidine, green spheres) viewed approximately perpendicular to the five-fold axis. The (+) subunit contributes the principal agonist-binding surface located on the inside of loop C, and the (−) subunit contributes the complementary binding surface. The right-hand panel shows an exploded view to highlight surface residues of the (−) subunit within van der Waals contact of agonist (yellow). (C) Same view as in panel (B), but showing the complex with α-conotoxin ImI. In the exploded view, N-terminal and C-terminal residues of the α-conotoxin are labeled. Residues colored yellow on the (−) subunit correspond with α7 residues that are known to contribute to molecular specificity. The residues at homologous positions within AChBP make direct contact with ImI. Direct contact with surfaces on the complementary face is one mode of molecular recognition leading to subtype-specific block by α-conotoxin. (D) Complex with α-conotoxin PnIA[A10L, D14K]. Residues colored blue on the (+) subunit and yellow on the (−) subunit correspond with α3 residues that contribute to enhanced sensitivity to PnIA. Neither set of residues makes direct contact with toxin. In this case, specificity-determinants most likely influence hinge-like motions of loop C to encourage or prevent association with PnIA, dependent on receptor subtype. This type of effect on loop C motion is a second, allosteric mode of molecular recognition that serves to establish subtype-specificity for α-conotoxins. (E) Complex with α-conotoxin BuIA. Both direct-contact and allosteric mechanisms are apparently at work to establish molecular specificity for this α-conotoxin. Residues colored blue on the (+) subunit and yellow on the (−) subunit correspond with α6 residues that enhance sensitivity 60 000 fold compared with α4* receptors (* indicates receptors include other types of subunits). As is the case for the complex with PnIA[A10L, D14K], neither set of residues makes direct contact with toxin, and the blue set impinges on hinge-positions of loop C, supporting the idea that α6 *versus* α4 discrimination is established through an allosteric mechanism. Residues colored red on the (−) subunit correspond with β4 residues that strongly influence toxin dissociation kinetics. The residues at homologous positions of AChBP make van der Waals contact with BuIA, supporting the idea that β4 identity is established by a direct-contact mechanism. Structural representations were built using coordinates from the following pdb entries: *2byn* apo structure and *2byq* agonist-bound structure;[42] *2br8* αPnIA[A10L, D14K] structure;[40] *2c9t* αImI structure;[43] *4ez1* αBuIA structure.[50]

Table 6.4 Crystal structures of α-conotoxin in complex with AChBP from *Aplasia californica*

α-conotoxin	*pdb id*	*Subtype Specificity*	*Degree of Specificity*	*Specificity Determinants (corresponding sites in AChBP)*
α-ImI	*2byp, 2c9t*	α7	x50 α7/α1	α7: 55, 59, 77 (53, 57, 75)
α-PnIA [A10L, D14K]	*2br8*	α3 (native toxin); α7 ([A10L])	x1000 α3/α2 (native); x10 α7/α3 ([A10L])	α3: 182, 188, 198 (178, 184, 194)
α-BuIA	*4ez1*	α6; β4	x60,000 α6/α4; x60 *k*-off β4/β2	α6: 185, 187, 188, 198, 205 (181, 183, 184, 194, 201); β4: 59, 111, 119 (54, 106, 114)

Specificity determinants that distinguish nAChR subunits have been mapped by mutational analysis. In this approach, two receptor subunits that differ in sensitivity to a particular α-conotoxin are employed to construct subunit chimeras that are then assembled into functional receptors. Residues that contribute to subtype identity (from the toxin's point of view) are found by monitoring changes in toxin sensitivity as residues from one subunit are swapped for the corresponding residues in the other subunit. In the case of α-conotoxin ImI from *C. imperialis*,[44] acetylcholine-responsive receptors were constructed by linking the acetylcholine-binding domain of α7 and α1 onto the transmembrane channel-forming domain of the 5-HT_3 receptor.[45] The α7- and α1-subunits assemble into homomeric assemblies so no other subunits were required. ImI inhibited the α7-5-HT_3 and α1-5-HT_3 channels but with different IC_{50} values, indicating that ImI preferentially inhibits α7 over α1 subtypes by a factor of 50-fold.[45] The identity of the amino acid residue at positions 55, 59, and 77 within the α sequence was found to be deterministic with respect to ImI sensitivity.[45] In an α7 background, changing these three positions to the residue found in the α1 sequence made the chimera relatively resistant to ImI. Conversely, importing the α7 residues at any of these three positions increased sensitivity to ImI in an otherwise α1 background.

Structural models allow interpretation of these mutational analyses to arrive at a mechanistic understanding of molecular specificity.[42,43] Figure 6.4C shows ImI as found in co-crystals of *Aplysia* AChBP. The structure explains competitive inhibition since the site occupied by ImI completely overlaps with the site of agonist binding. The structure also suggests how the toxin can differentiate between homologous α-subunits. The peptide makes close contact with residues at positions corresponding with those implicated as specificity determinants by the residue swapping experiments. The toxin-contacting surface for these specificity determinants is on the complementary (−) face. The ImI peptide apparently distinguishes α7 and α1 subtypes by direct contact with residues on the complementary (−) surface. Direct contact with variable regions of the receptor is one modality for establishing molecular recognition that fits into the lock-and-key paradigm for specificity.

As will be described for α-conotoxin PnIA and α-conotoxin BuIA, a second modality that involves moving parts and allosteric interactions also effectively establishes subtype specificity.

The α-conotoxins, PnIA and PnIB, discovered in *C. penneceus*,[46] demonstrated differential inhibition of α3β2 and α7 receptors expressed in *Xenopus* oocytes, with PnIA preferring α3β2 and PnIB preferring α7.[47] Leucine at position 10 in the peptide sequence was found to be important for potency towards the α7 receptor, since the single amino-acid substitution variant, PnIA[A10L], switched receptor preference from α3β2 to α7.[47] Neuronal α3β2 receptors are at least 1000-fold more sensitive to PnIA compared with α2β2 receptors.[48] To map specificity determinants on the α3-subunit, Everhart and co-workers compared PnIA sensitivity for a series of receptors comprising α3/α2 chimeric subunits and β2-subunits.[48] Positions 182, 188, and 198 were identified as subtype-specificity determinants used by PnIA to distinguish α3 from α2 since replacement with the α2 residue in an α3 background significantly diminished potency.[48] These subtype determinants map to loop C of the principal (+) subunit. Loop C contains several residues that construct the (+) portion of the agonist and α-conotoxin binding pocket. However, the particular residues implicated in α3 *versus* α2 discrimination map to corresponding residues in the AChBP that do not make direct contact with peptide in a PnIA[A10L, D14K]-AChBP crystal structure (Figure 6.4D). Homology models of α3β2 (which were constructed before the toxin-AChBP structure became available) also highlighted the disposition of subtype-specificity determinants as distant from the agonist-binding pocket and suggested that structural changes in loop C or changes in the dynamic properties of loop C communicate residue identity at these distant sites.[48]

The distant location of specificity determinants poses an enigma with respect to the mechanism of achieving subtype specificity. Of course, a trivial explanation is that structural inferences based on the AChBP are flawed and that specificity-determining residues in authentic α3β2 receptors are in fact making close contact with inhibitor. However, many other structural inferences derived from AChBP crystal structures are supported by biochemical and mutagenesis data, lending a high degree of credibility to the use of AChBP structure as an accurate surrogate for the ligand-binding domain of acetylcholine receptors.[41] The prevalent explanation for understanding subtype specificity is consistent with currently available structures and attributes the influence of distant specificity determinants to changes in the structural and dynamic properties of loop C.

Conceptually, this idea is related to the fundamental principles that apply to any allosteric system. Toxin and agonist binding are coupled to structural changes focused on loop C.[43] The acetylcholine-binding domain is therefore an allosteric system with different structural states, and the distribution of these states clearly depends on ligand occupancy. For allosteric systems, the converse is also true. Namely, a change in the structural state distribution (*e.g.* as affected by residue substitutions) will influence ligand-binding affinity. The specificity determinants uncovered for PnIA, although not close to the toxin binding pocket, impinge on positions that seem well placed with respect to any type of hinge-like motion for loop C. It is reasonable to think that the physiochemical nature of residues at hinge positions will increase flexibility in some cases and prevent certain conformations in other cases and that the outcome of these effects will be to increase or decrease the likelihood of toxin association (*i.e.* increase or decrease toxin potency). We will refer to this molecular explanation, illustrated by PnIA, as the allosteric mechanism for establishing subtype selectivity, which is to be distinguished from the direct-contact mechanism illustrated by ImI.

The two mechanisms for subtype selectivity are not mutually exclusive, and they may be synergistic. The α-conotoxin BuIA, isolated from *Conus bullatus*,[49] appears to utilize both direct-contact and allosteric mechanisms for subtype specificity. BuIA differentiates between α4 and α6 extracellular acetylcholine-binding domains, preferring the latter by a factor of ~60 000.[49,50] BuIA also discriminates among β-subunits on the basis of dissociation kinetics with release from α3β4 receptors about 60-fold slower than from α3β2 receptors.[51] The determinants used by BuIA to discern α- and β-subunit identity have been mapped by the chimera receptor approach to five residue positions within loop C for α6 recognition[50] and to two residues residing close to the agonist (−) binding surface for slow β4 dissociation.[51] Figure 6.4E shows a crystal structure of BuIA bound with AChBP with these specificity determinants color-coded. Similar to the situation found for PnIA, the residues corresponding to those that confer high sensitivity to α4/α6* chimeras are far away from the (+) agonist binding site[50] (* indicates assembled receptors contain additional subunits besides the chimeric α4/α6-subunit). The distant location of these specificity determinants speaks for an allosteric mechanism for subtype specificity in discriminating α6* and α4* receptors. For β4 specificity determinants, the corresponding positions in AChBP are making van der Waals contact with BuIA, meaning the direct-contact mechanism is the simplest explanation for differential dissociation kinetics.

The two modalities for subtype selectivity are expected to apply generally for other α-conotoxins and for other families of conotoxins that target voltage-gated and ligand-gated ion channels. The direct-contact mechanism is akin to lock-and-key descriptions of substrate specificity, which seems an engraved dogma of biochemical lore and intuition. The hypothesis that the allosteric modality will be a general mechanism needs to be tested with experiments, including structure determination of toxin-receptor complexes for other systems. *In lieu* of these, some logical arguments may be suggested. All of the neuromuscular signaling molecules involve moving parts and allostery. The necessity of moving parts stems from the requirement to couple some extra-cellular event (neurotransmitter binding or membrane depolarization) to channel opening in the transmembrane region. Examples of bioactive peptides can be found within conoidean venoms that target several possible structural states of the signaling molecules (*e.g.* closed, open, and desensitized). Given these commonly observed traits, it seems unavoidable that the transitions between structural states will provide opportunities for inhibition that can drive the evolution of toxins with some selective advantage for a given conoidean species.

6.4 Diversity of Conoidean Venom Components

6.4.1 Molecular Genetic Framework: Venom Peptide Superfamilies

Most biologically active components in conoidean venoms are disulfide-rich peptides (or small proteins). Typically, 50–200 different venom peptides make up the repertoire of each species. The molecular genetics of cone snail venoms has been reviewed.[52] Although other conoidean lineages have not been studied as extensively, the basic molecular genetic principles are consistent with the genetic framework elucidated for venoms of cone snails.

Each venom peptide is encoded by a different gene (see Figure 6.5). In most cases, the initial translation product is a prepropeptide precursor with an N-terminal signal sequence, an intervening propeptide region and a mature toxin region at the C-terminal end, which is released by proteolytic cleavage. A conventional proteolytic signal (−KR or −R in most cases) is found between the propeptide and mature toxin regions. The disulfide-rich peptides in cone snail venoms are encoded by only a few gene superfamilies. Hypermutation of these superfamilies has been largely responsible for the impressive chemical diversity that has been evolved in cone snail venoms.[10,53] Each species has its own distinct complement of disulfide-rich venom peptides. A notable feature of all peptides in the same gene superfamily is the extreme conservation of the signal sequence, and the characteristic arrangement of cysteine residues within the mature toxin region. Examples of two gene superfamilies of *Conus* venom peptides are illustrated in Figure 6.5. Each gene superfamily is indicated by a capital letter (*e.g.* A-, M-, O-, P-, S-, T-superfamilies).

In any particular cone snail venom, the majority of the individual peptides are encoded by only three to five gene superfamilies. Some gene superfamilies are conserved across the entire *Conus* genus (such as the O-superfamily), while other superfamilies are more lineage-specific. A single cone snail species may express dozens of different peptides that belong to the same gene superfamily, and these may diverge substantially in sequence (except that the cysteine residues are conserved). Within a venom peptide gene superfamily, the propeptide regions are generally more conserved than the mature toxin regions, but less conserved than the signal sequence regions. The size of the propeptide region varies considerably. In a small minority of gene superfamilies, the propeptide region is absent altogether (the fraction of superfamilies that encode precursors without a pro-region may be greater in some of the other conoidean lineages). Thus, a typical cone snail peptide precursor has a mature-toxin region that ranges in size from 10 to 35 residues, with signal sequences 20–25 residues in length. The propeptide regions vary between 15 and 50 residues, depending on the superfamily. The A and

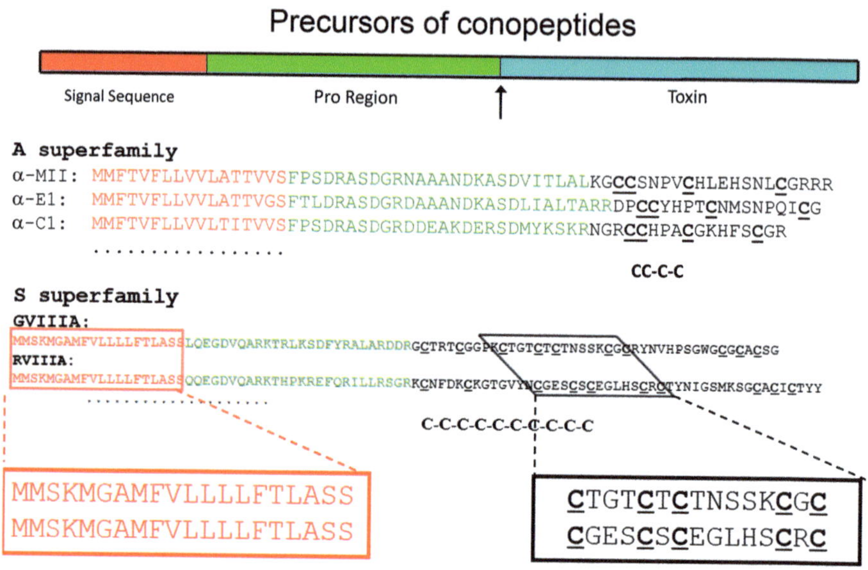

Figure 6.5 Sequences of conopeptide precursors. The sequences shown are conopeptide precursors from five different species of *Conus*; the entire predicted open reading frame is shown before post-translational modification. Three of the sequences belong to the A-gene superfamily, and two belong to the S-gene superfamily. The signal sequences are shown in red, and the propeptide region in green. The mature peptides are always encoded at the C-terminal end in single copy. The three peptides that belong to the A-superfamily have a predicted length of 15–17 residues after post-translational processing, and contain two disulfide linkages formed from the four cysteine residues. The two S-superfamily sequences encode mature peptides that have ten cysteine residues, predicted to form five disulfide bonds. A comparison (boxed sequences) is made between the two S-superfamily sequences with regard to the signal sequences (which are identical) and a segment of the mature peptide toxin region. Note that in the latter, while the cysteine residues align perfectly, no other amino-acid residues are conserved. The two S-superfamily sequences diverge in their molecular targeting specificity: GVIIIA is a specific antagonist of the 5-HT_3 receptor, while RVIIIA is a broad spectrum nAChR antagonist.

S-superfamilies shown in Figure 6.5 illustrate a stark contrast with respect to the size of the mature peptides. The three examples shown for the A superfamily are 15–17 residues in length for the mature peptides with two disulfide bonds after post-translational processing, while the two peptides shown for the S-superfamily are much larger, over 40 residues in length for the mature peptides, with five disulfide cross-links.

Conus venom peptides are much more highly modified post-translationally compared with peptides from other animal venoms,[54,55] and in this respect they resemble peptidic natural products secreted by bacteria. In some gene families that encode *Conus* venom peptides, a specific type of post-translational modification might be particularly common. Such post-translational modifications include the γ-carboxylation of glutamate, bromination of tryptophan, and hydroxylation of proline residues. For example, γ-carboxyglutamate is found at a high frequency and in a characteristic arrangement in peptides belonging to the conantokin family. The enzymes and mechanisms responsible for post-translational modification are discussed below.

6.4.2 Generation of Peptide Diversity

The evolution of the gene superfamilies that encode *Conus* venom peptides occurs at an unprecedented rate. The striking juxtaposition of conserved and hypervariable regions within these rapidly evolving genes (see Figure 6.5) and the extreme degree of conservation of signal sequences within the superfamilies of venom peptides are fairly unique in biological systems. In cone snails, there may be special genetic mechanisms in place to conserve signal sequences, since even silent mutations typically are not observed. In contrast, in the mature toxin region, hypermutation occurs at non-cysteine positions (see the example of S-superfamily peptides in Figure 6.5). The biological consequence is that each *Conus* species has its own distinct complement of venom peptides, and even closely related *Conus* species have venom peptides with divergent amino-acid sequences. The highly divergent members of a gene superfamily can be confidently identified in different *Conus* species on the basis of signal sequences, which are extremely well conserved even at the nucleotide level. Additionally, there are conserved nucleotide sequences in un-translated regions that are also characteristic of each gene superfamily.

Conus venom peptides evolved and diversified concurrently with the successive adaptive radiations that have made *Conus* one of the most species-rich genera of marine animals. The unique juxtaposition of almost totally conserved signal sequences and hypervariable mature-toxin sequences is reminiscent of the constant and hypervariable regions observed in antibodies, where the conservation of a structural framework is coupled to the rapid generation of new binding functions. Different from antibodies, however, the mechanism that has produced hyper-diversification in the mature-toxin region of conopeptides is unknown. Conticello, Fainzilber and co-workers have suggested a detailed hypothetical mechanism based on their analysis of the patterns of mutation observed.[56] A rationale for the unprecedented signal-sequence conservation observed in *Conus* venom peptide precursors has been discussed, with alternative explanations considered.[57]

The number of venom peptide toxins produced by a single cone snail varies. Estimates of 50–200 toxins per snail venom were traditionally based on reverse-phase high-performance liquid chromatography (HPLC) fractionation. However, highly sensitive mass spectrometric methods have led to the detection of a far larger number of peptides in some cone snail venoms. Estimates range from several hundreds to thousands.[58–61] Intra-species variation in venom composition has been reported for *Conus consors*,[62] *C. erminius*,[63] and *C. vexillum* collected from two locations in the Red Sea.[64] Combined transcriptomic and mass spectrometric analysis of the milked venom from *C. marmoreus* revealed over 2000 unique peptides. However, all of these estimates include the multiple isoforms of peptides generated by heterogenous modification and processing of precursors encoded by the same gene.[61] Naturally occurring folding isoforms of conotoxins may also occur.[65] The diverse peptide isoforms produced from the same gene through variations in post-translational processing (and possibly through alternative protein folding) may have physiological significance in some cases, but this possibility has not been directly evaluated. An examination of the N-terminal sequences of the mature-toxin region often reveals multiple sites where proteolytic cleavage may occur, suggesting that a series of peptides with different N-terminal sequences may be produced, which may differ in their physiological activity *in vivo*. A similar heterogeneity can be generated at the C-terminal end. A precursor that has the X-GRRR sequence (where X is the conopeptide) at its C-terminus is generally considered to mature, *via* processing by a known amidating enzyme,[66] to yield X–NH_2, a peptide with the terminal carboxyl group protected by amidation. However, X-G, X-GR and X-GRR are all possible post-translationally produced intermediates that may accumulate in the venom and which differ in biological activity.

The structural diversity of conotoxins has been a subject of multiple reviews, and there is a dedicated database, ConoServer, that summarizes the known *Conus* venom peptide structures.[67,68] The structural diversity has been generated in part through a diversification of disulfide scaffolds (see Chapter 2 for more details). While most conotoxins contain multiple disulfide bridges, there are conopeptide families that lack Cys residues altogether. These include the conantokins, contulakin-G, and the conolysins.[52] A striking feature of conantokins, antagonists of *N*-methyl-D-aspartate receptors, is the presence of multiple γ-carboxyglutamate residues that stabilize the helical conformation of the peptide through chelated divalent metal ions.[69] In a glycopeptide known as contulakin-G from *C. geographus* (an agonist of neurotensin receptors), interactions between sugar and peptide backbone were reported,[70] suggesting that post-translational modifications (other than disulfide bridges) can be recruited to stabilize the conformation of these disulfide-poor conopeptides. Various post-translational modifications, including L- to D-isomerization, bromination of Trp or sulfation of Tyr, were summarized by Buczek and co-workers.[71]

Most conotoxins comprise less than 40 amino-acid residues, with multiple disulfide bridges. More than two dozen distinct disulfide-rich scaffolds have been described.[52] One mechanism by which snails diversify scaffolds is rapid gene duplication.[72,73] Although disulfide bridges stabilize bioactive conformations of conotoxins, a recent study showed that even noncritical bridges play an important role in ensuring their proper folding.[74] It is noteworthy that cone snails "experiment" with engineering disulfide bridges even within well-known gene families encoding highly conserved disulfide scaffolds, such as Kunitz domains (a protein family in which the disulfide scaffold is highly conserved across biological systems). Conotoxins constructed with a Kunitz domain are called conkunitzins, and several members of the conkunitzin family either lack one of the canonical disulfide bridges or contain an extra disulfide bridge. Structure determination by nuclear magnetic resonance (NMR) and X-ray crystallography revealed how one disulfide bridge could be replaced with close van der Waals contacts made possible by a Cys→Gly substitution encountered in conkunitzin-S1.[75,76] Dimers and trimers of the Kunitz domains were also identified in this family.[77] Indeed, diversification of Kunitz-domain scaffolds discovered in cone snail venoms is, so far as we know, much greater than has been reported for all other biological systems combined.

Most conotoxins are relatively small (under 5 kDa), but there are venom components larger than 9 kDa, for example con-ikot-ikot from *Conus striatus* that targets α-amino-3-hydroxy-5-methyl-4-isoxazolepropionic acid (AMPA) receptors[78] or p21a from *C. purpurascens*.[79] Enzyme components of cone snail venoms include phospholipase A2-like conodipine-M[80] and glycosyl hydrolases (hyaluronidases).[81,82] It is noteworthy that neuroactive metabolites have recently been demonstrated to be produced by bacteria living in symbiosis with cone snails.[83] Overall estimates of unique conotoxins and additional venom components produced by all cone snail species currently exceed 100 000. In all venomous marine snails, there are literally millions of unique pharmacologically active compounds.

Although earlier studies on the biosynthesis of conotoxins were sparse (reviewed in ref. 84), recent proteomic studies have provided more detail about *Conus* enzymes involved in venom peptide biosynthesis and folding. The discovery of a multitude of protein disulfide isomerase isoforms[59] suggests that these may play a role in the oxidative folding of conotoxin precursors containing a propeptide sequence.[85] Additional enzymes involved in the biosynthesis and folding of conotoxins include the *Conus* γ-glutamyl carboxylase,[86] peptidyl-prolyl *cis*-trans isomerase (PPI),[87] and a multi-enzyme complex of ER-based chaperones hsp70, hsp40 and protein disulfide isomerase (PDI) recently detected in the venom glands of *C. victoria*.[63] Our understanding of the putative mechanisms of enzyme-catalyzed oxidative folding of disulfide-rich peptides will undoubtedly improve as these enzymes and enzyme complexes are further explored.

6.5 Conoidean Venom Peptides: Some Future Directions

6.5.1 K_V-Channel Pharmacology: The Next Great Challenge

In the preceding sections, we summarized the targeting selectivity of different α-conotoxins that have evolved to match the diversity of homomeric and heteromeric nAChR subtypes in various prey, predators and competitors of cone snails. The resulting molecular diversity of α-conotoxins has been studied and developed systematically, providing selective ligands for various nAChR subtypes over three decades. The development of subtype-selective pharmacology for nAChRs has been and continues to be a formidable challenge, because many different nAChR pentameric isoforms can be formed from 16 different nAChR subunits encoded in mammalian genomes. By comparison, the on-going development of subtype-selective pharmacology for potassium channels is even more challenging because, in mammals, there are approximately 70 genes that encode K channel subunits, including 40 genes that encode subunits of voltage-gated K_V channels.[88]

The potential complexity of K_V channels is illustrated with just one example in Figure 6.6A, which shows the possible K_V channel subtypes that can be formed by only two subunits, $K_V1.1$ and $K_V1.2$ (from the "Shaker" or K_V1 subfamily), and in Figure 6.6C, which shows the striking increase in complexity generated if triheteromeric complexes are considered. The importance of considering the complexity is highlighted by recent results from the Terlau group obtained by testing precisely defined K_V channel architectures with the κM-conotoxin RIIIJ (RIIIJ), shown in Figure 6.6B. Although it was known that among homomeric Kv1 channels, RIIIJ blocked channels with four $K_V1.2$ subunits most potently, these new results show that there are striking differences in sensitivity among the possible isoforms of heteromeric K_V channels with both $K_V1.1$ and $K_V1.2$ subunits. Evolution of a toxin that strongly prefers channels constructed with one $K_V1.1$ subunit and three $K_V1.2$ subunits strongly suggests that this particular isoform plays a critical role in neuron function. The number of different heteromeric K_V channel subtypes that may theoretically be assembled is staggeringly large, but it is not presently clear which of these homomeric or heteromeric K_V channels actually exist *in vivo*. Venom peptides have the potential to become invaluable resources for identifying the specific K_V channel subtypes that are actually expressed in native cells and for elucidating the physiological roles of each.

There are multiple families of conopeptides that inhibit K_V channels. These have not yet been investigated in the systematic way that the α-conotoxins have been analyzed. However, there is compelling initial data that these block specific K_V channel subtypes with a high degree of selectivity. κM-conotoxin RIIIJ (RIIIJ) again provides a poignant example. RIIIJ selectively inhibits homomeric $K_V1.2$ channels over other K_V1 homomers, *i.e.* $K_V1.1$ and $K_V1.3$–1.8.[89] However, when tested against native K_V channels expressed in neurons, RIIIJ elicited a diversity of effects,[90] suggesting that RIIIJ may inhibit multiple $K_V1.2/1.X$ heteromers that are differentially expressed in different neuronal cell types. In Figure 6.7, we show examples of a neuron that was resistant to RIIIJ and others that were sensitive to RIIIJ, with a diversity of response phenotypes. Notably, when a different conopeptide (known as Pl14a) that is selective for $K_V1.6$ was applied to the same neuronal population, a subset of RIIIJ-resistant neurons responded, suggesting that K_V channels that contain $K_V1.2$ and $K_V1.6$ subunits are expressed in different subsets of sensory neurons.[90] RIIIJ and Pl14a, in conjunction with other

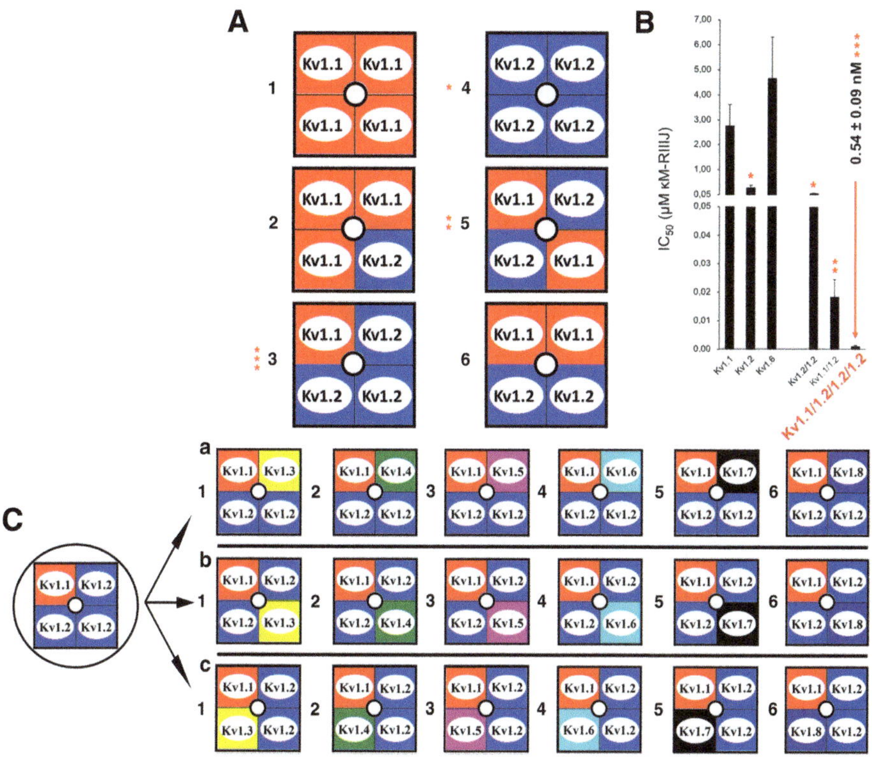

Figure 6.6 The potential complexity of heteromeric K_V channel subtypes. (A) Heteromeric K_V channel subtypes comprising two different K_V1 subfamily subunits, $K_V1.1$ and $K_V1.2$. (B) Affinity of κM-conotoxin RIIIJ, a peptide from *Conus radiatus* venom for K_V channels containing $K_V1.1$ and $K_V1.2$ subunits. The highest affinity was found when a single $K_V1.1$ and three $K_V1.2$ subunits were combined. The heteromeric channels were made by covalently linking the appropriate $K_V1.1$ and $K_V1.2$ subunits. Asterisks identify the presumed channel shown in panel (A) that corresponds to the IC_{50} value in panel (B). (C) Potential tri-heteromeric subunits derived from the highest affinity target for κM-conotoxin in RIIIJ shown in panels (A) and (B). One of the other K_V1 subunits is substituted for one of the $K_V1.2$ subunits, to yield a tri-heteromeric K_V channel. The possibility that these are also high affinity targets for κM-RIIIJ cannot be eliminated.

subtype-selective K_V channel venom peptides, are the type of pharmacological tools that can be used to identify the specific K_V channel subtypes expressed in each neuronal cell type and elucidate the physiological roles of each. The challenge is to differentiate between the types of K_V channel isoforms shown in Figure 6.6. Clearly a much larger array of subtype-selective peptides with highly specific targeting selectivity will be required to adequately address this challenge.

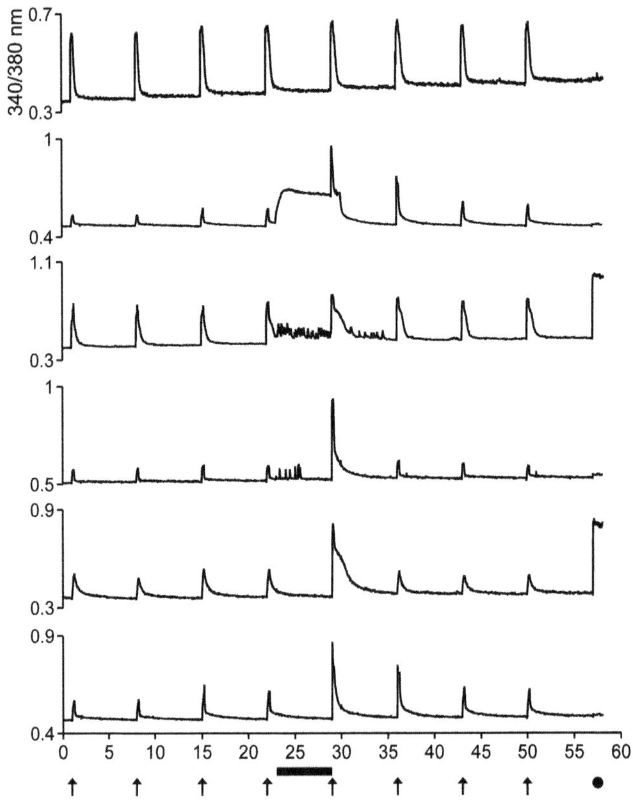

Figure 6.7 Calcium-imaging traces from a culture of heterogeneous DRG neurons, demonstrating different response phenotypes to κM-conotoxin RIIIJ. Each trace represents the response of a single neuron. In this experiment, more than 100 neuronal responses were monitored simultaneously, and seven of these are highlighted here. The *X*-axis is the same for all traces (time in minutes). The *Y*-axis is a measure of cytoplasmic calcium concentration as measured by the ratio of excitation with 340 nm and 380 nm light, while monitoring fluorescence emission at 510 nm (cells loaded with Fura-2-AM dye). The arrows indicate when the cells were depolarized with 30 mM extracellular K^+. An upward deflection of the trace indicates that the cytoplasmic calcium concentration increased transiently when the neuron was depolarized. The black horizontal bar indicates when 1 μM (RIIIJ) was present in the bath. Some neurons were apparently unaffected by the application of RIIIJ (top trace). In some neurons, the calcium baseline increased during the application of RIIIJ (trace second from top). In other neurons, the calcium baseline exhibited repetitive calcium spikes (third and fourth traces from top). In some neurons, the calcium transient was amplified substantially by the application of RIIIJ in either peak height, width or both (all traces but top trace). The black circle indicates when capsaicin was added at the end of the experiments. Notably, cells affected by RIIIJ included both capsaicin-sensitive and resistant neurons.

The therapeutic potential of selective K_V channel ligands cannot be overstated. Even non-selective K_V channel antagonists, such as 4-aminopyridine, are used clinically for multiple sclerosis and several other neurological diseases[91,92] (reviewed in ref. 93). The subtype-selective K_V channel antagonists found in venoms are expected to provide greater therapeutic benefits with fewer side effects. Indeed, a derivative of the sea anemone toxin ShK, which selectively targets the $K_V1.3$ isoform, is currently in human clinical trials for treatment of multiple sclerosis.[94–96] Preclinical studies for κ-conopeptide PVIIA, which selectively targets Shaker K_V channels, demonstrated that this conopeptide is cardioprotective in animal models. Terlau and co-workers showed that this peptide is state selective, and that channels inhibited with κ-PVIIA were also inhibited in the progression to a slow inactivated state.[97] These few examples of preclinical therapeutic development illustrate the vast and largely untapped potential of subtype-selective K_V channel-targeting peptides from venoms.

6.5.2 Lessons Learned and Translating these to Biomedical Science

By current estimates we have surveyed and characterized only a small portion of the structural and functional diversity of venom peptides from marine snails. Nevertheless, there are several important lessons to be learned from the pharmacological strategies that evolved in marine snails—lessons that can be fruitfully applied to the biomedical sciences.

The first lesson is that Nature has already provided subtype-selective ligands for therapeutically relevant receptor and ion channel targets. Consideration of the vast number of unstudied or understudied conoidean species indicates that the majority of these subtype-selective ligands are as yet undiscovered or uncharacterized. A second major insight is that a specific physiological end-point can be effectively achieved by combining structurally distinct ligands that bind different molecular targets and/or bind different sites on a common molecular target. Such flexibility is underappreciated in pharmacology, where drug-discovery programs typically focus on a single molecular target and mechanism of action.

These points deserve special attention when considering how to translate advances in conoidean peptide research to biomedical science. The cabal strategies employed by fish-hunting cone snails has revealed that a partial block of a few functionally coupled signaling proteins achieves a physiological end-point more effectively compared with complete block of one signaling protein in a physiological circuit. Drug-discovery programs would do well to follow this lesson, taken from the cone snail playbook, for two reasons, both of which relate to specificity. First, molecular targeting specificity is always concentration dependent. At high doses, spillover effects will be greater, and even a highly selective drug will uncover off-target binding, leading to unwanted side effects when taken in excess. Second, a cell type or circuit may be targeted more selectively with a therapeutic cabal. Any one signaling molecule (*e.g.* a particular subunit of the K_V channel) is generally expressed in several circuits, in different tissues, and in many cell types. But a particular combination of signaling molecules is likely to be uniquely expressed in a single target circuit, tissue and cell type. Thus, if two functionally linked proteins are inhibited only partially in the targeted circuit, the effects may be synergistic to achieve a desired therapeutic benefit, but the side effects potentially will be less severe in other circuits or cells where only one of the signaling proteins is expressed and functionally coupled to a different set of signaling proteins.

The potential for cell-specific outcomes triggered by perturbation of signaling molecules is corroborated by the study of a mutation in one Na_V channel subtype, $Na_V1.7$, that results in opposite phenotypes in different cell types. The gain-of-function mutation causes a depolarizing shift in the resting membrane potential leading to hyper-excitability in dorsal root ganglion (DRG) neurons, where $Na_V1.7$ is coupled functionally to $Na_V1.8$. However, in sympathetic neurons, the same mutation results in hypo-excitability due to channel inactivation because $Na_V1.8$ is not expressed in these neurons.[98] Thus, the same gain-of-function mutation has different phenotypes dependent on the co-expression of other Na_V channels. Cell type, as defined by the repertoire of signaling molecules functionally co-expressed, is clearly important in determining how the cell will respond to some event, in this case to a mutation, or by analogy, to a pharmacological perturbation. The same principle forms the foundation for a new paradigm, described in the final section, which aims to define cell types by monitoring the way cells respond to a judiciously designed set of diverse pharmacological challenge events.

6.5.3 A New Paradigm for Discovery of Novel Pharmacological Agents

A set of daunting challenges faces the field of conoidean venom peptide research. The lessons learned from past research indicate the existence of highly specific pharmacologies that could precisely target different cell types defined, for example, by the exact isoforms of functionally expressed K_V channels. Finding the necessary ligands with current methodologies that rely on intuition and exhaustive tests on a one-signaling-molecule-per-trial basis seems woefully inadequate, given the enormity of the task. We desire to efficiently uncover the most valuable of these pharmacologies within a reasonable time span, and therefore have been motivated to find a faster and "wider-net" approach inspired, appropriately, from the strategies that naturally evolved in marine snails. This section attempts to sketch how this approach works, provide some examples of recent results obtained, and define realistic goals for future work.

The approach for discovery of novel pharmacologies we are currently developing tackles two difficult and related challenges at the same time. The first challenge is to define cell types of the nervous system more precisely. For example, we want to replace the low-resolution demarcation of cells on the basis of tetrodotoxin-sensitivity with high-resolution characters based on the precise isoforms of Na_V channels present on each cell. The μ-conotoxins and other conopeptide families targeting different isoforms of Na_V channels should make a higher resolution demarcation of cell types feasible, especially if additional subtype-selective agents that target another signal molecule family are included. The expectation is that with a sufficiently expanded arsenal of subtype-selective ligands we should be able to precisely distinguish all of the cell types found within a particular anatomical locus.

The second challenge is to obtain and analyze results looking for new pharmacologies in a massively parallel manner. To meet this challenge, we employ heterogeneous mixtures of cells and simultaneously monitor the response of each cell during application of a series of stimulatory and inhibitory agents. Since all neuronal signaling events ultimately transduce to changes in intra-cellular calcium ion concentration, cell responses are recorded *via* calcium imaging (see Figure 6.7). Part of the protocol is designed to identify the multiple cell types present (and the constellation of signaling molecules present in each of those cell types). In this phase, the response (or lack of response) to multiple events provides a functional fingerprint that can differentiate between the different cell types present. The cells are then challenged with novel ligands or mixtures of ligands (*e.g.* venom fractions) and the resulting responses are obtained for all cells in the population (100–200 cells can be simultaneously analyzed). This is expected to quickly identify sources (*e.g.* venoms or venom fractions) with new potentially valuable pharmacologies. For want of a better name, we are calling the platform described here "constellation pharmacology". The key to making the constellation pharmacology platform work is that the solutions to each of the two challenges reinforce each other. Definition of cell types will increasingly attain better resolution as new ligands with improved specificity are discovered, which will better enable discovery of novel ligands with unique targeting selectivity.

As proof-of-principle, we have characterized two types of cold-sensitive neurons from rodent DRG: cold thermosensors and cold nociceptors.[12] The constellations of receptors and ion channels in the two DRG subclasses differ substantially, as determined by the presently available toolkit of pharmacological agents, including several subtype-selective venom peptides. These differences were exploited to pharmacologically manipulate the cold threshold required to activate cold nociceptors. Ordinarily the cold nociceptors only responded to noxious-cold temperatures (*i.e.* $<15\,°C$). However, by blocking the K_V channels present in those neurons, they began to respond to innocuous cool temperatures.[12]

The changes in response thresholds elicited by pharmacology are relevant to pathological conditions. The same K_V channel subtype that we blocked using pharmacology was also downregulated in a mouse model of cold allodynia (produced by a chemotherapy drug).[99] Thus, the cold threshold can be lowered either by gene expression changes or through pharmacology. Conversely, knowledge of the constellation of ion channels and receptors expressed in the cold nociceptors raises the possibility of novel pharmacological strategies for specifically suppressing cold allodynia, which would alleviate a major source of suffering experienced by cancer patients undergoing chemotherapy treatment. Our ability to discriminate between different neuronal cell types through their cell-specific constellations as reflected by functional fingerprinting, should increase significantly as more subtype-selective ligands are characterized. In turn, this creates an opportunity to design more sophisticated and targeted therapeutic interventions.

The continuing development of constellation pharmacology should provide a scientific bridge between molecular and systems neuroscience. There currently exists a formidable barrier separating these two fields: molecular neuroscience focuses on the properties of individual signaling molecules, and systems neuroscience aims to understand the behavior of neuronal circuits. Essentially, this barrier results from a "zone of ignorance" at the cellular level. At present, except for morphologically distinct neurons (*e.g.* rods and cones, and cerebellar Purkinje cells), the exact subclass of a neuron being studied is generally difficult to define. An additional layer of confusion is contributed by the vast number of possible subtypes of signaling molecules, such as K_V channels or nicotinic receptors. It is currently difficult to answer the seemingly simple question of which K_V channel subtypes are expressed in which particular neuronal subclasses. In order to follow the emergence of new neuronal subclasses during brain development or to compare neurons within or among species, a better definition of cell types is required. Constellation pharmacology should breach the significant gap that exists between molecular and systems neuroscience by uncovering the functionally integrated molecular components within each neuronal subclass.

The great promise of conoidean venom peptides for future biomedical applications is that they will provide major tools for linking together molecular and systems neuroscience into a seamlessly integrated, cohesive discipline. Such integration is essential for accelerating progress in understanding and treating pathologies of the nervous system. To take specific examples, there is an urgent need for early biomarkers in many of the intractable neurodegenerative diseases. Constellation pharmacology can be used to investigate which cell-specific constellations change with the disease progression, potentially detecting early cell-specific changes that occur long before any clinical symptoms appear. Subtype-selective conopeptides have the potential to monitor such changes, thereby becoming diagnostic biomarkers. Understanding what cell-specific changes occur may be the best hope we have of understanding the presently intractable progression of neurodegenerative diseases, and of blocking this progression through a targeted pharmacological intervention.

Acknowledgements

The work of the authors has been supported by National Institute of General Medical Sciences Grant GM48677. We are grateful to Maren Watkins for organizing the sequences presented, and to Terry Merritt and My Huynh for support in preparing the Chapter.

References

1. B. M. Olivera, *Annu. Rev. Ecol. Evol. Syst.*, 2002, **33**, 25–47.
2. B. M. Olivera, P. Showers-Corneli, M. Watkins and A. Fedosov, *Annu. Rev. Animal Biosci.*, 2014, **2**, 487–513.
3. N. Puillandre, P. Bouchet, T. F. Duda, Jr., S. Kauferstein, A. J. Kohn, B. M. Olivera, M. Watkins and C. Meyer, *Mol. Phylogenet. Evol.*, 2014, **78**, 290–303.
4. N. Puillandre, T. F. Duda, Jr., C. Meyer, B. M. Olivera and P. Bouchet, *J. Mollus. Stud.*, 2014, advance access published online 5 September 2014 doi:10.1093/mollus/eyu055.
5. P. Bouchet, Y. I. Kantor, A. V. Sysoev and N. Puillandre, *J. Mollus. Stud.*, 2011, 77, 273–308.
6. J. K. Tucker and M. J. Tenorio, *Systematic Classification of Recent and Fossil Conoidean Gastropods: With Keys to the Genera of Cone Shells*. ConchBooks, 2009.
7. J. D. Taylor, Y. I. Kantor and A. V. Sysoev, *Bulletin of the Natural History Museum (London), Zoology Series,* 1993, **59**, 125–170.
8. N. Puillandre, C. P. Meyer, P. Bouchet and B. M. Olivera, *Zool. Scr.*, 2011, **40**, 350–363.
9. A. B. Cabang, J. S. Imperial, J. Gajewiak, M. Watkins, P. S. Corneli, B. M. Olivera and G. P. Concepcion, *Toxicon,* 2011, **58**, 672–680.
10. B. M. Olivera, *Mol. Biol. Cell,* 1997, **8**, 2101–2109.
11. H. Terlau, K. J. Shon, M. Grilley, M. Stocker, W. Stuhmer and B. M. Olivera, *Nature,* 1996, **381**, 148–151.
12. R. W. Teichert, S. Raghuraman, T. Memon, J. L. Cox, T. Foulkes, J. E. Rivier and B. M. Olivera, *Proc. Natl. Acad. Sci. U.S.A.*, 2012, **109**, 12758–12763.
13. P. K. Pallaghy, K. J. Nielsen, D. J. Craik and R. S. Norton, *Protein Sci.*, 1994, **3**, 1833–1839.
14. O. Buczek, D. Wei, J. J. Babon, X. Yang, B. Fiedler, P. Chen, D. Yoshikami, B. M. Olivera, G. Bulaj and R. S. Norton, *Biochemistry,* 2007, **46**, 9929–9940.
15. B. M. Olivera and R. W. Teichert, *Mol. Interv.*, 2007, **7**, 251–260.
16. D. R. Witcher, M. De Waard, J. Sakamoto, C. Franzini-Armstrong, M. Pragnell, S. D. Kahl and K. P. Campbell, *Science,* 1993, **261**, 486–489.
17. L. J. Cruz and B. M. Olivera, *J. Biol. Chem.*, 1986, **261**, 6230–6233.
18. M. M. Zhang, M. J. Wilson, L. Azam, J. Gajewiak, J. E. Rivier, G. Bulaj, B. M. Olivera and D. Yoshikami, *Br. J. Pharmacol.*, 2013, **168**, 1597–1610.

19. M. M. Zhang, M. J. Wilson, J. Gajewiak, J. E. Rivier, G. Bulaj, B. M. Olivera and D. Yoshikami, *Br. J. Pharmacol.*, 2013, **169**, 102–114.
20. E. Leipold, R. Markgraf, A. Miloslavina, M. Kijas, J. Schirmeyer, D. Imhof and S. H. Heinemann, *Neuropharmacology*, 2011, **61**, 105–111.
21. M. J. Wilson, D. Yoshikami, L. Azam, J. Gajewiak, B. M. Olivera, G. Bulaj and M. M. Zhang, *Proc. Natl. Acad. Sci. U.S.A.*, 2011, **108**, 10302–10307.
22. G. F. King, P. Escoubas and G. M. Nicholson, *Channels (Austin)*, 2008, **2**, 100–116.
23. F. Bosmans and K. J. Swartz, *Trends Pharmacol. Sci.*, 2010, **31**, 175–182.
24. B. B. Carstens, R. J. Clark, N. L. Daly, P. J. Harvey, Q. Kaas and D. J. Craik, *Curr. Pharm. Des.*, 2011, **17**, 4242–4253.
25. K. K. Khoo, M. J. Wilson, B. J. Smith, M. M. Zhang, J. Gulyas, D. Yoshikami, J. E. Rivier, G. Bulaj and R. S. Norton, *J. Med. Chem.*, 2011, **54**, 7558–7566.
26. R. J. French, D. Yoshikami, M. F. Sheets and B. M. Olivera, *Mar. Drugs*, 2010, **8**, 2153–2161.
27. A. Van Der Haegen, S. Peigneur and J. Tytgat, *FEBS J.*, 2011, **278**, 3408–3418.
28. N. Unwin, *J. Mol. Biol.*, 1993, **229**, 1101–1124.
29. N. P. Mongan, A. K. Jones, G. R. Smith, M. S. Sansom and D. B. Sattelle, *Protein Sci.*, 2002, **11**, 1162–1171.
30. M. Quik, *Trends Neurosci.*, 2004, **27**, 561–568.
31. H. Chen, X. Huang, X. Guo, R. B. Mailman, Y. Park, F. Kamel, D. M. Umbach, Q. Xu, A. Hollenbeck, A. Schatzkin and A. Blair, *Neurology*, 2010, **74**, 878–884.
32. M. Quik, X. A. Perez and T. Bordia, *Mov. Disord.*, 2012, **27**, 947–957.
33. M. Quik and S. Wonnacott, *Pharmacol. Rev.*, 2011, **63**, 938–966.
34. M. Quik and J. M. McIntosh, *J. Pharmacol. Exp. Ther.*, 2006, **316**, 481–489.
35. A. R. Tapper, S. L. McKinney, R. Nashmi, J. Schwarz, P. Deshpande, C. Labarca, P. Whiteaker, M. J. Marks, A. C. Collins and H. A. Lester, *Science*, 2004, **306**, 1029–1032.
36. T. Bordia, S. R. Grady, J. M. McIntosh and M. Quik, *Mol. Pharmacol.*, 2007, **72**, 52–61.
37. R. Exley, N. Maubourguet, V. David, R. Eddine, A. Evrard, S. Pons, F. Marti, S. Threlfell, P. Cazala, J. M. McIntosh, J. P. Changeux, U. Maskos, S. J. Cragg and P. Faure, *Proc. Natl. Acad. Sci. U.S.A.*, 2011, **108**, 7577–7582.
38. K. Brejc, W. J. van Dijk, R. V. Klaassen, M. Schuurmans, J. van Der Oost, A. B. Smit and T. K. Sixma, *Nature*, 2001, **411**, 269–276.
39. A. B. Smit, N. I. Syed, D. Schaap, J. van Minnen, J. Klumperman, K. S. Kits, H. Lodder, R. C. van der Schors, R. van Elk, B. Sorgedrager, K. Brejc, T. K. Sixma and W. P. Geraerts, *Nature*, 2001, **411**, 261–268.
40. P. H. Celie, I. E. Kasheverov, D. Y. Mordvintsev, R. C. Hogg, P. van Nierop, R. van Elk, S. E. van Rossum-Fikkert, M. N. Zhmak, D. Bertrand, V. Tsetlin, T. K. Sixma and A. B. Smit, *Nat. Struct. Mol. Biol.*, 2005, **12**, 582–588.

41. P. Rucktooa, A. B. Smit and T. K. Sixma, *Biochem. Pharmacol.*, 2009, **78**, 777–787.
42. S. B. Hansen, G. Sulzenbacher, T. Huxford, P. Marchot, P. Taylor and Y. Bourne, *EMBO J.*, 2005, **24**, 3635–3636.
43. C. Ulens, R. C. Hogg, P. H. Celie, D. Bertrand, V. Tsetlin, A. B. Smit and T. K. Sixma, *Proc. Natl. Acad. Sci. U.S.A.*, 2006, **103**, 3615–3620.
44. J. M. McIntosh, D. Yoshikami, E. Mahe, D. B. Nielsen, J. E. Rivier, W. R. Gray and B. M. Olivera, *J. Biol. Chem.*, 1994, **269**, 16733–16739.
45. P. A. Quiram and S. M. Sine, *J. Biol. Chem.*, 1998, **273**, 11001–11006.
46. M. Fainzilber, A. Hasson, R. Oren, A. L. Burlingame, D. Gordon, M. E. Spira and E. Zlotkin, *Biochemistry*, 1994, **33**, 9523–9529.
47. S. Luo, T. A. Nguyen, G. E. Cartier, B. M. Olivera, D. Yoshikami and J. M. McIntosh, *Biochemistry*, 1999, **38**, 14542–14548.
48. D. Everhart, E. Reiller, A. Mirzoian, J. M. McIntosh, A. Malhotra and C. W. Luetje, *J. Pharmacol. Exp. Ther.*, 2003, **306**, 664–670.
49. L. Azam, C. Dowell, M. Watkins, J. A. Stitzel, B. M. Olivera and J. M. McIntosh, *J. Biol. Chem.*, 2005, **280**, 80–87.
50. H. W. Kim and J. M. McIntosh, *FASEB J.*, 2012, **26**, 4102–4110.
51. D. L. Shiembob, R. L. Roberts, C. W. Luetje and J. M. McIntosh, *Biochemistry*, 2006, **45**, 11200–11207.
52. H. Terlau and B. M. Olivera, *Physiol. Rev.*, 2004, **84**, 41–68.
53. A. O. Lluisma, B. A. Milash, B. Moore, B. M. Olivera and P. K. Bandyopadhyay, *Mar. Genom.*, 2012, **5**, 43–51.
54. P. G. Arnison, M. J. Bibb, G. Bierbaum, A. A. Bowers, T. S. Bugni, G. Bulaj, J. A. Camarero, D. J. Campopiano, G. L. Challis, J. Clardy, P. D. Cotter, D. J. Craik, M. Dawson, E. Dittmann, S. Donadio, P. C. Dorrestein, K. D. Entian, M. A. Fischbach, J. S. Garavelli, U. Goransson, C. W. Gruber, D. H. Haft, T. K. Hemscheidt, C. Hertweck, C. Hill, A. R. Horswill, M. Jaspars, W. L. Kelly, J. P. Klinman, O. P. Kuipers, A. J. Link, W. Liu, M. A. Marahiel, D. A. Mitchell, G. N. Moll, B. S. Moore, R. Muller, S. K. Nair, I. F. Nes, G. E. Norris, B. M. Olivera, H. Onaka, M. L. Patchett, J. Piel, M. J. Reaney, S. Rebuffat, R. P. Ross, H. G. Sahl, E. W. Schmidt, M. E. Selsted, K. Severinov, B. Shen, K. Sivonen, L. Smith, T. Stein, R. D. Sussmuth, J. R. Tagg, G. L. Tang, A. W. Truman, J. C. Vederas, C. T. Walsh, J. D. Walton, S. C. Wenzel, J. M. Willey and W. A. van der Donk, *Nat. Prod. Rep.*, 2013, **30**, 108–160.
55. A. G. Craig, P. Bandyopadhyay and B. M. Olivera, *Eur. J. Biochem.*, 1999, **264**, 271–275.
56. S. G. Conticello, Y. Gilad, N. Avidan, E. Ben-Asher, Z. Levy and M. Fainzilber, *Mol. Biol. Evol.*, 2001, **18**, 120–131.
57. B. M. Olivera, M. Watkins, P. Bandyopadhyay, J. S. Imperial, E. P. de la Cotera, M. B. Aguilar, E. L. Vera, G. P. Concepcion and A. Lluisma, *Ann. N. Y. Acad. Sci.*, 2012, **1267**, 61–70.
58. D. Biass, S. Dutertre, A. Gerbault, J. L. Menou, R. Offord, P. Favreau and R. Stocklin, *J. Proteomics*, 2009, **72**, 210–218.

59. H. Safavi-Hemami, W. A. Siero, D. G. Gorasia, N. D. Young, D. Macmillan, N. A. Williamson and A. W. Purcell, *J. Proteome Res.*, 2011, **10**, 3904–3919.
60. A. Violette, D. Biass, S. Dutertre, D. Koua, D. Piquemal, F. Pierrat, R. Stocklin and P. Favreau, *J. Proteomics*, 2012, **75**, 5215–5225.
61. S. Dutertre, A. H. Jin, Q. Kaas, A. Jones, P. F. Alewood and R. J. Lewis, *Mol. Cell. Proteomics*, 2013, **12**, 312–329.
62. S. Dutertre, D. Biass, R. Stocklin and P. Favreau, *Toxicon*, 2010, **55**, 1453–1462.
63. J. A. Rivera-Ortiz, H. Cano and F. Mari, *Peptides*, 2011, **32**, 306–316.
64. M. A. Abdel-Rahman, I. M. Abdel-Nabi, M. S. El-Naggar, O. A. Abbas and P. N. Strong, *Comp. Biochem. Physiol. C Toxicol. Pharmacol.*, 2011, **154**, 318–325.
65. H. Safavi-Hemami, D. G. Gorasia, A. M. Steiner, N. A. Williamson, J. A. Karas, J. Gajewiak, B. M. Olivera, G. Bulaj and A. W. Purcell, *J. Biol. Chem.*, 2012, **287**, 34288–34303.
66. S. Ul-Hasan, D. M. Burgess, J. Gajewiak, Q. Li, H. Hu, M. Yandell, B. M. Olivera and P. K. Bandyopadhyay, *Toxicon*, 2013, **74**, 215–224.
67. Q. Kaas, J. C. Westermann and D. J. Craik, *Toxicon*, 2010, **55**, 1491–1509.
68. Q. Kaas, R. Yu, A. H. Jin, S. Dutertre and D. J. Craik, *Nucleic Acids Res.*, 2012, **40**, D325–D330.
69. M. Prorok and F. J. Castellino, *Curr. Drug Targets*, 2007, **8**, 633–642.
70. L. Kindahl, C. Sandström, A. G. Craig, T. Norberg and L. Kenne, *Can. J. Chem.*, 2002, **80**, 1022–1031.
71. O. Buczek, G. Bulaj and B. M. Olivera, *Cell. Mol. Life Sci.*, 2005, **62**, 3067–3079.
72. A. D. Santos, J. M. McIntosh, D. R. Hillyard, L. J. Cruz and B. M. Olivera, *J. Biol. Chem.*, 2004, **279**, 17596–17606.
73. D. Chang and T. F. Duda, Jr., *Mol. Biol. Evol.*, 2012, **29**, 2019–2029.
74. K. H. Gowd, K. D. Blais, K. S. Elmslie, A. M. Steiner, B. M. Olivera and G. Bulaj, *Biopolymers*, 2012, **98**, 212–223.
75. M. Bayrhuber, V. Vijayan, M. Ferber, R. Graf, J. Korukottu, J. Imperial, J. E. Garrett, B. M. Olivera, H. Terlau, M. Zweckstetter and S. Becker, *J. Biol. Chem.*, 2005, **280**, 23766–23770.
76. C. Y. Dy, P. Buczek, J. S. Imperial, G. Bulaj and M. P. Horvath, *Acta Crystallogr. D Biol. Crystallogr.*, 2006, **62**, 980–990.
77. B. M. Olivera, G. Bulaj, J. E. Garrett, H. Terlau and J. S. Imperial, Peptide toxins from the venoms of cone snails and other toxoglossan gastropods. In *Animal Toxins – State of the Art: Perspectives in Health and Biotechnology*, De Lima, M. E., Ed. Editora UFMG: Terreo: 2009, pp. 25–48.
78. C. S. Walker, S. Jensen, M. Ellison, J. A. Matta, W. Y. Lee, J. S. Imperial, N. Duclos, P. J. Brockie, D. M. Madsen, J. T. Isaac, B. Olivera and A. V. Maricq, *Curr. Biol.*, 2009, **19**, 900–908.
79. C. Möller and F. Mari, *Biopolymers*, 2011, **96**, 158–165.

80. J. M. McIntosh, F. Ghomashchi, M. H. Gelb, D. J. Dooley, S. J. Stoehr, A. B. Giordani, S. R. Naisbitt and B. M. Olivera, *J. Biol. Chem.,* 1995, **270**, 3518–3526.
81. A. Violette, A. Leonardi, D. Piquemal, Y. Terrat, D. Biass, S. Dutertre, F. Noguier, F. Ducancel, R. Stocklin, I. Krizaj and P. Favreau, *Mar. Drugs,* 2012, **10**, 258–280.
82. A. Leonardi, D. Biass, D. Kordis, R. Stocklin, P. Favreau and I. Krizaj, *J. Proteome Res.,* 2012, **11**, 5046–5058.
83. Z. Lin, R. R. Antemano, R. W. Hughen, M. D. Tianero, O. Peraud, M. G. Haygood, G. P. Concepcion, B. M. Olivera, A. Light and E. W. Schmidt, *J. Nat. Prod.,* 2010, **73**, 1922–1926.
84. G. Bulaj and B. M. Olivera, *Antioxid. Redox. Signal.,* 2008, **10**, 141–155.
85. O. Buczek, B. M. Olivera and G. Bulaj, *Biochemistry,* 2004, **43**, 1093–1101.
86. P. K. Bandyopadhyay, J. E. Garrett, R. P. Shetty, T. Keate, C. S. Walker and B. M. Olivera, *Proc. Natl. Acad. Sci. U.S.A.,* 2002, **99**, 1264–1269.
87. H. Safavi-Hemami, G. Bulaj, B. M. Olivera, N. A. Williamson and A. W. Purcell, *J. Biol. Chem.,* 2010, **285**, 12735–12746.
88. G. A. Gutman, K. G. Chandy, S. Grissmer, M. Lazdunski, D. McKinnon, L. A. Pardo, G. A. Robertson, B. Rudy, M. C. Sanguinetti, W. Stuhmer and X. Wang, *Pharmacol. Rev.,* 2005, **57**, 473–508.
89. P. Chen, A. Dendorfer, R. K. Finol-Urdaneta, H. Terlau and B. M. Olivera, *J. Biol. Chem.,* 2010, **285**, 14882–14889.
90. R. W. Teichert, N. J. Smith, S. Raghuraman, D. Yoshikami, A. R. Light and B. M. Olivera, *Proc. Natl. Acad. Sci. U.S.A.,* 2012, **109**, 1388–1395.
91. R. M. Sherratt, H. Bostock and T. A. Sears, *Nature,* 1980, **283**, 570–572.
92. R. E. Jones, J. R. Heron, D. H. Foster, R. S. Snelgar and R. J. Mason, *J. Neurol. Sci.,* 1983, **60**, 353–362.
93. S. Sedehizadeh, M. Keogh and P. Maddison, *Clin. Neuropharmacol.,* 2012, **35**, 191–200.
94. O. Castaneda, V. Sotolongo, A. M. Amor, R. Stocklin, A. J. Anderson, A. L. Harvey, A. Engstrom, C. Wernstedt and E. Karlsson, *Toxicon,* 1995, **33**, 603–613.
95. M. W. Pennington, M. E. Byrnes, I. Zaydenberg, I. Khaytin, J. de Chastonay, D. S. Krafte, R. Hill, V. M. Mahnir, W. A. Volberg, W. Gorczyca and W. R. Kem, *Int. J. Pept. Protein Res.,* 1995, **46**, 354–358.
96. S. Rangaraju, V. Chi, M. W. Pennington and K. G. Chandy, *Expert Opin. Ther. Targets,* 2009, **13**, 909–924.
97. H. Terlau, A. Boccaccio, B. M. Olivera and F. Conti, *J. Gen. Physiol.,* 1999, **114**, 125–140.
98. A. M. Rush, S. D. Dib-Hajj, S. Liu, T. R. Cummins, J. A. Black and S. G. Waxman, *Proc. Natl. Acad. Sci. U.S.A.,* 2006, **103**, 8245–8250.
99. J. Descoeur, V. Pereira, A. Pizzoccaro, A. Francois, B. Ling, V. Maffre, B. Couette, J. Busserolles, C. Courteix, J. Noel, M. Lazdunski, A. Eschalier, N. Authier and E. Bourinet, *EMBO Mol. Med.,* 2011, **3**, 266–278.
100. I. Vetter and R. J. Lewis, *Curr. Top. Med. Chem.,* 2012, **12**, 1546–1552.

101. G. P. Miljanich, *Curr. Med. Chem.,* 2004, **11**, 3029–3040.
102. C. J. Armishaw and P. F. Alewood, *Curr. Protein Pept. Sci.,* 2005, **6**, 221–240.
103. J. M. McIntosh, N. Absalom, M. Chebib, A. B. Elgoyhen and M. Vincler, *Biochem. Pharmacol.,* 2009, **78**, 693–702.
104. T. S. Han, R. W. Teichert, B. M. Olivera and G. Bulaj, *Curr. Pharm. Des.,* 2008, **14**, 2462–2479.
105. V. D. Twede, G. Miljanich, B. M. Olivera and G. Bulaj, *Curr. Opin. Drug Discovery. Dev.,* 2009, **12**, 231–239.
106. E. D. Koch, B. M. Olivera, H. Terlau and F. Conti, *Biophys. J.,* 2004, **86**, 191–209.
107. M. Vincler, S. Wittenauer, R. Parker, M. Ellison, B. M. Olivera and J. M. McIntosh, *Proc. Natl. Acad. Sci. U.S.A.,* 2006, **103**, 17880–17884.
108. W. R. Gray, A. Luque, B. M. Olivera, J. Barrett and L. J. Cruz, *J. Biol. Chem.,* 1981, **256**, 4734–4740.
109. C. Hopkins, M. Grilley, C. Miller, K. J. Shon, L. J. Cruz, W. R. Gray, J. Dykert, J. Rivier, D. Yoshikami and B. M. Olivera, *J. Biol. Chem.,* 1995, **270**, 22361–22367.
110. E. C. Jimenez, B. M. Olivera and R. W. Teichert, *Biochemistry,* 2007, **46**, 8717–8724.
111. M. Loughnan, A. Nicke, A. Jones, C. I. Schroeder, S. T. Nevin, D. J. Adams, P. F. Alewood and R. J. Lewis, *J. Biol. Chem.,* 2006, **281**, 24745–24755.
112. K. J. Shon, S. C. Koerber, J. E. Rivier, B. M. Olivera and J. M. McIntosh, *Biochemistry,* 1997, **36**, 15693–15700.
113. L. J. Cruz, W. R. Gray, B. M. Olivera, R. D. Zeikus, L. Kerr, D. Yoshikami and E. Moczydlowski, *J. Biol. Chem.,* 1985, **260**, 9280–9288.
114. M. Fainzilber, D. Gordon, A. Hasson, M. E. Spira and E. Zlotkin, *Eur. J. Biochem.,* 1991, **202**, 589–595.
115. R. W. Teichert, R. Jacobsen, H. Terlau, D. Yoshikami and B. M. Olivera, *Toxicon,* 2007, **49**, 318–328.
116. M. Ferber, A. Sporning, G. Jeserich, R. DeLaCruz, M. Watkins, B. M. Olivera and H. Terlau, *J. Biol. Chem.,* 2003, **278**, 2177–2183.
117. J. S. Imperial, P. S. Bansal, P. F. Alewood, N. L. Daly, D. J. Craik, A. Sporning, H. Terlau, E. Lopez-Vera, P. K. Bandyopadhyay and B. M. Olivera, *Biochemistry,* 2006, **45**, 8331–8340.
118. L. M. Kerr and D. Yoshikami, *Nature,* 1984, **308**, 282–284.
119. J. Bernaldez, S. A. Roman-Gonzalez, O. Martinez, S. Jimenez, O. Vivas, I. Arenas, G. Corzo, R. Arreguin, D. E. Garcia, L. D. Possani and A. Licea, *Mar. Drugs,* 2013, **11**, 1188–1202.
120. L. J. England, J. Imperial, R. Jacobsen, A. G. Craig, J. Gulyas, M. Akhtar, J. Rivier, D. Julius and B. M. Olivera, *Science,* 1998, **281**, 575–578.
121. I. A. Sharpe, J. Gehrmann, M. L. Loughnan, L. Thomas, D. A. Adams, A. Atkins, E. Palant, D. J. Craik, D. J. Adams, P. F. Alewood and R. J. Lewis, *Nat. Neurosci.,* 2001, **4**, 902–907.

122. L. J. Cruz, V. de Santos, G. C. Zafaralla, C. A. Ramilo, R. Zeikus, W. R. Gray and B. M. Olivera, *J. Biol. Chem.*, 1987, **262**, 15821–15824.
123. M. Fainzilber, T. Nakamura, J. C. Lodder, E. Zlotkin, K. S. Kits and A. L. Burlingame, *Biochemistry*, 1998, **37**, 1470–1477.
124. J. M. McIntosh, B. M. Olivera, L. J. Cruz and W. R. Gray, *J. Biol. Chem.*, 1984, **259**, 14343–14346.

CHAPTER 7

Scorpion Venoms as a Platform for Drug Development

RICARDO C. RODRÍGUEZ DE LA VEGA*[a,b], GERARDO CORZO[c], AND LOURIVAL D. POSSANI[c]

[a]Ecologie, Systématique et Evolution, UMR 8079, Université Paris-Sud, F-91405 Orsay Cedex, France; [b]UMR 8079, CNRS, F-91405 Orsay Cedex, France; [c]Departamento de Medicina Molecular y Bioprocesos, Instituto de Biotecnología, Universidad Nacional Autónoma de México, Av. Universidad, 2001 Cuernavaca, Mor. 62210, México
*E-mail: ricardo.rodriguezdelavega@gmail.com

7.1 Introduction

Frightening, mesmerizing or whatever else it might be, the sight of a scorpion is something humans do not often dismiss. We know a scorpion sting can be dangerous, even life-threatening, due to the presence of neurotoxic components in the complex mixture that makes up venom. In many tropical and sub-tropical countries scorpionism (the clinical manifestation of scorpion envenoming) remains a public health problem,[1] and thus it is not surprising that scorpion venom research has been largely driven by interest in the understanding of human envenomation caused by the sting of these arachnids. Notwithstanding, some cultures have used scorpion venom to treat human pathologies, most of them mirroring the "like cures like" Paracelsian principle (*similia similibus curantur*) to treat scorpion envenomation,[2–4] but also other conditions such as pain, convulsions and diabetic neuropathy.[4–6] Whereas there is no, or only poor, scientific support for most

RSC Drug Discovery Series No. 42
Venoms to Drugs: Venom as a Source for the Development of Human Therapeutics
Edited by Glenn F. King

Published by the Royal Society of Chemistry, www.rsc.org

of the claimed beneficial effects of scorpion venom, the study of isolated venom components and their mechanism of action has provided some interesting drug candidates. This chapter reviews the potential use of scorpion-venom components to treat several human pathophysiological conditions, with special emphasis given to the frontrunners in the search for novel drug leads from scorpion-venom-derived peptides.

7.2 Overview of Bioactive Peptides in Scorpion Venoms

Scorpion venoms are complex mixtures consisting of dozens of different components, among which low molecular weight proteins contribute to the largest diversity.[7–11] Hundreds of amino acid sequences are available in public databases[12] and many more are being described, thanks to high-throughput sequencing; nonetheless, these sequences still make up a rather low proportion of the molecular diversity revealed by mass spectrometry (MS)-assisted proteomic analysis of the venoms (see, for example, ref. 7 and 11) and they account for only a tiny portion of the estimated biodiversity of scorpion-venom peptides.[9] A rough comparison of the molecular weight frequency distribution illustrates the large mismatch between published sequences and available venom proteomes (Figure 7.1). While almost 40% of all sequenced scorpion-venom peptides are sodium channel modulators or related peptides, only 8% of the determined masses in proteomic studies fall within the typical mass range for these toxins (6–7.5 kDa). On the other side of the spectrum, most proteomic studies have revealed a large contribution to the scorpion-venom proteome, around 20% on average, from very small peptides (1.5 kDa or less), but only 2% of the peptides in sequence databases are within this mass range.

Low molecular weight (less than 10 kDa) proteinaceous components from scorpion venoms can be roughly divided into reticulated (disulfide-bridged) and linear (non-disulfide-bridged) peptides.[9,13] Larger proteins are also found in scorpion venoms and in some cases constitute the main active components;[14] however, their contribution to the molecular diversity of scorpion venoms is limited. The largest group of described scorpion-venom peptides is constituted by disulfide-bridged peptides, 21–77 residues long, with the sequence signature of the cystine stabilized α/β (CSα/β) motif,[15] even if not all of them fold into the most common αββ three-dimensional topology.[16] To a large extent the size of CSα/β scorpion venom peptides serves as indication of their mechanism of action: short-chain peptides (smaller than 45 residues) are generally potassium channel blockers and long-chain peptides (55 residues or larger) mostly modulate the activity of sodium channels.[9] Some scorpion CSα/β peptides display a remarkable specificity for certain subtypes of ion channels, and thus great efforts have been made to identify specific modulators of ion channels involved in several channelopathies. Some of the most promising prospects are described in the sections to follow.

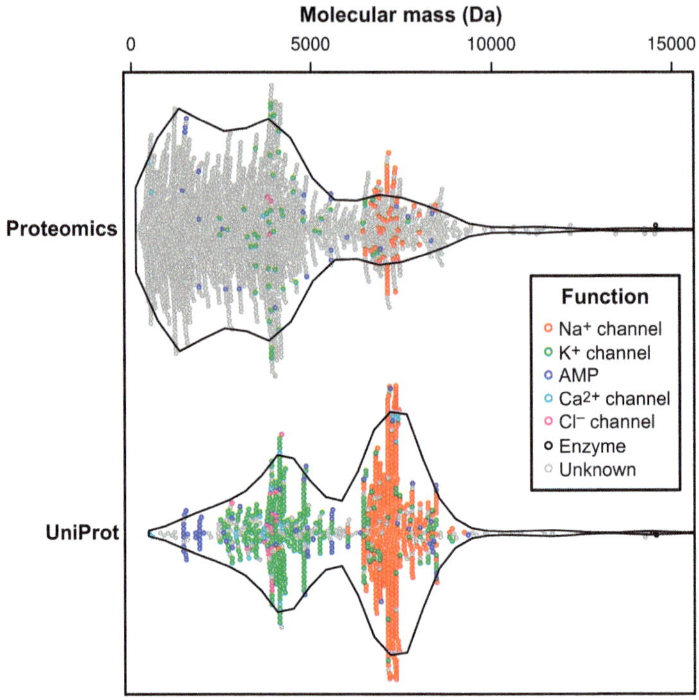

Figure 7.1 Violin plots of molecular weight distributions in scorpion venom based on UniProt entries (mature peptides of 780 non-redundant peptides as August 2012; see ref. 12 and 40) and MS-assisted venom profiling data (5552 values pooled from 21 venom profiles, see ref. 7–11). The graph was obtained using the R "violins" package, with the smoothening parameter $h = 444$. Overlaid dots are individual molecular weight values for 780 non-redundant peptides in UniProt and 1671 randomly picked frequency-weighted values from the MS data. Dots are colored according to their functional annotation and plotted with the R "beeswarm" package (method = "swarm", corral = "wrap"). Sodium channel toxins = red; potassium channel toxins = green; AMPs = blue; calcium channel toxins = light blue; Cltx-like toxins = pink; enzymatic activity = black; no functional data = gray.

Peptides folding into the inhibitor cystine knot (ICK) scaffold[17] constitute another widely distributed, though much less diversified, group of scorpion-venom components.[18–20] These peptides are high-affinity agonists of ryanodine receptors (RyRs, the calcium channels residing in the sarcoplasmic reticulum), and thus they are often called calcines. Although the activation of RyRs does not have clear relevance in the ecological context of venom use, the confirmation that calcines can reach intracellular RyRs even if applied to the extracellular milieu of intact cells[20–22] prompted the discovery of their cell-penetrating properties. Thus, as detailed below, these peptides are now being studied for intracellular targeting of attached drugs.

Whereas the ion channel modulating activities of scorpion-venom peptides with the sequence signature of the CSα/β or ICK motifs are relatively well understood, the heterogeneous collection of non-disulfide bridged peptides (NDPs) also found in scorpion venoms has received much less attention,[13] despite making the largest contribution to the molecular diversity of most scorpion venoms characterized so far using proteomic techniques.[7–11,23] Scorpion-venom linear peptides have no common biochemical or structural features, although they are typically enriched in charged and hydrophobic residues, and thus prone to behave as amphipathic molecules. A variety of different functions have been assigned to these NDPs, from cytolytic or antimicrobial activity to bradykinin potentiating and immunomodulatory activities.[13] The precise mechanism of action of most bioactive scorpion-venom linear peptides is only beginning to be elucidated; nonetheless, as detailed below, some of them are interesting candidates for therapeutic applications.

The spectrum of human pathophysiological conditions for which scorpion-venom peptides may offer a therapeutic alternative spans from controlling parasitic infections to treatment of autoimmune diseases and specific labeling of tumors (see Table 7.1). Here, special attention is given to the physiological basis of their potential therapeutic use in the context of selected pathophysiological conditions and to the use of toxin scaffolds for rational drug design; the reader is encouraged to consult the extensive literature regarding bioprospecting for bioactive peptides in animal venoms, including scorpions, and the molecular basis of their mechanism of action (see, for example, ref. 24–26).

7.3 Drug Candidates

7.3.1 Antimicrobial Peptides

The available therapeutic tool-kit to combat infections is becoming obsolete as rapid generation times, high mutation rates, and strong selection pressure combine to assure microbial pathogens a good lead in the arms-race against antibiotics.[27] This explains in part why the discovery of new small-molecule antimicrobials has slowed down dramatically in the last 50 years.[28] In the meantime, some researchers have begun to look for new antibiotic leads among naturally occurring peptides that participate in the innate immune response of multicellular organisms, the so-called (cationic) antimicrobial or host-defense peptides. These structurally diverse and multifunctional antimicrobial peptides (AMPs) are amongst the first lines of humoral defense in diverse organisms.[29] Some peptides from scorpion venoms resemble AMPs, both structurally and functionally. Although their precise role in the ecology of venom use is not well understood, it is assumed that they serve as spreading factors that help other venom components reach their targets.[30]

Table 7.1 Selected therapeutic leads from scorpion venoms

Pathophysiological condition	*Pharmaceutical principle*	*Scorpion-venom-derived peptide*	*Structural class*	*References*
Malaria	*In host* plasmodicidal activity	Scorpine	Mixed NDP-CSα/β	45,46
Autoimmune diseases	Kv1.3 blockade	Mokatoxin (engineered)	CSα/β	57,58,8,59
		ADWX-1 (engineered)	CSα/β	
		Vm24	CSα/β	
Cancer	Tumor-marker	Chlorotoxin-Cyc5.5	CSα/β	72,73
	Drug carrier	Chlorotoxin-^{131}I	CSα/β	
HIV	Inhibition of gp120 binding to CD4	Scyllatoxin (engineered)	CSα/β	109
Drug carrier	Cell-penetrating	Maurocalcine (engineered)	ICK	92

Regardless of their role in the ecology of venom use, most functionally characterized scorpion-venom NDPs display antibiotic activities, often concomitant with significant cytolytic action that limits their potential therapeutic application.[13,30,31] More recently, a number of synthetic variants of some of these peptides have been designed in which the cytolytic action against mammalian cells (mainly tested in hemolytic assays) has been reduced, while the antimicrobial effect of the native peptide has been preserved (see, for example, ref. 32 and 33). More promising candidates for therapeutic applications are NDPs with very low or non-recorded cytolytic activity against mammalian cells, such as Meucin-24 and Meucin-25, two peptides from *Mesobuthus eupeus* venom with plasmodicidal activity,[34] BmKbpp, a multi-functional AMP from *Mesobuthus martensii*,[35] and the antibacterial VmCT2 from *Vaejovis mexicanus*.[36] Moreover, some scorpion-venom AMPs can also act synergistically with conventional antibiotics.[37] An intriguing feature of some scorpion-venom AMPs is their ability to affect mammalian cells at non-lytic concentrations, as exemplified by the NDP Parbutoporin from *Parabuthus schlechteri*, which promotes neutrophil chemotaxis and degranulation while delaying spontaneous neutrophil apoptosis.[31] In this sense, scorpion-venom AMPs are not different from mammalian host defense peptides, now recognized as not "merely antimicrobials", but important signaling molecules of the innate immune defense system.[29] Such features broaden the scope of potential applications of scorpion-venom AMPs; however, much research is still needed to clarify their precise molecular targets and mechanism of action.

The potential application of scorpion-venom NDPs as antimicrobials is still an open question, and further research is required to obtain more specific AMPs and to explore their immunomodulatory effects. However, another class of scorpion-venom AMP is gaining significant attention with the aim of combating malaria. In 2000, a peptide found in *Pandinus imperator* venom was reported to inhibit several stages of the murine malaria parasite *Plasmodium berghei*.[38] The peptide, called scorpine, contains an extended N-terminal region of some 30 residues followed by a typical CSα/β C-terminal region; this molecular architecture resembles the previously described potassium channel blockers grouped within the β-KTx subfamily of scorpion-venom peptides.[39] Scorpine/β-KTx homologs have been found in several scorpion species (an updated list is available from UniProt's toxin annotation program).[40] Although scorpine-like peptides are cytolytic,[41,42] a potassium channel blocking activity can be unmasked by removing the N-terminal region.[41] While the cytolytic action of scorpine-like peptides limits their potential therapeutic application, their plasmocidal activity could still contribute towards the fight against malaria. Target expression of plasmodicidal peptides in vector insects has been long foreseen as a promising strategy to combat malaria,[43] but it faces important ecological and social concerns related to the release of transgenic mosquitoes into the field (see ref. 44). Partially overcoming this limitation, two groups recently engineered scorpine-producing transgenic entomopathogenic fungi[45] and bacteria[46] that can grow within mosquitoes, thus deploying the AMP directly into the vector organism where parasites spend half of their life cycle.[47]

7.3.2 Autoimmune Diseases

Years of research have firmly established that modulation of excitable cells is the main mechanism by which scorpion venoms exert their toxic function, impairing cellular communication by the concerted depolarizing action of potassium channel blockers and sodium channel activators. However, virtually all cells possess ion channels similar to the ones affected in excitable cells, if not one and the same. Thus, non-excitable cells can be affected by scorpion toxins inasmuch as the targeted channels are important for their function, as is the case for T lymphocytes, the white blood cells in charge of the cellular immune response in mammals.[48] Activation of cytotoxic T cells depends on a constant influx of calcium that needs to be counterbalanced by potassium efflux in order to maintain the membrane potential; different subsets of T cells achieve this by up-regulating the expression of voltage-gated or calcium-activated potassium channels (Kv1.3 and KCa3.1, respectively).[49] Notably, chronically activated effector memory T cells (T_{EM}) are responsible for the tissue damage characteristic of several autoimmune diseases such as multiple sclerosis, rheumatoid arthritis and type-I diabetes mellitus.[50] T_{EM} cells overexpress Kv1.3 channels and thus their activation can be inhibited by specific blockers of this channel,[51] while other subsets of T cells remain unaffected as their membrane potential is mainly maintained by KCa3.1 channels.[49] As several CSα/β toxins from scorpion venoms are high affinity blockers of these channels, they were readily recognized as potential drug leads for treating autoimmune diseases.[52]

The story of scorpion venoms and autoimmune diseases can be traced back to a clinical report in 1983, where a scorpion sting was suspected to be the cause of a temporal relapsing of the symptoms of a multiple sclerosis patient.[53] Later kalitoxin, a short-chain potassium channel blocking toxin from *Androctonus mauretanicus mauretanicus* venom, was shown to block T cell Kv1.3 channels in animals with experimental autoimmune encephalitis, a model for T cell mediated autoimmune disease, and improve their condition.[54] Since then, a number of very high affinity blockers of Kv1.3 channels have been isolated from scorpion venoms (summarized in ref. 51); however, most of these naturally occurring scorpion-venom peptides also block other potassium channels, including KCa3.1, with potencies that preclude their use as therapeutic agents (see ref. 55). Considerable effort has been dedicated to engineering molecules based on naturally occurring Kv1.3 peptide blockers; the biggest success so far has been achieved with derivatives of the sea anemone peptide ShK, which has just successfully completed Phase I clinical trials for treatment of multiple sclerosis.[56] Although scorpion-venom peptides lag behind in this respect, the panorama could change due to recent success in obtaining highly specific engineered peptides[57,58] and the discovery of highly selective and potent Kv1.3 blockers found in previously uncharted scorpion venoms.[59,60] Moreover, the pivotal role of Kv1.3 and KCa3.1 channels in other pathophysiological conditions, including a number of inflammatory diseases[50] and graft rejection,[61] is another open, though much less traveled, road for the use of the ever increasing number of highly selective and potent scorpion-venom-derived peptides. The use of specific scorpion-venom peptides as immunosupressors is protected by several patents.[62–64]

7.4 Pharmaceutical Tools

7.4.1 Tumor Labeling and Targeting

Uncontrolled cell growth is the landmark of various etiologically diverse pathologies commonly known as cancer. The therapeutic alternatives for cancer patients often involve invasive surgical procedures to remove the malignant neoplasm. In such cases, it is critical to identify with high precision the boundaries of the tumor in order to avoid further compromise of the affected organ and to improve the prognosis of the patient.[65] Thus, the search for tumor-marker specific molecular probes is not surprisingly an active field of research.[66] In the mid-1990s, glial cell tumors (gliomas) were shown to express a specific chloride current that can be blocked with a four disulfide-bridged CSα/β scorpion-venom-derived peptide called chlorotoxin (Cltx),[67] originally isolated from the venom of *Leiurus quinquiestratus quinquiestratus*.[68] Later it was shown that a radiolabeled derivative of Cltx specifically binds to glioma cells, thus revealing its potential as a tumor marker.[69] Interestingly, the high affinity receptor on glioma cells proved not to be the chloride channel itself, but either an associated metalloprotease (MMP-2), previously shown to be involved in the tumor invasiveness,[70] or annexin A2.[71] Although direct inhibition of MMP-2 or chloride channels seems not to be sufficient for arresting the growth of cancer cells, the conjugation of Cltx and the dye Cyc5.5 generates a fluorescent derivative that is being tested as a tumor marker for intra-operative visualization.[72] Moreover, the selective Cltx binding to tumor cells prompted the development of other conjugates with attached cancer cell-killing moieties (such as radioactive iodine[73] and methotrexate[74]) and even as carriers for gene delivery,[75] with the aim of directly and specifically delivering anti-cancer drugs and genes to the neoplastic cells. Iodine-labeled synthetic Cltx was approved for Phase I/II clinical trials by the United States Food and Drug Administration (FDA) in 2002 under an Orphan Drug and Fast Track Development Program status, and an initial report was published not much later.[76] However, no updates have been released since 2010. Several patents cover Cltx conjugates and their use as tumor-labeling agents and drug carriers.[77–79]

At least 20 other scorpion-venom peptides are close homologs of Cltx (>60% sequence identity); some of them bind glioma cells,[80,81] another specifically inhibits the cystic fibrosis transmembrane conductance regulator, an important therapeutic target itself,[82] and yet others are reported to be insecticidal *via* an unknown mechanism.[83] Several other scorpion venoms and venom components have been reported to inhibit proliferation of a number of cancer cell lines (reviewed in ref. 84), thus expanding the number of potential drug leads. In both cases, however, much research is still needed to establish their therapeutic potential, if any. It is important to note that there is little scientific evidence supporting the recent outcry about the use of scorpion venom preparations as complementary cancer treatment. In particular, the venom of the Cuban scorpion *Rhopalurus junceus* has been repeatedly reported in the lay media as a "cancer cure" due to its analgesic, anti-inflammatory and antiproliferative effects.[5] The few reports available dealing with the venom of this species point to the usual composition of Buthidae venoms, with a single sodium channel toxin-like peptide accounting for 25% of the total protein content of the soluble venom.[85] Although an international patent dealing with the anti-tumor use of a combination of seven low molecular weight peptides isolated from *R. junceus* venom was granted in 2012,[86] no information about the putative mechanism of cancer inhibition by *R. junceus* venom or venom components is available in the scientific literature.

7.4.2 Drug Carriers

A common drawback of peptide- and protein-based drug leads is their poor bioavailability, as these (relatively) large and (at least partially) hydrophilic molecules are unlikely to traverse the many biological barriers of the body. Important efforts have thus been made to develop drug carriers that can take therapeutic compounds as cargoes to their molecular targets, a particularly important issue when the targets are localized inside the cell. One class of such potential drug carriers is constituted by cell-penetrating peptides (CPPs), a molecularly diverse group of peptides that can translocate across membranes without damaging the cell.[87] By the beginning of 2000 the puzzling observation that maurocalcine, a scorpion-venom ICK peptide from *Scorpio maurus*, induces intracellular calcium release from intact cells[88] fueled further investigations in order to clarify how could maurocalcine reach its intracellular target. Over subsequent years it became clear that maurocalcine is a bona fide CPP[21] that can carry large cargoes to the interior of the cell.[89] Interestingly, cell penetration can be decoupled from pharmacological activity on RyR by replacement of the toxin's cysteine residues with isosteric 2-aminobutyrate residues;[90] this disulfide-less variant served as basis for development of shorter carrier-competent CPP, only nine residues long, based on maurocalcine.[91] As a proof of concept, disulfide-less maurocalcine was shown to efficiently deliver a conjugated anti-cancer drug (doxorubicin) into cancer cell lines.[92] Scorpion venoms from at least 10 other species contain ICK peptides, and two of them have been confirmed as CPPs (imperatoxin A from *P.imperator*[22] and hadrurocalcine from *Hadrurus gertschi*[20]). The confirmed diversity of scorpion venom CPPs could provide leads for much needed cell-specific CPPs.[66,93] Maurocalcine and its CPP analogs are covered by patents.[94,95]

7.5 "Toxin" Scaffolds as Platform for Drug Design

The therapeutic potential of scorpion-venom peptides is not exhausted by their pharmacological targets or their capabilities as drug carriers. Rational engineering of scorpion-venom-derived peptides has provided variants with improved potency and selectivity against Kv1.3[54,96–98] or KCa2.2 channels.[99] Likewise, a 10-fold improvement in the selectivity of Css4 (from *Centruroides suffusus suffusus*) for muscle Nav1.4 channels was achieved by mutating three residues in the functional site.[100] More recently, the cytolytic activity of scorpion NDPs was successfully modulated by site-directed mutagenesis in order to diminish their cytotoxic effect against mammalian cells,[33,101,102,] to improve their antibiotic function,[32] and to develop antiviral activities.[103,104]

Moreover, the large sequence diversity and high stability of many structurally constrained small disulfide-rich proteins make them attractive scaffolds for molecular engineering of novel biological functions.[105–107] Often a therapeutically interesting function is found in pharmacologically intractable proteins (whether due to their complexity, poor *in vivo* stability, or off-target effects). Pioneering work on the scorpion-venom peptides charybdotoxin and scyllatoxin (from *Leiurus quinquestriatus hebraeus* venom) demonstrated that functional sites of larger proteins could be grafted onto the highly stable CSα/β scaffold.[108] In an impressive *tour de force*, an engineered peptide mimicking the HIV gp120 receptor, CD4, was obtained after a combination of functional site grafting, site-directed mutagenesis and molecular design using a scyllatoxin template (thoroughly recounted in ref. 109). The most modified CD4 mimic (named CD4M33) retained 18 residues out of the 31 in the template; it effectively inhibits cell-free infection by several HIV strains and binds to the CD4 binding site of gp120 while retaining the CSα/β scaffold of the parental scyllatoxin.[109] The CD4 mimics based on scyllatoxin are covered by an international patent.[110] Further demonstrating the permissiveness of the CSα/β to incorporate grafted functional sites, an effective inhibitor of the interaction between the tumor suppressor p53 and the oncoproteins MDM2 and MDMX was obtained by grafting the MDM2/MDMX binding site of p53 onto BmBKTx1, a potassium channel blocker from the venom of *M. martensii*.[111] Finally, the uncommon helix-hairpin fold of some scorpion-venom peptides has been also been used to graft the androgen receptor binding motif of several coactivators such as p160.[112]

7.6 Future Prospects and Concluding Remarks

In the previous sections we singled out no more than a dozen scorpion-venom-derived peptides from the three broad structural classes that constitute the vast majority of known scorpion-venom peptide sequences (CSα/β, ICK and NDP; see Figure 7.2). Although only five pharmaceutical principles were covered, the potential applications of these few peptides span a wide

Malaria

Scorpine	GWINEEKIQKKIDERMGNTVLGGMAKAIVHKMAKNEFQCMANMDMLGNCEKHCQTSGEKGYCHGTKCKCGTPLSY		
	[----------NDP-----------][------------------CSα/β---------------------]		

Autoimmune diseases

Mokatoxin	..INVKCSLPQQCIKPCKDAGMRFGKCMNKKCRCYS.	2kir	CSα/β
ADWX-1	VGINVKCKHSRQCLKPCKDAGMRFGKCTNGKCHCTPK	2k4u	CSα/β
Vm24	.AAAISCVGSPECPPKCRAQGCKNGKCMNRKCKCYYC	2k9o	CSα/β

Cancer

Chlorotoxin	MCMPCFTTDHQMARKCDDCCGGKGRGKCYGPQCLCR	1chl	CSα/β

HIV

CD4M33	tpaNLHFCQLRCKSLGLLGKCAGSbfCAC	1yyl_S	CSα/β

Drug carrier

Maurocalcine	GDCLPHLKLCKENKDCCSKKCKRRGTNIEKRCR	1c6w	ICK

Figure 7.2 Amino acid sequences of selected scorpion-venom-derived candidates for therapeutic applications. The two domains of scorpine are indicated below the sequence. Kv1.3 blockers are aligned in order to highlight their structural similarity. In the scyllatoxin-based sequence CD4M33, "tpa" and "bf" are the unnatural amino acids tiopropionic acid and bis-fenilalanine, respectively. Protein Data Bank codes, where available, are indicated at the right of the sequences. Cysteine residues involved CSα/β and ICK motifs are highlighted in white on a black background.

diversity of pathophysiological conditions. The complete absence of neuroactive peptides in the above summary may be puzzling, inasmuch as excitable cells are quite obviously the primary targets of scorpion venom. Reports of analgesic or anticonvulsant effects of scorpion toxins that act on neuronal sodium channels are not rare in the scientific literature (for reviews, see ref. 113 and 114); however, the complex etiology of pain and epileptogenic conditions, the lack of unambiguously identified receptors, and the often promiscuous action of sodium channel scorpion toxins have hampered development of these peptides as therapeutic leads.

Recent advances in the systematic survey of the molecular diversity of scorpion venoms have revealed a large contribution from non-canonical venom components. Although their functional characterization has just begun, novel functions have been already described (*e.g.*, the protease inhibitory effect of recently described Kunitz-type scorpion-venom peptides[115]). Thus, it is expected that the large diversity of novel peptides revealed by proteomic and transcriptomic analyses (*e.g.*, see ref. 116 and 117), together with the use of high-throughput functional assays,[118–120] should lead to the discovery of other interesting activities in the near future. Scorpion venoms remain a largely uncharted source of bioactive compounds with potential therapeutic applications.

Acknowledgements

RCRV research is funded by the European Union Seventh Framework Programme (FP7/2007–2013) under Grant 246556. GC receives funding by DGAPA-UNAM IN220809 and CONACyT 153606. Work at LDP group is partially funded by grants DGAPA-UNAM IN204110 and CONACyT-SEP153496.

References

1. J. P. Chippaux, *Drug Des. Dev. Ther.,* 2012, **6**, 165.
2. J. O. Dewey, *Lancet,* 1949, **253**, 757.
3. F. J. Monzon Muñoz and R. M. Blasco Gil, *Bol. SEA,* 1997, **12**, 47.
4. J. A. Gonzalez and J. R. Vallejo, *J. Ethnopharmacol.*, 2013, **146**, 62–74.
5. L. F. de Armas, *Rev. Iberica Arachnol.,* 2001, **4**, 99.
6. W. Xie, Y. Zhao, D. Gu, L. Du, G. Cai and Y. Zhang, *Evid. Based Complement. Alternat. Med.,* 2011, **2011**, 683561.
7. E. Diego-García, S. Peigneur, E. Clynen, T. Marien, L. Czech, L. Schoofs and J. Tytgat, *Proteomics,* 2012, **12**, 313.
8. G. B. Gurrola, R. Hernández-López, R. C. Rodríguez de la Vega, Z. Varga, C. V. Batista, S. P. Salas-Castillo, G. Panyi, F. del Río-Portilla and L. D. Possani, *Biochemistry,* 2012, **51**, 4049.
9. R. C. Rodríguez de la Vega, N. Vidal and L. D. Possani in *Handbook of Biologically Active Peptides* (2nd edn), ed., A. Kastin, Academic Press, San Diego, 2013, p. 423.
10. J. J. Smith, A. Jones and P. F. Alewood, *J. Venom Res.,* 2012, **3**, 7.
11. K. Luna-Ramírez, V. Quintero-Hernández, L. Vargas-Jaimes, C. V. Batista, K. D. Winkel KD and L. D. Possani, *Toxicon,* 2013, **63**, 44.
12. F. Jungo, L. Bougueleret, I. Xenarios and S. Poux, *Toxicon,* 2012, **60**, 551.
13. X. C. Zeng, G. Corzo and R. Hahin, *IUBMB Life,* 2005, **57**, 13.
14. L. Borchani, A. Sassi, R. Ben Yekhlef, I. Safra and M. El Ayeb, *Toxicon,* 2011, **58**, 681.
15. F. Bontems, C. Roumestand, B. Gilquin, A. Ménez and F. Toma, *Science,* 1991, **254**, 1521.
16. A. L. Saucedo, D. Flores-Soli, R. C. Rodríguez de la Vega, B. Ramírez-Cordero, R. Hernández-López, P. Cano-Sánchez, R. Noriega Navarro, J. García-Valdés, F. Coronas-Valderrama, A. de Roodt, L. G. Brieba, L. D. Possani and F. del Río-Portilla, *J. Biol. Chem.,* 2012, **287**, 12321.
17. P. K. Pallaghy, K. J. Nielsen, D. J. Craik and R. S. Norton, *Protein Sci.,* 1994, **3**, 1833.
18. F. Z. Zamudio, G. B. Gurrola, C. Arévalo, R. Sreekumar, J. W. Walker, H. H. Valdivia and L. D. Possani, *FEBS Lett.,* 1997, **405**, 385.
19. A. Mosbah, R. Kharrat, Z. Fajloun, J. G. Renisio, E. Blanc, J. M. Sabatier, M. El Ayeb and H. Darbon, *Proteins,* 2000, **40**, 436.

20. E. F. Schwartz, E. M. Capes, E. Diego-García, F. Z. Zamudio, O. Fuentes, L. D. Possani and H. H. Valdivia, *Br., J. Pharmacol.,* 2009, **157**, 392.
21. E. Estève, K. Mabrouk, A. Dupuis, S. Smida-Rezgui, X. Altafaj, D. Grunwald, J. C. Platel, N. Andreotti, I. Marty, J. M. Sabatier, M. Ronjat and M. De Waard, *J. Biol. Chem.,* 2005, **280**, 12833.
22. G. B. Gurrola, E. M. Capes, F. Z. Zamudio, L. D. Possani and H. H. Valdivia, *Pharmaceuticals,* 2010, **3**, 1093.
23. B. Rates, K. K. Ferraz, M. H. Borges, M. Richardson, M. E. De Lima and A. M. Pimenta, *Toxicon,* 2008, **52**, 611.
24. R. J. Lewis and M. L. Garcia, *Nat. Rev. Drug Discovery,* 2003, **2**, 790.
25. E. Beraud and K. G. Chandy, *Inflamm. Allergy Drug Targets,* 2011, **10**, 322.
26. G. F. King, *Expert Opin. Biol. Ther.,* 2011, **11**, 1469.
27. M. N. Alekshun and S. B. Levy, *Cell,* 2007, **128**, 1037.
28. K. Lewis, *Nature,* 2012, **485**, 439.
29. A. T. Yeung, S. L. Gellatly and R. E. Hancock, *Cell. Mol. Life Sci.,* 2011, **68**, 2161.
30. L. Kuhn-Nentwig, *Cell. Mol. Life Sci.,* 2003, **60**, 2651.
31. Q. Remijsen, F. Verdonck and J. Willems, *Toxicon,* 2010, **55**, 180.
32. L. Cao, C. Dai, Z. Li, Z. Fan, Y. Song, Y. Wu, Z. Cao and W. Li, *PLoS One,* 2012, 7, e40135.
33. L. Sánchez-Vásquez, J. Silva-Sánchez, J. M. Jiménez-Vargas, A. Rodríguez-Romero, C. Muñoz-Garay, M. C. Rodríguez, G. B. Gurrola and L. D. Possani, *Biochim. Biophys. Acta*, 2013, **1830**, 3427–3436.
34. B. Gao, J. Xu, M. del, C. Rodriguez, H. Lanz-Mendoza, R. Hernández-Rivas, W. Du and S. Zhu, *Biochimie,* 2010, **92**, 350.
35. X. C. Zeng, S. Wang, Y. Nie, L. Zhang and X. Luo, *Peptides,* 2012, **33**, 44.
36. S. Ramírez-Carreto, V. Quintero-Hernández, J. M. Jiménez-Vargas, G. Corzo, L. D. Possani, B. Becerril and E. Ortiz, *Peptides,* 2012, **34**, 290.
37. F. García, E. Villegas, G. P. Espino-Solis, A. Rodríguez, J. Paniagua-Solis, G. Sandoval, L. D. Possani and G. Corzo, *J. Antibiot.,* 2013, **66**, 3.
38. R. Conde, F. Z. Zamudio, M. H. Rodríguez and L. D. Possani, *FEBS Lett.,* 2000, **471**, 165.
39. C. Legros, B. Céard, P. E. Bougis and M. F. Martin-Eauclaire, *FEBS Lett.,* 1998, **431**, 375.
40. http://www.uniprot.org/program/Toxins (last accessed September 2012).
41. E. Diego-García, Y. Abdel-Mottaleb, E. F. Schwartz, R. C. Rodríguez de la Vega, J. Tytgat and L. D. Possani, *Cell. Mol. Life Sci.,* 2008, **65**, 187.
42. S. Zhu, B. Gao, A. Aumelas, M. del, C. Rodríguez, H. Lanz-Mendoza, S. Peigneur, E. Diego-García, M. F. Martin-Eauclaire, J. Tytgat and L. D. Possani, *Biochim. Biophys. Acta,* 2010, **1804**, 872.
43. L. D. Possani, M. Corona, M. Zurita and M. H. Rodríguez, *Arch. Med. Res.,* 2002, **33**, 398.
44. T. W. Scott, W. Takken, B. G. Knols and C. S. Boëte, *Science,* 2002, **298**, 117.
45. W. Fang, J. Vega-Rodríguez, A. K. Ghosh, M. Jacobs-Lorena, A. Kang and R. J. St Leger, *Science,* 2011, **331**, 1074.

46. S. Wang, A. K. Ghosh, N. Bongio, K. A. Stebbings, D. J. Lampe and M. Jacobs-Lorena, *Proc. Natl. Acad. Sci. U.S.A.*, 2012, **109**, 12734.
47. S. Wang and M. Jacobs-Lorena, *Trends Biotechnol.*, 2013, **31**, 185–193.
48. S. Feske, E. Y. Skolnik and M. Prakriya, *Nat. Rev. Immunol.*, 2012, **12**, 532.
49. M. D. Cahalan and K. G. Chandy, *Immunol. Rev.*, 2009, **231**, 59.
50. J. Lam and H. Wulff, *Drug Dev. Res.*, 2011, **72**, 573.
51. G. Panyi, L. D. Possani, R. C. Rodríguez de la Vega, R. Gáspár and Z. Varga, *Curr. Pharm. Des.*, 2006, **12**, 2199.
52. R. J. Leonard, M. L. Garcia, R. S. Slaughter and J. P. Reuben, *Proc. Natl. Acad. Sci. U.S.A.*, 1992, **89**, 10094.
53. A. E. Breland and R. D. Currier, *Lancet*, 1983, **322**, 1021.
54. C. Beeton, J. Barbaria, P. Giraud, J. Devaux, A. M. Benoliel, M. Gola, J. M. Sabatier, D. Bernard, M. Crest and E. Béraud, *J. Immunol.*, 2001, **166**, 936.
55. K. M. Giangiacomo, Y. Ceralde and T. J. Mullmann, *Toxicon*, 2004, **43**, 877.
56. V. Chi, M. W. Pennington, R. S. Norton, E. J. Tarcha, L. M. Londono, B. Sims-Fahey, S. K. Upadhyay, J. T. Lakey, S. Iadonato, H. Wulff, C. Beeton and K. G. Chandy, *Toxicon*, 2012, **59**, 529.
57. Z. Takacs, M. Toups, A. Kollewe, E. Johnson, L. G. Cuello, G. Driessens, M. Biancalana, A. Koide, G. G. Ponte, E. Perozo, T. F. Gajewski, G. Suarez-Kurtz, S. Koide and S. A. Goldstein, *Proc. Natl. Acad. Sci. U.S.A.*, 2009, **106**, 22211.
58. Z. Li, W. H. Liu, S. Han, B. W. Peng, J. Yin, Y. L. Wu, X. H. He and W. X. Li, *J. Biol. Chem.*, 2012, **287**, 29479.
59. Z. Varga, G. Gurrola-Briones, F. Papp, R. C. Rodríguez de la Vega, G. Pedraza-Alva, R. B. Tajhya, R. Gaspar, L. Cardenas, Y. Rosenstein, C. Beeton, L. D. Possani and G. Panyi, *Mol. Pharmacol.*, 2012, **82**, 372.
60. S. Xie, J. Feng, C. Yu, Z. Li, Y. Wu, Z. Cao, W. Li, X. He, M. Xiang and S. Han, *Peptides*, 2012, **36**, 94.
61. I. Grgic, H. Wulff, I. Eichler, C. Flothmann, R. Köhler and J. Hoyer, *Transplant. Proc.*, 2009, **41**, 2601.
62. Patent number: WO 1995/003065 A1.
63. Patent number: WO 2006/002850 A3.
64. Patent number: WO 2008/139243 A1.
65. http://www.ncbi.nlm.nih.gov/books/NBK23357/ (last accessed September 2012).
66. N. Svensen, J. G. Walton and M. Bradley, *Trends Pharmacol. Sci.*, 2012, **33**, 186.
67. N. Ullrich, A. Bordey, G. Y. Gillespie and H. Sontheimer, *Neuroscience*, 1996, **83**, 1161.
68. J. A. DeBin, J. E. Maggio and G. R. Strichartz, *Am. J. Physiol.*, 1993, **264**, C361.
69. L. Soroceanu, Y. Gillespie and M. B. Khazaeli, *Cancer Res.*, 1998, **58**, 4871.
70. J. Deshane, C. C. Garner and H. Sontheimer, *J. Biol. Chem.*, 2003, **278**, 4135.

71. K. Kesavan, J. Ratliff, E. W. Johnson, W. Dahlberg, J. M. Asara, P. Misra, J. V. Frangioni and D. B. Jacoby, *J. Biol. Chem.*, 2010, **285**, 4366.
72. O. Veiseh, C. Sun, J. Gunn, N. Kohler, P. Gabikian, D. Lee, N. Bhattarai, R. Ellenbogen, R. Sze, A. Hallahan, J. Olson and M. Zhang, *Nano Lett.*, 2005, **5**, 1003.
73. S. Shen, M. B. Khazaeli, G. Y. Gillespie and V. L. Alvarez, *J. Neurooncol.*, 2005, **71**, 113.
74. C. Sun, C. Fang, Z. Stephen, O. Veiseh, S. Hansen, D. Lee, R. G. Ellenbogen, J. Olson and M. Zhang, *Nanomedicine,* 2008, **3**, 495.
75. F. M. Kievit, O. Veiseh, C. Fang, N. Bhattarai, D. Lee, R. G. Ellenbogen and M. Zhang, *ACS Nano,* 2010, **4**, 4587.
76. A. N. Mamelak, S. Rosenfeld, R. Bucholz, A. Raubitschek, L. B. Nabors, J. B. Fiveash, S. Shen, M. B. Khazaeli, D. Colcher, A. Liu, M. Osman, B. Guthrie, S. Schade-Bijur, D. M. Hablitz, V. L. Alvarez and M. A. Gonda, *J. Clin. Oncol.*, 2006, **24**, 3644.
77. Patent number: WO 2003/101474 A1.
78. Patent number: WO 2009/021136 A1.
79. Patent number: WO 2011/142858 A9.
80. Y. J. Fu, L. T. Yin, A. H. Liang, C. F. Zhang, W. Wang, B. F. Chai, J. Y. Yang and X. J. Fan, *Neurosci. Lett.*, 2007, **412**, 62.
81. I. Rjeibi, K. Mabrouk, H. Mosrati, C. Berenguer, H. Mejdoub, C. Villard, D. Laffitte, D. Bertin, L. Ouafik, J. Luis, M. Elayeb and N. Srairi-Abid, *Peptides,* 2011, **32**, 656.
82. M. D. Fuller, C. H. Thompson, Z. R. Zhang, C. S. Freeman, E. Schay, G. Szakács, E. Bakos, B. Sarkadi, D. McMaster, R. J. French, J. Pohl, J. Kubanek and N. A. McCarty, *J. Biol. Chem.*, 2007, **282**, 37545.
83. R. Wudayagiri, B. Inceoglu, R. Herrmann, M. Derbel, P. V. Choudary and B. D. Hammock, *BMC Biochem.*, 2001, **2**, 16.
84. T. A. Heinen and A. B. da Veiga, *Toxicon,* 2011, **57**, 497.
85. B. I. García-Gómez, F. I. Coronas, R. Restano-Cassulini, R. R. Rodríguez and L. D. Possani, *Toxicon,* 2011, **58**, 18.
86. Patent number: WO 2012/041261 A2.
87. R. M. Johnson, S. D. Harrison and D. Maclean, *Methods Mol. Biol.,* 2011, **683**, 535.
88. E. Estève, S. Smida-Rezgui, S. Sarkozi, C. Szegedi, I. Regaya, L. Chen, X. Altafaj, H. Rochat, P. Allen, I. N. Pessah, I. Marty, J. M. Sabatier, I. Jona, M. De Waard and M. Ronjat, *J. Biol. Chem.*, 2003, **278**, 37822.
89. M. Boisseau, K. Mabrouk, N. Ram, N. Garmy, V. Collin, A. Tadmouri, M. Mikati, J. M. Sabatier, M. Ronjat, J. Fantini and M. De Waard, *Biochim. Biophys. Acta,* 2006, **1758**, 308.
90. N. Ram, S. Aroui, E. Jaumain, H. Bichraoui, K. Mabrouk, M. Ronjat, H. Lortat-Jacob and M. De Waard, *J. Biol. Chem.*, 2008, **283**, 24274.
91. C. Poillot, H. Bichraoui, C. Tisseyre, E. Bahemberae, N. Andreotti, J. M. Sabatier, M. Ronjat and M. De Waard, *J. Biol. Chem.*, 2012, **287**, 17331.

92. S. Aroui, N. Ram, F. Appaix, M. Ronjat, A. Kenani, F. Pirollet and M. De Waard, *Pharm. Res.*, 2009, **26**, 836.
93. J. M. Sabatier, and M. De Waard, in *Handbook of Biologically Active Peptides* (2nd edn), ed., A. Kastin, Academic Press, San Diego, 2013, p. 407.
94. Patent number: WO 2001/064724 A2.
95. Patent number: WO 2006/051224 A1.
96. S. Mouhat, G. Teodorescu, D. Homerick, V. Visan, H. Wulff, Y. Wu, S. Grissmer, H. Darbon, M. De Waard and J. M. Sabatier, *Mol. Pharmacol.*, 2006, **69**, 354.
97. S. Han, H. Yi, S. J. Yin, Z. Y. Chen, H. Liu, Z. J. Cao, Y. L. Wu and W. X. Li, *J. Biol. Chem.*, 2008, **283**, 19058.
98. S. Han, S. Yin, H. Yi, S. Mouhat, S. Qiu, Z. Cao, J. M. Sabatier, Y. Wu and W. Li, *J. Proteome Res.*, 2010, **9**, 3118.
99. V. G. Shakkottai, I. Regaya, H. Wulff, Z. Fajloun, H. Tomita, M. Fathallah, M. D. Cahalan, J. J. Gargus, J. M. Sabatier and K. G. Chandy, *J. Biol. Chem.*, 2001, **276**, 43145.
100. L. Cohen, N. Ilan, M. Gur, W. Stühmer, D. Gordon and M. Gurevitz, *J. Biol. Chem.*, 2007, **282**, 29424.
101. A. Rodríguez, E. Villegas, H. Satake, L. D. Possani and G. Corzo, *Amino Acids,* 2011, **40**, 61.
102. A. Almaaytah, M. Zhou, L. Wang, T. Chen, B. Walker and C. Shaw, *Peptides,* 2012, **35**, 291.
103. Y. Chen, L. Cao, M. Zhong, Y. Zhang, C. Han, Q. Li, J. Yang, D. Zhou, W. Shi, B. He, F. Liu, J. Yu, Y. Sun, Y. Cao, Y. Li, W. Li, D. Guo, Z. Cao and H. Yan, *PLoS One,* 2012, 7, e34947.
104. Z. Zhao, W. Hong, Z. Zeng, Y. Wu, K. Hu, X. Tian, W. Li and Z. Cao, *J. Biol. Chem.*, 2012, **287**, 30181.
105. D. J. Craik, N. L. Daly and C. Waine, *Toxicon,* 2001, **39**, 43.
106. H. Kolmar, *Curr. Opin. Pharmacol.*, 2009, **9**, 608.
107. S. J. Moore and J. R. Cochran, *Methods Enzymol.*, 2012, **503**, 223.
108. C. Vita, *Curr. Opin. Biotechnol.*, 1997, **8**, 429.
109. F. Stricher, L. Martin and C. Vita, *Methods Mol. Biol.*, 2006, **340**, 113.
110. Patent number: WO 2002/059146 A2.
111. C. Li, M. Liu, J. Monbo, G. Zou, C. Li, W. Yuan, D. Zella, W. Y. Lu and W. Lu, *J. Am. Chem. Soc.*, 2008, **130**, 13546.
112. B. Vaz, S. Möcklinghoff, S. Folkertsma, S. Lusher, J. de Vlieg and L. Brunsveld, *Chem. Commun.*, 2009, **2009**, 5377.
113. M. R. Mortari, A. O. Cunha, L. B. Ferreira and W. F. dos Santos, *Pharmacol. Ther.*, 2007, **114**, 171.
114. E. P. Cherniack, *Altern. Med. Rev.*, 2011, **16**, 50.
115. R. Zhao, H. Dai, S. Qiu, T. Li, Y. He, Y. Ma, Z. Chen, Y. Wu, W. Li and Z. Cao, *PLoS One,* 2011, **6**, e27548.
116. M. Rendón-Anaya, L. Delaye, L. D. Possani and A. Herrera-Estrella, *PLoS One,* 2012, 7, e43331.

117. J. Xu, X. Zhang, Z. Guo, J. Yan, L. Yu, X. Li, X. Xue and X. Liang, *Proteomics,* 2012, **12**, 3076.
118. C. Legros, C. Schulze, M. L. Garcia, P. E. Bougis, M. F. Martin-Eauclaire and O. Pongs, *Biochemistry,* 2002, **41**, 15369.
119. F. J. Vonk, K. Jackson, R. Doley, F. Madaras, P. J. Mirtschin and N. Vidal, *BioEssays,* 2011, **33**, 269.
120. K. S. Kudryashova, O. V. Nekrasova, A. I. Kuzmenkov, A. A. Vassilevski, A. A. Ignatova, Y. V. Korolkova, E. V. Grishin, M. P. Kirpichnikov and A. V. Feofanov, *Anal. Bioanal. Chem.*, 2013, **405**, 2379–2389.

CHAPTER 8

Therapeutic Applications of Spider-Venom Peptides

JENNIFER J. SMITH, CARUS HO YEE LAU, VOLKER HERZIG, MARIA P. IKONOMOPOULOU, LACHLAN D. RASH*, AND GLENN F. KING*

Institute for Molecular Bioscience, The University of Queensland, St Lucia, Queensland 4072, Australia
*E-mail: l.rash@uq.edu.au; glenn.king@imb.uq.edu.au

8.1 Introduction

As a result of competition between species, a multitude of physical adaptations have emerged that confer an evolutionary benefit to the animals that possess them. One advantageous strategy that has arisen independently in numerous phyla is the development of venom, which serves to subdue prey, defend against predation and deter competitors.[1,2] Spiders are the most successful venomous taxa in terms of diversity, with an estimated 100 000 extant species occupying a multitude of ecological niches, including marine and desert environments.[3] As such, spiders encounter a wide variety of predators and prey, and their venoms have evolved to affect a broad range of species, including vertebrates. Spider venoms are a heterogeneous mixture of salts, low molecular weight organic molecules, nucleotides, peptides and proteins.[4–6] The principal mode of action of spider venoms is incapacitation of prey *via* paralysis. In most species this is effected through the activity of venom peptides on receptors or ion channels within the central or peripheral nervous system.[6]

RSC Drug Discovery Series No. 42
Venoms to Drugs: Venom as a Source for the Development of Human Therapeutics
Edited by Glenn F. King

Published by the Royal Society of Chemistry, www.rsc.org

Peptides are the major components of most spider venoms, with some species containing over 1000 unique peptides with masses ranging from 2 to 8 kDa.[6–8] Currently, only ~1000 spider venom peptides have been described from 86 species, which represents ~0.01% of the estimated 10 million novel venom peptides present in the total number of spider species.[9] Less than half of the reported spider-venom peptides have been functionally characterised. Nevertheless the range of molecular targets known to be affected by spider toxins is considerable (Figure 8.1). Most of the characterised peptides have high affinity for particular molecular targets and they are often selective for specific receptor subtypes. It is this selectivity, functional diversity and biological stability (discussed below) that makes spider-venom peptides attractive drug leads.[8] In addition to peptides, the comparatively understudied small molecules and proteins of spider venoms may provide a source of novel drug leads.

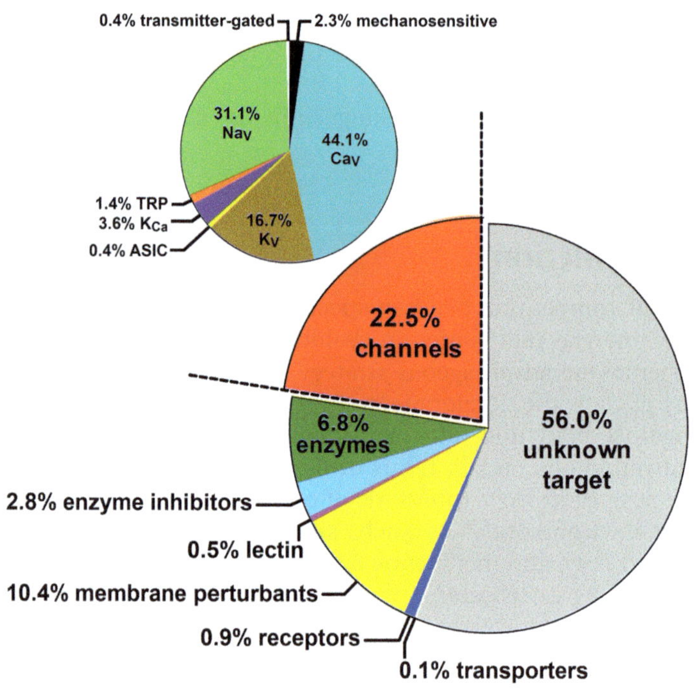

Figure 8.1 Pharmacology of spider-venom peptides and proteins. Summary of the known targets/mechanism of action of spider-venom peptides and proteins. The majority of spider-venom peptides target ion channels (60% of peptides with a known target), and of these 75% target Na_V or Ca_V channels. Targeting of receptors and transporters remains a relatively unexplored area for spider-venom peptides, with only 1% of described toxins known to target these membrane proteins.

8.2 Non-ICK Spider-Venom Components

Spider venoms are incredibly complex, containing components ranging from small molecules and salts to moderately sized proteins comprised of several hundred amino acid residues. The most studied components of spider venoms are the disulfide-rich neurotoxic peptides. These peptides are typically <40 residues in length and they adopt a three-dimensional fold known as the inhibitor cystine knot (ICK) motif (see Chapter 2), which provides these toxins with high levels of chemical, thermal, and biological stability. Additionally, some spider venoms contain small peptides with no disulfide bonds that have cytolytic activity due to their ability to interact with, and form pores in, lipid membranes.[10–14] These peptides often have antimicrobial activity due to this generalised cytolytic action. Some of the larger spider-venom proteins such as sphingomyelinase also have cytotoxic activity; in combination with the linear cytolytic peptides, this can lead to localised tissue damage at the site of envenomation, which may lead to enhanced diffusion of the disulfide-rich neurotoxic venom peptides.[15]

Spider venoms also contain non-peptidic compounds, of which the acylpolyamines are of pharmacological interest. Spider-venom acylpolyamines are low molecular weight compounds that primarily inhibit a class of ligand-gated cation channels known as ionotropic glutamate receptors. The ionotropic glutamate family of receptors is comprised of *N*-methyl-D-aspartate (NMDA), α-amino-3-hydroxy-5-methylisoxasole-4-propionic acid (AMPA), and kainate receptors. Glutamate is the primary excitatory neurotransmitter in the central nervous system (CNS). Abnormal glutamate signalling plays a role in neurological and psychiatric diseases, and therefore glutamate receptors are considered valid drug targets.[16] At least 100 spider acylpolyamines have been isolated and many have been chemically synthesised. Most of these compounds are non-selective,[17] but several recent studies have endeavoured to modify acylpolyamines to chemically tune their subtype preferences, with varying degrees of success.[18–20] Further study of spider-venom acylpolyamines may be fruitful in providing novel lead compounds that target ionotropic glutamate receptors.

8.3 Spider-Venom ICK Peptides as Drug Leads

Spider-venom peptides display a diverse array of pharmacological activities directed towards modulating the activity of ion channels and receptors in the nervous system of prey and predators (Figure 8.1). As such, spider-venom ICK peptides have received considerable attention as pharmacological tools and as leads for the development of drugs to treat pathophysiological conditions involving the central and/or peripheral nervous systems, such as chronic pain and neurodegenerative diseases.[2,8,21] In the following sections we examine some of the most promising potential therapeutic applications of ICK spider-venom peptides.

8.3.1 Spider-Venom Peptides for Treatment of Chronic Pain

Chronic pain is a global health problem that affects 15–20% of the world's population.[22] The annual economic cost of chronic pain in the USA was recently estimated to be ~\$600 billion, which is more than the combined economic burden of cancer, heart disease, and diabetes.[23] Thus, there is considerable worldwide interest in the development of new analgesic drugs as the therapeutic utility of current-generation analgesics is restricted by problems such as dose-limiting side-effects, tolerance, and the potential for addiction.[24]

Noxious stimuli that induce pain are sensed by specialised afferent nerve fibres known as primary sensory neurons or nociceptors. The peripheral terminals of these neurons are enriched in specialised channels and receptors that transduce chemical, mechanical, and thermal stimuli into a depolarisation of the cell membrane.[24] Many of these sensory ion channels and receptors, such as acid-sensing ion channels (ASICs) and transient receptor potential (TRP) channels, are potential analgesic targets. The small cell depolarisations generated by these sensory transducers can be transformed into an action potential by nearby voltage-gated ion channels. Arrival of this action potential at the central projection of the sensory neuron leads to activation of second order sensory neurons that carry the encoded signal to the brain, where it is perceived as pain.[24] Voltage-gated sodium (Na_V) channels and voltage-gated calcium (Ca_V) channels play key roles in action potential generation in nociceptors and neurotransmitter release at the central terminals of these neurons, respectively, and consequently these channels have become important analgesic targets.

8.3.1.1 Spider-Venom Peptides Targeting Voltage-Gated Calcium Channels

Humans contain 10 subtypes of Ca_V channel denoted $Ca_V1.1$–1.4 (L-type), $Ca_V2.1$ (P/Q-type), $Ca_V2.2$ (N-type), $Ca_V2.3$ (R-type) and $Ca_V3.1$–3.3 (T-type) based on their electrophysiological and pharmacological properties.[25] Of these, $Ca_V2.1$, $Ca_V2.2$, and Ca_V3 are the best validated analgesic targets[26] (see Chapter 9). For example, the FDA-approved analgesic Prialt®, an ICK peptide from the venom of the aquatic cone snail *Conus magus*, targets $Ca_V2.2$ channels (see Chapters 6 and 9 for more details).

The dominant pharmacology of spider-venom peptides that target ion channels is blockade of Ca_V channels (see Figure 8.1), but very few of these peptides display selectivity for particular Ca_V channel subtypes. Activity against a range of Ca_V channel subtypes may not necessarily exclude a toxin from development as an analgesic so long as the targeted subtypes are involved in pain pathways and they do not play other critical physiological roles. For example, although the peptide ω-ctenitoxin-Pn2a from the venom of *Phoneutria nigriventer* is a non-selective blocker of Ca_V1, $Ca_V2.1$, $Ca_V2.2$, and $Ca_V2.3$ channels, it was shown to be analgesic without causing adverse motor effects in mice models of neuropathic pain.[27,28] Similarly, ω-ctenitoxin-Pn4a from the same spider targets $Ca_V1.2$, $Ca_V2.2$, $Ca_V2.1$, and $Ca_V2.3$ but was shown to be efficacious in rat and mice models of inflammatory and neuropathic pain without causing alterations in gross behaviour of the rodents at doses that induced maximal antinociceptive effects.[29,30] ω-Ctenitoxin-Pn4a also has the potential to be a useful adjuvant in opioid treatment, since it potentiates the analgesic effect of morphine and reduces the adverse effects of repeated doses of morphine, including tolerance, constipation, hyperalgesia and withdrawal symptoms.[31]

Other spider-venom peptides that display selectivity for Ca_V channels involved in pain signalling include the $Ca_V2.2$ blockers ω-agatoxin-Aa2a,[32,33] ω-theraphotoxin-Hh1a,[34] and ω-segestritoxin Sf1a,[35] and the Ca_V1 blockers ω-theraphotoxin-Cc1a[36] and ω-ctenitoxin-Cs1a.[37] The possible analgesic effects of these toxins await assessment in animal models of pain.

$Ca_V2.1$ is highly expressed in the CNS and it is believed to play a key role in neurological disorders such as familiar hemiplegic migraine, epilepsy, ataxia, and neurodegeneration in Alzheimer's disease.[38] The prototypic blocker of this channel ($IC_{50} \sim 2$ nM) is ω-agatoxin-Aa4a (ω-agatoxin IVA), a 48-residue peptide from the venom of the American funnel-web spider *Agelenopsis aperta.*[39] Indeed, ω-agatoxin-Aa4a and the closely related paralogues ω-agatoxin-Aa4b ($IC_{50} \sim 3$ nM) and ω-agatoxin-Aa4c ($IC_{50} \sim 36$ nM) are the most selective $Ca_V2.1$ blockers described to date. Further studies of these venom peptides are warranted due to their potency and high level of selectivity. However, it should be noted that peptides generally have poor brain permeability, and therefore chemical modifications to improve the CNS permeability of these peptides and/or the use of non-conventional routes of administration such as intranasal will be necessary for these peptides to access the central population of $Ca_V2.1$. The search for selective $Ca_V2.1$ antagonists is ongoing; the global healthcare company Abbott recently developed a high-throughput fluorescence-based screen for rapid identification of $Ca_V2.1$ blockers.[40,41] Screening of spider venoms using this assay might provide a source of molecules to satisfy the unmet need for selective modulators of $Ca_V2.1$.

8.3.1.2 Spider-Venom Peptides Targeting Voltage-Gated Sodium Channels

Insect genomes encode only a single Na_V channel, which is a common target of peptide toxins from venomous predators.[42] In contrast, the mammalian Na_V channel family is comprised of nine subtypes designated $Na_V1.1$–$Na_V1.9$, and spider-venom peptides have been isolated that target each of these subtypes (Figure 8.2). $Na_V1.7$, $Na_V1.8$, and $Na_V1.9$ are expressed predominantly in nociceptive neurons and are thought to play important roles in pain signalling.[24] $Na_V1.7$ in particular is thought to be crucial in setting the threshold for action potential generation in nociceptive neurons due to its ability to generate substantial ramp currents in response to weak

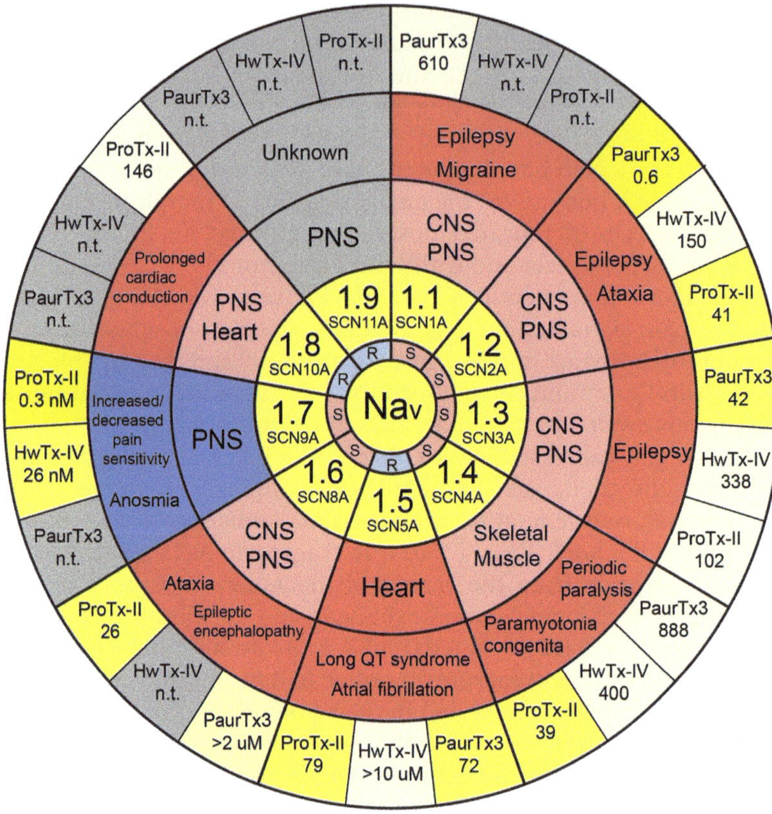

Figure 8.2 Summary of Na_V channel subtypes. Each of the nine Na_V channel subtypes is expressed in different anatomical locations. Mutations of each of these subtypes are associated with a variety of diseases (channelopathies). The innermost ring indicates if the subtype is tetrodotoxin sensitive (S, in red) or resistant (R, in blue). The primary anatomical location and diseases caused by mutations in each subtype are highlighted as unfavourable (red) and acceptable (blue). The outermost ring shows the IC_{50} for inhibition of each subtype by the listed spider toxins (bright yellow being most potent). Data taken from ArachnoServer database (www.arachnoserver.org).[9,124]

depolarisations.[24,43,44] Human genetic studies have shown that this channel is a promising target for development of analgesics to treat chronic pain.[44] Gain-of-function mutations in the gene *SCN9A* that encodes Na_V1.7 can lead to inherited erythromelalgia and paroxysmal extreme pain disorder,[45] which are characterised by extreme episodic pain. In contrast, loss-of-function mutations in *SCN9*A result in a congenital inability to sense pain.[46] Interestingly, people with non-functional Na_V1.7 are completely healthy without any sensory deficits except loss of smell (anosmia).[47] Thus, drugs that inhibit Na_V1.7 function should be useful analgesics for treating a wide range of chronic pain conditions.

The ArachnoServer database[9] currently lists six spider-venom peptides that target $Na_V1.7$. These toxins were all isolated from theraphosid spiders, commonly known as tarantulas. The most potent reported blocker of $Na_V1.7$ is β-theraphotoxin-Tp2a (ProTx-II), a 30-residue ICK peptide isolated from the venom of the Peruvian green velvet tarantula (*Thrixopelma pruriens*). ProTx-II potently inhibits $Na_V1.7$ ($IC_{50} = 0.3$ nM) with >80-fold selectivity over other subtypes *in vitro*, including $Na_V1.2$, $Na_V1.3$, $Na_V1.5$, $Na_V1.6$, and $Na_V1.8$.[48] However, ProTx-II failed to elicit analgesia after intravenous or intrathecal dosing in rodent models of acute and inflammatory pain. The lack of analgesia was attributed to its poor pharmacodynamic profile, which includes a very slow on-rate and poor penetration into the perineurium.[48] Moreover, ProTx-II was lethal to rats following intrathecal injection at a dose of 0.1 mg kg^{-1} or intravenous injection at a dose of 1.0 mg kg^{-1},[48] presumably due to off-target inhibition of $Na_V1.5$, which is critical for the rising phase of the cardiac action potential, and $Na_V1.6$, which is essential for action potential generation at nodes of Ranvier in myelinated motor neurons. Interestingly, β-theraphotoxin-Gr1b (GsAF 1), a peptide isolated from the venom of the Chilean rose tarantula (*Grammostola rosea*), was reported to be analgesic in several rat pain models without causing confounding side-effects, despite the fact that it shares ~90% sequence identity with ProTx-II.[49] Intrathecal β-theraphotoxin-Gr1b induced non-opioid receptor-mediated analgesia in rodent models of thermal, chemical, and mechanical pain.[49]

Another potent blocker of $Na_V1.7$ is μ-theraphotoxin-Hs2a (huwentoxin-IV; HWTX-IV), a 35-residue peptide isolated from the venom of the Chinese bird spider *Haplopelma schmidti.* Native HWTX-IV inhibits $hNa_V1.7$ with an IC_{50} of 26 nM.[50] A comprehensive scanning mutagenesis and analoguing study by the pharmaceutical company MedImmune (a subsidiary of AstraZeneca) led to definition of the pharmacophore that mediates interaction of HWTX-IV with $hNa_V1.7$ and development of an analogue with much higher potency against this channel ($IC_{50} \sim 0.4$ nM).[51] A contemporaneous study by the pharmaceutical company Janssen (a subsidiary of Johnson & Johnson) confirmed the peptide's pharmacophore for $hNa_V1.7$ and led to a homology/docking model of the toxin bound to the domain II voltage sensor of $hNa_V1.7$.[52] Native HWTX-IV is moderately selective for $hNa_V1.7$, with IC_{50} values of ~150 nM, ~350 nM and >10 μM for $hNa_V1.2$, $hNa_V1.3$, and $hNa_V1.4$/$hNa_V1.5$, respectively.[51] However, the activity of HWTX-IV and optimised analogues against the key off-target subtype $hNa_V1.6$ remains to be determined.

In summary, although spider venoms appear to be an excellent natural source of $Na_V1.7$ inhibitors, none of the spider-venom peptides isolated to date have sufficient selectivity over key off-target Na_V subtypes to be therapeutically useful analgesics. Nevertheless, the development of detailed structure–activity relationships for these peptides, as exemplified by the recent Janssen and MedImmune studies, should facilitate the rational engineering of analogues with therapeutic utility.

8.3.1.3 Spider-Venom Peptides Targeting Purinergic Receptors

P2X receptors are ATP-gated non-selective cation channels that belong to the purinergic receptor family. Seven subunits have been described that associate to form homo- or heteromeric trimers.[53] The channel subtypes involved in nociception are $P2X_3$, $P2X_4$, and $P2X_7$, of which $P2X_3$ is the best studied.[54] Several small molecule antagonists of $P2X_3$ and $P2X_7$ have entered clinical trials for the treatment of pain associated with osteoarthritis, cystitis, and inflammation.[54]

Purotoxin-1 and -2 (PT1 and PT2), two structurally unrelated spider-venom peptides isolated from the venom of a burrowing wolf spider (*Geolycosa* spp.), act with nanomolar potency against the $P2X_3$ receptor.[55] They both display a complex mode of action resulting in a concentration-dependent prolongation of channel desensitisation.[56] PT1 was analgesic in rat models of inflammatory pain, exhibiting a similar analgesic effect to the $P2X_3$ antagonist A-317491 at doses three orders of magnitude lower.[55] Thus, PT1 appears to be a promising drug lead, and it will be interesting to see if other spider species yield additional modulators of P2X receptors with therapeutic potential.

8.3.1.4 Spider-Venom Peptides Targeting Acid-Sensing Ion Channels

The concentration of protons, or pH, is one of the most strictly controlled physiological parameters. Perturbations to physiological pH are most often associated with pathological conditions, so it is unsurprising that mammals are equipped with a system to detect acidosis. Acid-sensing ion channels were discovered in the late 1990s, almost 20 years after the observation that sensory neurons depolarise in response to a sudden drop in pH.[57,58] Although they belong to the epithelial sodium channel/degenerin (ENaC/DEG) family of channels, ASICs are distinguished by their restriction to chordates, predominantly neuronal distribution, and activation by protons. Alternative splicing of five ASIC-encoding genes leads to the expression of seven subunits (ASIC1a, ASIC1b, ASIC2a, ASIC2b, ASIC3, ASIC4, and ASIC5) that combine to form hetero- or homotrimeric channels that differ in their pH sensitivity, kinetics, and tissue distribution. ASICs appear to be the primary acid sensors in human nociceptors.[59] ASIC3, along with ASIC1a, is the most pH sensitive of the ASICs and is highly expressed in sensory neurons that innervate skin, muscle and many organs.[60] It plays a key role in inflammatory pain,[61,62] arthritis,[63] post-operative pain[64] and has recently been implicated as a key target in migraine.[65] Recent data from rodent models of pain suggest that ASIC1b, which is found exclusively in peripheral sensory neurons, is also a valid analgesic target.[66] ASIC1a is the most abundant ASIC subunit in the CNS and has been implicated as a therapeutic target for a broad range of pathophysiological conditions, including pain, ischemic stroke,[67] depression, high grade glioblastomas,[68] and autoimmune and neurodegenerative diseases such as multiple sclerosis,[69] Huntington's disease, and Parkinson's disease.[70,71] Thus, ASIC inhibitors are of significant interest as potential analgesics and as drugs for treating stroke and a variety of CNS disorders.

The most potent and selective blocker of ASIC1a is π-theraphotoxin-Pc1a (Pc1a), also known as PcTx1, a 40-residue ICK peptide isolated from the venom of the Trinidad chevron tarantula (*Psalmopoeus cambridgei*). PcTx1 inhibits ASIC1a with an IC_{50} of ~1 nM[72,73] and heteromeric ASIC1a/2b channels with an IC_{50} of 3 nM,[74] but it does not inhibit any other ASIC subtypes at concentrations up to 50 nM.[72] The peptide acts as a gating modifier by stabilising the desensitised state of ASIC1a.[75] At higher concentrations, PcTx1 potentiates rat ASIC1b ($EC_{50} \sim 100$ nM),[76] human ASIC1b,[77] and opens chicken ASIC1[78,79] presumably by stabilising the open state of these subtypes. Intrathecal or intracerebroventricular (i.c.v.) administration of PcTx1 in mice is potently analgesic, indicating that the CNS population of ASIC1a is involved in central pain processing and is a viable analgesic target.[80] In contrast, blockade of ASIC1a with PcTx1 or genetic ablation of ASIC1a in mice failed to reveal a role for ASIC1a in peripheral pain.[61,66]

8.3.2 Spider-Venom Peptides for Treatment of Stroke

Severe oxygen depletion during cerebral ischemia compels the brain to switch from oxidative phosphorylation to anaerobic glycolysis, with the resulting lactic acidosis causing the extracellular pH to fall from ~7.3 to 6.0–6.5 in the ischemic core. ASIC1a, which is the dominant ASIC subtype in human and rodent brain, is robustly activated by this decrease in extracellular pH. Homomeric ASIC1a mediates uptake of Ca^{2+} in addition to Na^{+} and protons, and consequently brain ASIC1a contributes to the intracellular Ca^{2+} overload that occurs during stroke and leads to neuronal death. In rodent models of cerebral ischemia, infarct size and neurological deficits are greatly reduced by knockout or pharmacological blockade of ASIC1a,[67,81] indicating that this channel is a novel target for anti-stroke therapeutics.

In a rat model of transient focal ischemia (middle cerebral artery occlusion; MCAO), i.c.v. injection of 'PcTx1 venom' 30 min before and after induction of ischemia reduced infarct size by 60%. Consistent with this being an effect mediated by ASIC1a, infarct size was similarly reduced by 61% in ASIC1 knockout mice.[67] Even more exciting from a therapeutic perspective was the observation that 'PcTx1 venom' reduced infarct size by ~30% when delivered i.c.v. as late as 5 h *after* MCAO. PcTx1 was even neuroprotective when administered intranasally as late as 4 h *after* MCAO.[81] Thus, PcTx1 is a promising lead for development of novel anti-stroke therapeutics. This is an exciting development as the use of recombinant tissue plasminogen activator to help restore blood flow to the ischemic region is currently the only approved agent for treatment of acute stroke and it is rarely used due to its narrow therapeutic window and the risk of inducing intracranial hemorrhage.[82,83]

8.3.3 Spider-Venom Peptides for Treatment of Autoimmune Diseases

The voltage-gated potassium (K_V) channel $K_V1.3$ is an integral part of the network of ion channels responsible for initiation and maintenance of the calcium signalling cascade in T lymphocytes.[84] In humans, $K_V1.3$ levels are upregulated when quiescent effector memory T (T_{EM}) cells become terminally differentiated in autoimmune diseases such as multiple sclerosis and rheumatoid arthritis.[85] In contrast, $K_V1.3$ expression is virtually unchanged during activation of naïve and long-lived central memory T cells. Thus, selective inhibition of $K_V1.3$ provides a therapeutic mechanism for treating T cell-mediated autoimmune diseases without inducing generalised immunosuppression.[86–88] ShK-186, a sea anemone venom peptide that potently and selectively inhibits $K_V1.3$, is currently in clinical trials for treatment of multiple sclerosis and psoriatic arthritis (see Chapter 10 for details).

Modulation of K_V channels is a moderately common pharmacology in spider venoms (Figure 8.1). Spider-venom peptides have been isolated that target a range of K_V channel subtypes, including K_V1, K_V2, and K_V4 channels. While non-selective $K_V1.x$ blockers have been isolated from spider venoms, no specific blockers of $K_V1.3$ have been reported to date. Nevertheless, focussed screening of spider venoms against $K_V1.3$ is likely to uncover novel modulators of this therapeutic target.

8.3.4 Other Spider-Venom Peptides with Therapeutic Potential

8.3.4.1 Blockers of Large-Conductance Calcium-Activated Potassium Channels

Epileptic pathophysiologies are characterised by irregular neuronal excitability that leads to recurrent seizures. Malfunctioning large-conductance Ca^{2+}-activated K^+ (BK_{Ca}) channels have been implicated in epileptic aetiology. Depending on the phenotype of the aberrant channels, it has been suggested that either enhancement or suppression of BK_{Ca} channel activity may be remedial.[89] No *specific* blockers of vertebrate K_{Ca} channels have yet been isolated from spider venoms. κ-Hexatoxin-Hv1c, an unusual spider-venom peptide containing a functionally critical vicinal disulfide bond,[90–92] blocks insect K_{Ca} channels (Slo-type) with high potency ($IC_{50} \sim 3$ nM) but is much less potent on mammalian K_{Ca} channels ($IC_{50} \sim 10$ μM for block of mouse $K_{Ca}1.1$).[93] Thus, it remains to be seen whether focussed screening of spider venoms will uncover novel modulators of human K_{Ca} channels.

8.3.4.2 Modulators of Transient Receptor Potential Channels

The TRP family of ion channels is an emergent target for the treatment of a myriad of diseases including neuropathic pain, inflammatory bowel disease, and migraine.[94–97] TRP channels are responsible for mediating sensations such as warmth, cold, and pressure, and they respond to some pungent chemicals in foods such as capsaicin from chillies and menthol from mint.[97,98] TRPV1, TRPV3 and TRPA1 are of particular interest, as modulation of these receptors has an analgesic effect. Topical capsaicin, an agonist of TRPV1, is already used for pain management,[99] and several TRPV1 modulators are in preclinical and clinical trials for treatment of various types of pain including migraine, postoperative and neuropathic pain.[97,98] TRP agonism is a recently reported pharmacology for spider toxins, with discovery of the 34/35-residue vanillotoxins in 2006[100] and the much larger 75-residue double-knot toxin (DkTx) in 2010.[101] The promiscuous spider-venom peptide protoxin-1 (β/ω-theraphotoxin-Tp1a), which inhibits Na_V, Ca_V, and K_V channels, was also recently found to antagonise TRPA1.[102] These spider toxins have proved to be extremely useful pharmacological tools for exploring the structure and function of TRPV1 channels; for example, DkTx was recently used to capture the open-state structure of TRPV1.[103] Further studies of spider venoms are likely to reveal novel TRP channel modulators that may prove to be useful therapeutic leads.

8.4 Patents Detailing Therapeutic Applications of Spider-Venom Peptides

Table 8.1 provides a summary of patents from the period 1989–2014 that describe potential therapeutic applications of spider-venom peptides. Approximately 30–35 such patents were filed over this 25-year period (note that we excluded patents from the same inventors/assignees that detail similar use of the same peptides as well as patents focussed on venom components smaller than 1 kDa). Sixty per cent of these patents describe peptides that act on Na_V or Ca_V channels, while another 30% detail peptides that act on K_V channels, mechanosensitive channels, or ASIC1a (Figure 8.3). Most of the claimed applications are for treatment of neurological diseases, including pain, epilepsy, and neuronal damage following stroke, although several patents describe treatments for erectile dysfunction and cardiac arrhythmia. Notably, all six applications filed in the period 2012–2014 describe Na_V or Ca_V channel modulators for treatment of pain, highlighting the growing importance of these ion channel targets for analgesic drug development.[26,44] A small number of patents (10%) describe non-reticulated cytolytic toxins with application in the fields of cancer, drug delivery, and immunotherapy.

It is notable that patents focussed on spider-venom peptides have gradually become more specific, both with respect to the molecular target of the peptide and the claimed therapeutic applications. This is consistent with both our improved understanding of the pharmacology of spider-venom peptides and our increased awareness of the role played by their most common targets (*i.e.*, ion channels) in a variety of important diseases.

Table 8.1 Patents detailing potential therapeutic applications of spider-venom peptides filed during the period 1989–2014.

Patent number	*Year*	*Assignee*	*Peptide name*	*Spider species*	*No. of residues*	*SS bonds*	*Target/mode of action*	*Disease/ application*
US 8 664 179	2014	CNRS & Universite de Nice Sophia Antipolis	Psp3Tx1	*Paraphysa* spp.	28	3	Inhibitor of T-type voltage-gated calcium channel $Ca_V3.2$	Pain, hypertension, and prostate cancer
WO2014016673 A1	2014	Purdue Pharma L.P.	C-terminally modified protoxin-II	*Thrixopelma pruriens*	30	3	Inhibitor of voltage-gated sodium channels, particularly $Na_V1.7$	Na_V channel disorders, especially pain
WO 2013173706 A2	2013	Janssen Biotech Inc.	Huwentoxin-IV analogs	*Haploplema schmidti*	39	3	Inhibitor of voltage-gated sodium channel $Na_V1.7$	Pain
US 8 399 026	2013	Alomone Preclinical Ltd	Peptides A, B	*Pterinochilus* spp.	34, 29	3, 3	Inhibitor of voltage-gated sodium channels $Na_V1.3$ and $Na_V1.8$	Pain
			Peptides C, D	*Haplopelma lividum*	35, 33	3, 3		
			Peptide E, VSTX3	*Grammostola* spp.	34, 34	3, 3		
US 8 383 162 WO2008061329 A2	2013	Universidade Federal de Minas Gerais	Phα1β	*Phoneutria nigriventer*	55	6	Inhibitor of N-type voltage-gated calcium channel $Ca_V2.2$	Pain
WO2012125973 A2	2012	Amgen Inc.	GpTx-1 (ω-TRTX-Gr2a)	*Grammostola porteri*	34	3	Potent inhibitor of $Na_V1.3$ and $Na_V1.7$. Weak inhibitor of $Ca_V3.1$	Pain
US 7 892 764	2011	Legacy Emmanuel Hospital & Health Center	PcTx1 (π-TRTX-Pc1a)	*Psalmopoeus cambridgei*	40	3	Selective blocker of ASIC1a	Epilepsy (seizure suppression)
WO2009094742 A1	2009	Universidade Federal de Minas Gerais	Tx2-6 (δ-CNTX-Pn2a)	*Phoneutria nigriventer*	48	5	Inhibits inactivation of Na_V channels and increases NO release in the corpus cavernosum	Erectile dysfunction

US 7 396 816	2008	Pharmadesign Inc.	TVP003	*Grammostola rosea*	10	1	Engineered blockers of mechanosensitive channels based on pharmacophore of GsMTx4, 34-residue toxin from *G. rosea*.	Atrial fibrillation
			TVP004		8	1		
			TVP005		17	1		
US 7 259 145	2007	Research Foundation of State University of New York at Buffalo	D-GsMTx4	*Grammostola rosea*	34	3	All D-enantiomer of native peptide. Blocks stretch-activated ion channels.	Cardiac arrhythmia
US 7 125 847	2006	Research Foundation of State University of New York at Buffalo	GsMTx4 (M-TRTX-Gr1a)	*Grammostola rosea*	34	3	Blocks stretch-activated ion channels	Cardiac arrhythmia
US 7 132 505	2006	CNRS	PcTx1 (π-TRTX-Pc1a)	*Psalmopoeus cambridgei*	40	3	Selective blocker of ASIC1a	Block of ASIC1a (disease not defined)
US 6 998 389	2006	Toximed GmbH	Mixture of peptide toxins and enzymes	Spiders in the family Sicariidae	N.A.	N.A.	Tumour cell lysis	Cancer (tumours)
WO2003035677 A2	2003	University of Bern	Cupiennin 1a	*Cupiennius salei*	35	0	Lysis of bacterial and tumour cells	Tumours and infections caused by bacterial pathogens
			Cupiennin 1b		35	0		
			Cupiennin 1c		35	0		
			Cupiennin 1d		35	0		
US 6 670 329	2003	Xiamen Bioway Biotech Co.	HWAP-I	*Haplopelma schmidti*	33	3	Inhibition of presynaptic activity and high-voltage-activated Ca_V channels	Pain
US 6 156 726	2000	Elan Pharmaceuticals, Inc.	HG-1 (SNX-482)	*Hysterocrates gigas*	41	3	Inhibitor of E-type voltage-gated calcium channel $Ca_V2.3$	Premature labour, lacation let-down, epileptic seizures

(continued)

Table 8.1 (*continued*)

Patent number	*Year*	*Assignee*	*Peptide name*	*Spider species*	*No. of residues*	*SS bonds*	*Target/mode of action*	*Disease/ application*
US 5 968 838	1999	Zeneca Limited	GsAF II (κ-TRTX-Gr2c)	*Grammostola rosea*	31	3	Blocks stretch-activated ion channels	Cardiac arrhythmia
US 5 877 026 WO1999042480 A1	1999	Zeneca Limited	GsAF I (β-TRTX-Gr1b)	*Grammostola rosea*	29	3	Unknown for GsAF I; probably blocks Na_V channels based on homology with other spider toxins. GsAF II blocks mechano-sensitive channels and Kv channels.	Pain
			GsAF II (κ-TRTX-Gr2c)		31	3		
US 5 804 554	1998	Pfizer Inc.	Pept. 12 (ω-FLTX-Kh1b)	*Kukulcania hibernalis*	74	6	Inhibitor of voltage-gated calcium channels	Calcium-channel-mediated diseases
			Pept. 13-1 (ω-FLTX-Kh1c)		74	6		
			Pept. 13-2 (ω-FLTX-Kh1d)		74	6		
			Pept. 13-3 (ω-FLTX-Kh1e)		74	6		
			Pept. 10 (ω-FLTX-Kh2a)		72	6		
			Pept. 13-4 (ω-FLTX-Kh2b)		72	6		
			Pept. 14-1 (ω-FLTX-Kh3a)		73	6		
US 5 756 462	1998	Pfizer Inc.	Pept. 17 (M-LCTX-Gri1b)	*Geolycosa riograndae*	25	0	Cytolytic toxins that form pores in the plasma membrane	Drug delivery, cancer chemotherapy, and immunotherapy
			Pept. 15a (M-LCTX-Gri2c)		27	0		
			Pept. 15b (M-LCTX-Gri2d)		27	0		

US 5 627 154	1997	Pfizer Inc.	κ-SPRTX-Hv1a	*Heteropoda venatoria*	33	3	Block Ca_V and Kv4.x channels	Calcium-channel-mediated diseases
			κ-SPRTX-Hv1b		30	3		
			κ-SPRTX-Hv1c		31	3		
			κ-SPRTX-Hv1d		30	3		
			κ-SPRTX-Hv1e		29	3		
			ω-SPRTX-Hv1a		37	3		
			ω-SPRTX-Hv1b		36	3		
US 5 599 559	1997	Pfizer Inc.	Pept. J2 (ω-AGTX-Aa5a)	*Agelenopsis aperta*	48	5	Blocks calcium channels in rat cerebellar granule cells	Calcium-channel-mediated diseases
US 5 281 693	1994	University of Utah Research Foundation	AG1 (also known as ω-AGTX-Aa4c & ω-Aga-IVA)	*Agelenopsis aperta*	48	4	Selective blocker of inhibitor of voltage-gated calcium channel $Ca_V2.1$	Epilepsy; neurodegenerative disorders
WO1994021278 A1	1994	NPS Pharmaceuticals Inc.	κ-SPRTX-Hv1b	*Heteropoda venatoria*	30	3	Toxins block voltage-dependent transient outward K^+ current in rat ventricular myocytes, with no effect on delayed-rectifier or inward rectifier K^+ currents	Diseases mediated by transient outward K^+ channels
			κ-SPRTX-Hv1c	*Heteropoda venatoria*	31	3		
			κ-SPRTX-Ofl2a	*Olios fasciculatus*	34	3		
WO1994010196 A1	1994	NPS Pharmaceuticals Inc.	ω-TRTX-Asp1a	*Aphonopelma* sp.	39	3	Block calcium channels in rat cerebellar granule cells	Calcium-channel-mediated diseases
			ω-TRTX-Asp1b		39	3		
			ω-TRTX-Asp1c		39	3		
			ω-TRTX-Asp1d		39	3		
			ω-TRTX-Asp1e		35	2		
			ω-TRTX-Asp2a		33	3		
			ω-TRTX-Asp2b		34	3		
			ω-TRTX-Asp3a		42	3		

(*continued*)

Table 8.1 *(continued)*

Patent number	*Year*	*Assignee*	*Peptide name*	*Spider species*	*No. of residues*	*SS bonds*	*Target/mode of action*	*Disease/ application*
US 5 196 204 (Very similar to US 5 064 657 in 1991 & 4 925 664 in 1990)	1993	University of Utah Research Foundation	AG1 (ω-Aga-IVA)	*Agelenopsis aperta*	N.D.	N.D.	AG1 122 and HOI (both MW ~6 kDa) were not sequenced in this study. AG1 blocks synaptic transmission in chick cochlear nucleus neurons. HOI irreversibly blocks excitatory amino acid receptors.	Heart conditions (arrhythmia, angina, hypertension, and congestive heart failure) and seizures
			HOI	*Hololena curta*	N.D.	N.D.		
US 5 122 596	1992	Pfizer Inc.	Fraction Q (very likely to be AG1/ω-AGTX-Aa4c/ω-Aga-IVA)	*Agelenopsis aperta*	45+	4	Blocks calcium channels in rat cerebellar granule cells	Calcium-channel-mediated diseases
US Pat. App. Ser. No. 07/346 181	1989	Pfizer Inc.	Fraction G	*Agelenopsis aperta*	64+	4	Estimated masses for fractions G, H_2, L_1, and M are 7267 Da, 5494 Da, ~20 000 Da, and ~80 000 Da respectively.	Calcium-channel-mediated diseases
			Fraction H_1 (ω-AGTX-Aa1d)		37	4		
			Fraction H_2		37+	3		
			Fraction I (ω-AGTX-Aa1f)		37	4		
			Fraction J		27+	2		
			Fraction K (ω-AGTX-Aa4c)		48	4		
			Fraction L_1		28++	N.D.		
			Fraction L_2		N.D.	N.D.		
			Fraction M		18++	N.D.		

N.D. = not determined; N.A. = not applicable; "+" indicates that the peptide was only partially sequenced and is larger than indicated.

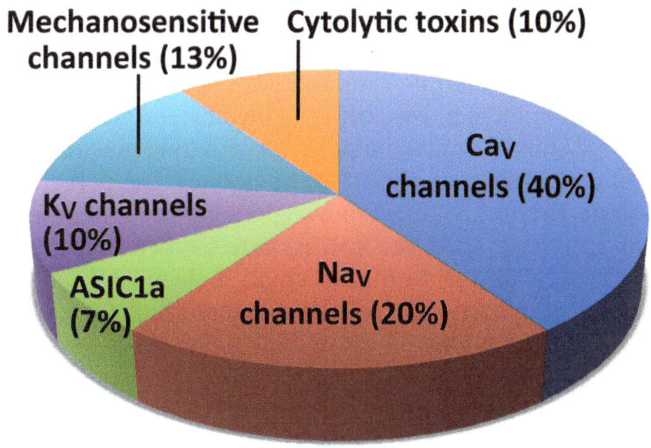

Figure 8.3 Distribution of patents describing therapeutic uses of spider-venom peptides based on molecular target/mechanism of action of the peptides.

8.5 Future Challenges

The majority of therapeutic leads derived from spider venoms are peptides, which presents significant challenges for drug development. Most peptides are readily degraded in the gastrointestinal tract, which makes oral delivery a challenge.[104] While disulfide-rich venom peptides are typically much more resistant to proteases than non-reticulated peptides,[105] their poor cell permeability still makes oral delivery difficult. However, a variety of chemical and physical modifications can be made to improve oral bioavailability, including N- to C-terminal cyclisation, which aims to prevent exopeptidase cleavage, attachment of lipophilic groups to improve cell permeability, formulations such as enteric coatings, or incorporation of the peptide into nanoparticles.[106–109] Moreover, oral drug delivery is not a critical requirement for treatment of chronic diseases. Synthetic insulin has been prescribed for subcutaneous (s.c.) injection for more than 30 years, and the anti-diabetic drug exenatide (a 39-residue, non-reticulated venom peptide) was prescribed as a twice-daily s.c. injection (Byetta™) for 7 years prior to approval of a once-weekly injectable formulation (Bydureon™) by the United States Food and Drug Administration (FDA) in 2012.[110] Peptide drugs can also be delivered *via* the intranasal route, which provides access to the brain *via* trigeminal neurons that innervate the nose and olfactory neurons that terminate in the nasal cavity.[111,112]

Another often-raised objection to the therapeutic use of peptides is that they are rapidly eliminated through the kidneys.[113] For example, exenatide has a plasma half-life of only 2.4 h.[110] However, whole-body blood circulation occurs in less than 1 min in humans,[109] and consequently venom-peptides with nanomolar or higher binding affinities for their cognate receptors or ion channels would be expected to reach and bind to their cell surface targets well before degradation and elimination became an issue. Moreover, the *in vivo* half-life of peptides can be extended *via* chemical modifications or *via* conjugation to either large polymers (*e.g.*, polyethylene glycol, polysialic acid, and hydroxyethyl starch) or plasma proteins with long plasma half-lives such as albumin and immunoglobulin.[114–117] In essence, the perceived disadvantages of peptide drugs are often outweighed by their potency and target selectivity. As a consequence, the success rate for peptides and proteins in clinical trials is very similar to that of small molecules,[118] and the average internal rate of return (IRR) for recently marketed biologics (~13%) significantly exceeds the IRR of ~7.5% for small-molecule drugs developed over the same period of time.[119]

In addition to their stability, potency, and target selectivity, low-cost mass production of spider-venom peptides is possible *via* heterologous expression in yeast or bacteria.[120] For example, Vestaron Corporation have scaled up production of an insecticidal spider-venom peptide to the point where has been approved by the US Environmental Protection Agency for sale as a foliar spray.[121] Thus, the challenges involved in developing spider-venom peptides as therapeutics are arguably less than for other peptide drugs due to their high level of stability and low-cost production.

8.6 Conclusions

Ion channels and receptors are ubiquitous membrane proteins that together represent about half of all major drug targets.[122] Our ability to manipulate these targets for therapeutic purposes relies on understanding their structure and relating it to their biological function. In this regard, venom peptides have proved to be critical tools for understanding the structure and function of ion channels and receptors.[123] Animal toxins have enabled the purification of ion channels, facilitated determination of their stoichiometry, structure and localisation, contributed to our understanding of their (patho)physiological roles, and not only aided in the discovery of ion channel subtypes but also defined many channel subtypes through their exquisite selectivity. Spider venoms are one of the richest known sources of bioactive peptides, and while this chapter has focussed on their therapeutic potential, we should not ignore the value of spider-venom peptides as pharmacological tools, including their application to the validation and characterisation of new therapeutic targets.

Acknowledgements

Work in our laboratory on the therapeutic application of venom peptides is supported by the Australian National Health and Medical Research Council (Project Grants APP1034958, APP1063798 and APP1067940, and Principal Research Fellowship APP1044414 to GFK).

References

1. B. G. Fry, K. Roelants, D. E. Champagne, H. Scheib, J. D. Tyndall, G. F. King, T. J. Nevalainen, J. A. Norman, R. J. Lewis, R. S. Norton, C. Renjifo and R. C. de la Vega, *Annu. Rev. Genomics Hum. Genet.*, 2009, **10**, 483–511.
2. G. F. King, *Expert Opin. Biol. Ther.*, 2011, **11**, 1469–1484.
3. J. A. Coddington and H. W. Levi, *Annu. Rev. Ecol. Syst.*, 1991, **22**, 565–592.
4. L. D. Rash and W. C. Hodgson, *Toxicon*, 2002, **40**, 225–254.
5. P. Escoubas and L. Rash, *Toxicon*, 2004, **43**, 555–574.
6. G. F. King and M. C. Hardy, *Annu. Rev. Entomol.*, 2013, **58**, 475–496.
7. P. Escoubas, B. Sollod and G. F. King, *Toxicon*, 2006, **47**, 650–663.
8. N. J. Saez, S. Senff, J. E. Jensen, S. Y. Er, V. Herzig, L. D. Rash and G. F. King, *Toxins*, 2010, **2**, 2851–2871.
9. V. Herzig, D. L. Wood, F. Newell, P. A. Chaumeil, Q. Kaas, G. J. Binford, G. M. Nicholson, D. Gorse and G. F. King, *Nucl. Acids Res.*, 2011, **39**, D653–D657.
10. C. W. Lee, S. Kim, S. H. Roh, H. Endoh, Y. Kodera, T. Maeda, T. Kohno, J. M. Wang, K. J. Swartz and J. I. Kim, *Biochemistry*, 2004, **43**, 890–897.
11. S. A. Kozlov, A. A. Vassilevski, A. V. Feofanov, A. Y. Surovoy, D. V. Karpunin and E. V. Grishin, *J. Biol. Chem.*, 2006, **281**, 20983–20992.
12. A. A. Vassilevski, S. A. Kozlov, O. V. Samsonova, N. S. Egorova, D. V. Karpunin, K. A. Pluzhnikov, A. V. Feofanov and E. V. Grishin, *Biochem. J.*, 2008, **411**, 687–696.
13. A. Won, A. Ruscito and A. Ianoul, *Biochim. Biophys. Acta*, 2012, **1818**, 3072–3080.
14. A. S. Kuznetsov, P. V. Dubovskii, O. V. Vorontsova, A. V. Feofanov and R. G. Efremov, *Biochemistry (Mosc)*, 2014, **79**, 459–468.
15. P. H. Lopes, R. Bertani, R. M. Goncalves-de-Andrade, R. H. Nagahama, C. W. van den Berg and D. V. Tambourgi, *PLoS Negl. Trop. Dis.*, 2013, **7**, e2394.
16. J. C. Watkins and D. E. Jane, *Br. J. Pharmacol.*, 2006, **147**(Suppl 1), S100–S108.
17. C. A. Olsen, A. S. Kristensen and K. Strømgaard, *Angew. Chem., Int. Ed.*, 2011, **50**, 11296–11311.
18. M. H. Poulsen, S. Lucas, T. B. Bach, A. F. Barslund, C. Wenzler, C. B. Jensen, A. S. Kristensen and K. Strømgaard, *J. Med. Chem.*, 2013, **56**, 1171–1181.
19. M. H. Poulsen, S. Lucas, K. Strømgaard and A. S. Kristensen, *Mol. Pharmacol.*, 2014, **85**, 261–268.
20. N. G. Norager, M. H. Poulsen, A. G. Jensen, N. S. Jeppesen, A. S. Kristensen and K. Strømgaard, *J. Med. Chem.*, 2014, **57**, 4940–4949.
21. J. K. Klint, S. Senff, D. B. Rupasinghe, S. Y. Er, V. Herzig, G. M. Nicholson and G. F. King, *Toxicon*, 2012, **60**, 478–491.
22. D. S. Goldberg and S. J. McGee, *BMC Public Health*, 2011, **11**, 770.

23. D. J. Gaskin and P. Richard, *J. Pain,* 2012, **13**, 715–724.
24. G. F. King and I. Vetter, *ACS Chem. Neurosci.*, 2014, **21**, 1087–1101.
25. G. F. King, *Toxicon,* 2007, **49**, 513–530.
26. E. Bourinet, C. Altier, M. E. Hildebrand, T. Trang, M. W. Salter and G. W. Zamponi, *Physiol. Rev.,* 2014, **94**, 81–140.
27. R. M. Leao, J. S. Cruz, C. R. Diniz, M. N. Cordeiro and P. S. Beirao, *Neuropharmacology,* 2000, **39**, 1756–1767.
28. G. D. Dalmolin, C. R. Silva, F. K. Rigo, G. M. Gomes, N. Cordeiro Mdo, M. Richardson, M. A. Silva, M. A. Prado, M. V. Gomez and J. Ferreira, *Pain,* 2011, **152**, 2224–2232.
29. L. B. Vieira, C. Kushmerick, M. E. Hildebrand, E. Garcia, A. Stea, M. N. Cordeiro, M. Richardson, M. V. Gomez and T. P. Snutch, *J. Pharmacol. Exp. Ther.,* 2005, **314**, 1370–1377.
30. A. H. Souza, J. Ferreira, N. Cordeiro Mdo, L. B. Vieira, C. J. De Castro, G. Trevisan, H. Reis, I. A. Souza, M. Richardson, M. A. Prado, V. F. Prado and M. V. Gomez, *Pain,* 2008, **140**, 115–126.
31. A. H. de Souza, M. C. Lima, C. C. Drewes, J. F. da Silva, K. C. Torres, E. M. Pereira, C. J. de Castro Junior, L. B. Vieira, M. N. Cordeiro, M. Richardson, R. S. Gomez, M. A. Romano-Silva, J. Ferreira and M. V. Gomez, *Toxicon,* 2011, **58**, 626–633.
32. V. P. Bindokas and M. E. Adams, *J. Neurobiol.,* 1989, **20**, 171–188.
33. M. E. Adams, V. P. Bindokas, L. Hasegawa and V. J. Venema, *J. Biol. Chem.,* 1990, **265**, 861–867.
34. Z. Liu, J. Dai, L. Dai, M. Deng, Z. Hu, W. Hu and S. Liang, *J. Biol. Chem.,* 2006, **281**, 8628–8635.
35. R. Newcomb, A. Palma, J. Fox, S. Gaur, K. Lau, D. Chung, R. Cong, J. R. Bell, B. Horne and L. Nadasdi, *Biochemistry,* 1995, **34**, 8341–8347.
36. J. K. Klint, G. Berecki, T. Durek, M. Mobli, O. Knapp, G. F. King, D. J. Adams, P. F. Alewood and L. D. Rash, *Biochem. Pharmacol.,* 2014, **89**, 276–286.
37. H. Kubista, R. A. Mafra, Y. Chong, G. M. Nicholson, P. S. Beirao, J. S. Cruz, S. Boehm, W. Nentwig and L. Kuhn-Nentwig, *Neuropharmacology,* 2007, **52**, 1650–1662.
38. V. Nimmrich and G. Gross, *Br. J. Pharmacol.,* 2012, **167**, 741–759.
39. M. E. Adams, *Toxicon,* 2004, **43**, 509–525.
40. M. Mezler, D. Hermann, A. M. Swensen, A. Draguhn, G. C. Terstappen, G. Gross, H. Schoemaker, G. Freiberg, S. Pratt, S. M. Gopalakrishnan and V. Nimmrich, *Comb. Chem. High Throughput Screening,* 2012, **15**, 372–385.
41. D. Hermann, M. Mezler, A. M. Swensen, C. Bruehl, A. Obergrusserger, K. Wicke, H. Schoemaker, G. Gross, A. Draguhn and V. Nimmrich, *Comb. Chem. High Throughput Screening,* 2013, **16**, 233–243.
42. G. F. King, P. Escoubas and G. M. Nicholson, *Channels,* 2008, **2**, 100–116.
43. J. W. Theile and T. R. Cummins, *Front. Pharmacol.,* 2011, **2**, 54.
44. S. D. Dib-Hajj, Y. Yang, J. A. Black and S. G. Waxman, *Nat. Rev. Neurosci.,* 2013, **14**, 49–62.

45. H. S. Ahn, S. D. Dib-Hajj, J. J. Cox, L. Tyrrell, F. V. Elmslie, A. A. Clarke, J. P. Drenth, C. G. Woods and S. G. Waxman, *Eur. J. Pain,* 2010, **14**, 944–950.
46. J. J. Cox, F. Reimann, A. K. Nicholas, G. Thornton, E. Roberts, K. Springell, G. Karbani, H. Jafri, J. Mannan, Y. Raashid, L. Al-Gazali, H. Hamamy, E. M. Valente, S. Gorman, R. Williams, D. P. McHale, J. N. Wood, F. M. Gribble and C. G. Woods, *Nature,* 2006, **444**, 894–898.
47. R. Staud, D. D. Price, D. Janicke, E. Andrade, A. G. Hadjipanayis, W. T. Eaton, L. Kaplan and M. R. Wallace, *Eur. J. Pain,* 2011, **15**, 223–230.
48. W. A. Schmalhofer, J. Calhoun, R. Burrows, T. Bailey, M. G. Kohler, A. B. Weinglass, G. J. Kaczorowski, M. L. Garcia, M. Koltzenburg and B. T. Priest, *Mol. Pharmacol.,* 2008, **74**, 1476–1484.
49. *United States Pat., 5,877,026*, 1998.
50. Y. Xiao, K. Blumenthal, J. O. Jackson, 2nd, S. Liang and T. R. Cummins, *Mol. Pharmacol.,* 2010, **78**, 1124–1134.
51. J. D. Revell, P. E. Lund, J. E. Linley, J. Metcalfe, N. Burmeister, S. Sridharan, C. Jones, L. Jermutus and M. A. Bednarek, *Peptides,* 2013, **44**, 40–46.
52. N. A. Minassian, A. Gibbs, A. Y. Shih, Y. Liu, R. A. Neff, S. W. Sutton, T. Mirzadegan, J. Connor, R. Fellows, M. Husovsky, S. Nelson, M. J. Hunter, M. Flinspach and A. D. Wickenden, *J. Biol. Chem.,* 2013, **288**, 22707–22720.
53. C. Coddou, Z. Yan, T. Obsil, J. P. Huidobro-Toro and S. S. Stojilkovic, *Pharmacol. Rev.,* 2011, **63**, 641–683.
54. R. A. North and M. F. Jarvis, *Mol. Pharmacol.,* 2013, **83**, 759–769.
55. E. V. Grishin, G. A. Savchenko, A. A. Vassilevski, Y. V. Korolkova, Y. A. Boychuk, V. Y. Viatchenko-Karpinski, K. D. Nadezhdin, A. S. Arseniev, K. A. Pluzhnikov, V. B. Kulyk, N. V. Voitenko and O. O. Krishtal, *Ann. Neurol.,* 2010, **67**, 680–683.
56. N. V. Kabanova, A. A. Vassilevski, O. A. Rogachevskaja, M. F. Bystrova, Y. V. Korolkova, K. A. Pluzhnikov, R. A. Romanov, E. V. Grishin and S. S. Kolesnikov, *Biochim. Biophys. Acta,* 2012, **1818**, 2868–2875.
57. O. A. Krishtal and V. I. Pidoplichko, *Neurosci. Lett.,* 1981, **24**, 243–246.
58. O. A. Krishtal and V. I. Pidoplichko, *Brain Res.,* 1981, **214**, 150–154.
59. S. Ugawa, T. Ueda, Y. Ishida, M. Nishigaki, Y. Shibata and S. Shimada, *J. Clin. Invest.,* 2002, **110**, 1185–1190.
60. E. Deval, X. Gasull, J. Noel, M. Salinas, A. Baron, S. Diochot and E. Lingueglia, *Pharmacol. Ther.,* 2010, **128**, 549–558.
61. E. Deval, J. Noel, N. Lay, A. Alloui, S. Diochot, V. Friend, M. Jodar, M. Lazdunski and E. Lingueglia, *EMBO J.,* 2008, **27**, 3047–3055.
62. J. Karczewski, R. H. Spencer, V. M. Garsky, A. Liang, M. D. Leitl, M. J. Cato, S. P. Cook, S. Kane and M. O. Urban, *Br. J. Pharmacol.,* 2010, **161**, 950–960.
63. M. Izumi, M. Ikeuchi, Q. Ji and T. Tani, *J. Biomed. Sci.,* 2012, **19**, 77.

64. E. Deval, J. Noel, X. Gasull, A. Delaunay, A. Alloui, V. Friend, A. Eschalier, M. Lazdunski and E. Lingueglia, *J. Neurosci.,* 2011, **31**, 6059–6066.
65. J. Yan, X. Wei, C. Bischoff, R. M. Edelmayer and G. Dussor, *Headache,* 2013, **53**, 1250–1261.
66. S. Diochot, A. Baron, M. Salinas, D. Douguet, S. Scarzello, A. S. Dabert-Gay, D. Debayle, V. Friend, A. Alloui, M. Lazdunski and E. Lingueglia, *Nature,* 2012, **490**, 552–555.
67. Z. G. Xiong, X. M. Zhu, X. P. Chu, M. Minami, J. Hey, W. L. Wei, J. F. MacDonald, J. A. Wemmie, M. P. Price, M. J. Welsh and R. P. Simon, *Cell,* 2004, **118**, 687–698.
68. A. K. Rooj, C. M. McNicholas, R. Bartoszewski, Z. Bebok, D. J. Benos and C. M. Fuller, *J. Biol. Chem.,* 2012, **287**, 4053–4065.
69. S. Vergo, M. J. Craner, R. Etzensperger, K. Attfield, M. A. Friese, J. Newcombe, M. Esiri and L. Fugger, *Brain,* 2011, **134**, 571–584.
70. R. L. Arias, M. L. Sung, D. Vasylyev, M. Y. Zhang, K. Albinson, K. Kubek, N. Kagan, C. Beyer, Q. Lin, J. M. Dwyer, M. M. Zaleska, M. R. Bowlby, J. Dunlop and M. Monaghan, *Neurobiol. Dis.,* 2008, **31**, 334–341.
71. H. K. Wong, P. O. Bauer, M. Kurosawa, A. Goswami, C. Washizu, Y. Machida, A. Tosaki, M. Yamada, T. Knopfel, T. Nakamura and N. Nukina, *Hum. Mol. Genet.,* 2008, **17**, 3223–3235.
72. P. Escoubas, J. R. De Weille, A. Lecoq, S. Diochot, R. Waldmann, G. Champigny, D. Moinier, A. Menez and M. Lazdunski, *J. Biol. Chem.,* 2000, **275**, 25116–25121.
73. N. J. Saez, M. Mobli, M. Bieri, I. R. Chassagnon, A. K. Malde, R. Gamsjaeger, A. E. Mark, P. R. Gooley, L. D. Rash and G. F. King, *Mol. Pharmacol.,* 2011, **80**, 796–808.
74. T. W. Sherwood, K. G. Lee, M. G. Gormley and C. C. Askwith, *J. Neurosci.,* 2011, **31**, 9723–9734.
75. X. Chen, H. Kalbacher and S. Gründer, *J. Gen. Physiol.,* 2005, **126**, 71–79.
76. X. Chen, H. Kalbacher and S. Gründer, *J. Gen. Physiol.,* 2006, **127**, 267–276.
77. E. N. Hoagland, T. W. Sherwood, K. G. Lee, C. J. Walker and C. C. Askwith, *J. Biol. Chem.,* 2010, **285**, 41852–41862.
78. I. Baconguis and E. Gouaux, *Nature,* 2012, **489**, 400–405.
79. D. S. Samways, A. B. Harkins and T. M. Egan, *Cell Calcium,* 2009, **45**, 319–325.
80. M. Mazzuca, C. Heurteaux, A. Alloui, S. Diochot, A. Baron, N. Voilley, N. Blondeau, P. Escoubas, A. Gelot, A. Cupo, A. Zimmer, A. M. Zimmer, A. Eschalier and M. Lazdunski, *Nat. Neurosci.,* 2007, **10**, 943–945.
81. G. Pignataro, R. P. Simon and Z. G. Xiong, *Brain,* 2007, **130**, 151–158.
82. S. N. Chapman, P. Mehndiratta, M. C. Johansen, T. L. McMurry, K. C. Johnston and A. M. Southerland, *Vasc. Health Risk Manage.,* 2014, **10**, 75–87.

83. M. A. Moskowitz, E. H. Lo and C. Iadecola, *Neuron,* 2010, **67**, 181–198.
84. M. D. Cahalan and K. G. Chandy, *Immunol. Rev.,* 2009, **231**, 59–87.
85. V. Chi, M. W. Pennington, R. S. Norton, E. J. Tarcha, L. M. Londono, B. Sims-Fahey, S. K. Upadhyay, J. T. Lakey, S. Iadonato, H. Wulff, C. Beeton and K. G. Chandy, *Toxicon,* 2012, **59**, 529–546.
86. C. Beeton, H. Wulff, J. Barbaria, O. Clot-Faybesse, M. Pennington, D. Bernard, M. D. Cahalan, K. G. Chandy and E. Béraud, *Proc. Natl. Acad. Sci. U.S.A.,* 2001, **98**, 13942–13947.
87. C. Beeton, H. Wulff, N. E. Standifer, P. Azam, K. M. Mullen, M. W. Pennington, A. Kolski-Andreaco, E. Wei, A. Grino, D. R. Counts, P. H. Wang, C. J. LeeHealey, S. A. B., A. Sankaranarayanan, D. Homerick, W. W. Roeck, J. Tehranzadeh, K. L. Stanhope, P. Zimin, P. J. Havel, S. Griffey, H. G. Knaus, G. T. Nepom, G. A. Gutman, P. A. Calabresi and K. G. Chandy, *Proc. Natl. Acad. Sci. U.S.A.,* 2006, **103**, 17414–17419.
88. C. Beeton, M. W. Pennington and R. S. Norton, *Inflamm. Allergy Drug Targets,* 2011, **10**, 313–321.
89. P. N'Gouemo, *Expert Opin. Ther. Targets,* 2011, **15**, 1283–1295.
90. X.-H. Wang, M. Connor, R. Smith, M. W. Maciejewski, M. E. H. Howden, G. M. Nicholson, M. J. Christie and G. F. King, *Nat. Struct. Biol.,* 2000, 7, 505–513.
91. F. Maggio and G. F. King, *J. Biol. Chem.,* 2002, **277**, 22806–22813.
92. A. D. de Araujo, V. Herzig, M. J. Windley, S. Dziemborowicz, M. Mobli, G. M. Nicholson, P. F. Alewood and G. F. King, *Antioxid. Redox Sign.,* 2013, **19**, 1976–1980.
93. S. J. Gunning, F. Maggio, M. J. Windley, S. M. Valenzuela, G. F. King and G. M. Nicholson, *FEBS J.,* 2008, **275**, 4045–4059.
94. P. Holzer, *Pharmacol. Ther.,* 2011, **131**, 142–170.
95. S. R. Eid, *Curr. Top. Med. Chem.,* 2011, **11**, 2118–2130.
96. S. Benemei, F. De Cesaris, C. Fusi, E. Rossi, C. Lupi and P. Geppetti, *J. Headache Pain,* 2013, **14**, 71.
97. D. Julius, *Annu. Rev. Cell Dev. Biol.,* 2013, **29**, 355–384.
98. L. Vay, C. Gu and P. A. McNaughton, *Br. J. Pharmacol.,* 2012, **165**, 787–801.
99. P. Anand and K. Bley, *Br. J. Anaesth.,* 2011, **107**, 490–502.
100. J. Siemens, S. Zhou, R. Piskorowski, T. Nikai, E. A. Lumpkin, A. I. Basbaum, D. King and D. Julius, *Nature,* 2006, **444**, 208–212.
101. C. J. Bohlen, A. Priel, S. Zhou, D. King, J. Siemens and D. Julius, *Cell,* 2010, **141**, 834–845.
102. J. Gui, B. Liu, G. Cao, A. M. Lipchik, M. Perez, Z. Dekan, M. Mobli, N. L. Daly, P. F. Alewood, L. L. Parker, G. F. King, Y. Zhou, S. E. Jordt and M. N. Nitabach, *Curr. Biol.,* 2014, **24**, 473–483.
103. E. Cao, M. Liao, Y. Cheng and D. Julius, *Nature,* 2013, **504**, 113–118.
104. J. H. Hamman, G. M. Enslin and A. F. Kotze, *BioDrugs,* 2005, **19**, 165–177.

105. S. Yang, Y. Xiao, D. Kang, J. Liu, Y. Li, E. A. B. Undheim, J. K. Klint, R. Mingqiang, R. Lai and G. F. King, *Proc. Natl. Acad. Sci. U.S.A.,* 2013, **110**, 17534–17539.
106. S. Gupta, A. Jain, M. Chakraborty, J. K. Sahni, J. Ali and S. Dang, *Drug Delivery,* 2013, **20**, 237–246.
107. V. Gupta, B. H. Hwang, J. Lee, A. C. Anselmo, N. Doshi and S. Mitragotri, *J. Control Release,* 2013, **172**, 753–762.
108. O. Cheneval, C. I. Schroeder, T. Durek, P. Walsh, Y. H. Huang, S. Liras, D. A. Price and D. J. Craik, *J. Org. Chem.,* 2014, **79**, 5538–5544.
109. L. J. Otvos, *Pharm. Outsourc.,* 2014, **15**, 40–43.
110. Y. Cai, L. Wei, L. Ma, X. Huang, A. Tao, Z. Liu and W. Yuan, *Drug Des. Devel. Ther.,* 2013, 7, 963–970.
111. W. Ying, *Future Neurol.,* 2008, **3**, 1–4.
112. L. Wallis, E. Kleynhans, T. D. Toit, C. Gouws, D. Steyn, J. Steenekamp, J. Viljoen and J. Hamman, *Protein Pept. Lett.*, 2014, **5**, 749–751.
113. L. Otvos, Jr and J. D. Wade, *Front. Chem.,* 2014, **2**, 62.
114. L. Pollaro and C. Heinis, *Medchemcomm,* 2010, **1**, 319–324.
115. R. Zhang, S. Jain, M. Rowland, N. Hussain, M. Agarwal and G. Gregoriadis, *J. Diabetes Sci. Technol.,* 2010, **4**, 532–539.
116. R. Liebner, R. Mathaes, M. Meyer, T. Hey, G. Winter and A. Besheer, *Eur. J. Pharm. Biopharm.,* 2014, **87**, 378–385.
117. B. Mukherjee, S. D. Karmakar, C. M. Hossain and S. Bhattacharya, *Protein Pept. Lett.,* 2014, in press.
118. L. J. Otvos and J. D. Wade, *Front. Chem.,* 2014, **2**, 62.
119. E. David, T. Tramontin and R. Zemmel, *Nat. Rev. Drug Discovery,* 2009, **8**, 609–610.
120. J. K. Klint, S. Senff, N. J. Saez, R. Seshadri, H. Y. Lau, N. S. Bende, E. A. Undheim, L. D. Rash, M. Mobli and G. F. King, *PLoS One,* 2013, **8**, e63865.
121. J. I. McIntyre, 4th Venoms to Drugs Meeting, *Heron Island, Australia,* 2011.
122. M. Rask-Andersen, M. S. Almén and H. B. Schiöth, *Nat. Rev. Drug Discovery,* 2011, **10**, 579–590.
123. J. Kalia, M. Milescu, J. Salvatierra, J. Wagner, J. K. Klint, G. F. King, B. M. Olivera and F. Bosmans, *J. Mol. Biol.*, doi: 10.1016/j.jmb.2014.07.027.
124. D. L. Wood, T. Miljenovic, S. Cai, R. J. Raven, Q. Kaas, P. Escoubas, V. Herzig, D. Wilson and G. F. King, *BMC Genomics,* 2009, **10**, 375.

CHAPTER 9

Case Study 1: Development of the Analgesic Drugs Prialt® and Xen2174 from Cone Snail Venoms

RICHARD J. LEWIS

Centre for Pain Research, Institute for Molecular Bioscience, The University of Queensland, St Lucia QLD 4072, Australia
E-mail: r.lewis@imb.uq.edu.au

9.1 Introduction

Cone snails comprise 500–700 recently evolved species of marine molluscs that capture prey of fish, molluscs or worms using a uniquely evolved and highly specialised envenomation machinery. Using a "needle and syringe" strategy, cone snails inject a complex mixture of venom peptides through a hollow, barbed harpoon using a muscular proboscis.[1] Interestingly, fish- and worm-hunting Conidae use a single injection strategy that paralyses prey within seconds, while mollusc-hunting species often inject prey multiple times before the prey can be easily removed from its shell. Each species of cone snail produces its own unique cocktail of mostly small, disulfide-bonded peptides. Individual venom peptides can be referred to as either conotoxins or conopeptides.

RSC Drug Discovery Series No. 42
Venoms to Drugs: Venom as a Source for the Development of Human Therapeutics
Edited by Glenn F. King

Published by the Royal Society of Chemistry, www.rsc.org

Conotoxins are genetically encoded as prepropeptides that are cleaved from their precursor protein by endoproteases in the venom.[2] Most conotoxins are small peptides (<4000 Da) stabilised through two to three disulfide bonds to generate rigid and well-defined globular structures that contribute to their high pharmacological specificity and chemical and enzymatic stability. The diverse range of membrane proteins that conotoxins selectively target make them valuable pharmacological probes for dissecting the role of receptors and ion channels in normal and disease processes. Many act at neuronal targets, especially those found in pain pathways, allowing highly specific dissection of the functional elements associated with normal and disease processes in the sensory system. Given this profile, it is not surprising that a number of conopeptides can reverse signs of pain in animal models, with several developed into potential novel analgesics (Table 9.1).

Conus spp. have evolved >50 000 unique conotoxin precursor sequences that generate many thousands of different venom peptides[3] through the recently described mechanism of variable peptide processing.[4] However, this broadly evolved class of venom peptides is mostly untapped as a source of bioactive peptides, with <1% of conopeptides having been pharmacologically characterised. This review will discuss two conopeptide classes with therapeutic potential in pain management, the ω-conotoxins that directly inhibit spinal $Ca_v2.2$ and the χ-conotoxins targeting the norepinephrine transporter that indirectly inhibit $Ca_v2.2$. A comparison of ω-MVIIA (Prialt®) marketed for the treatment of severe pain and χ-Xen2174 evaluated in Phase IIb clinical trials, reveals the strengths and weaknesses of these overlapping pharmacological approaches for targeting spinal pathways (Figure 9.1) for the management of severe pain.

Table 9.1 Conotoxins developed for the treatment of pain. The amino acid sequence, molecular target and current clinical status are indicated

Peptide	*Sequence*[a]	*Target*	*Current status*
ω-MVIIA	CKGKGAKCSRLMYDCCTGSCRSGKC[b]	$Ca_v2.2$ block	Marketed
Xen2174	ZGVCCGYKLCHOC[b]	NET block	Phase IIb (stalled)
ω-CVID	CKSKGAKCSKLMYDCCSGSCSGTVGC[b]	$Ca_v2.2$ block	Phase IIb (stalled)
Vc1.1	GCCSDPRCNYDHPEIC[b]	nAChR-$GABA_B$	Phase IIa (ceased)
Conantokin-G	GEγγLQγNQγLIRγKSN	NMDA-R block	Phase I (ceased)
Contulakin-G	pESEEGGSNAT$_g$KKPYIIL	Neurotensin agonist	Phase I (ceased)

[a]Abbreviations: γ = γ-carboxyglutamate; pE = pyroglutamate; T_g = glycosylated tyrosine; O = hydroxyproline.
[b]C-terminal amidation.

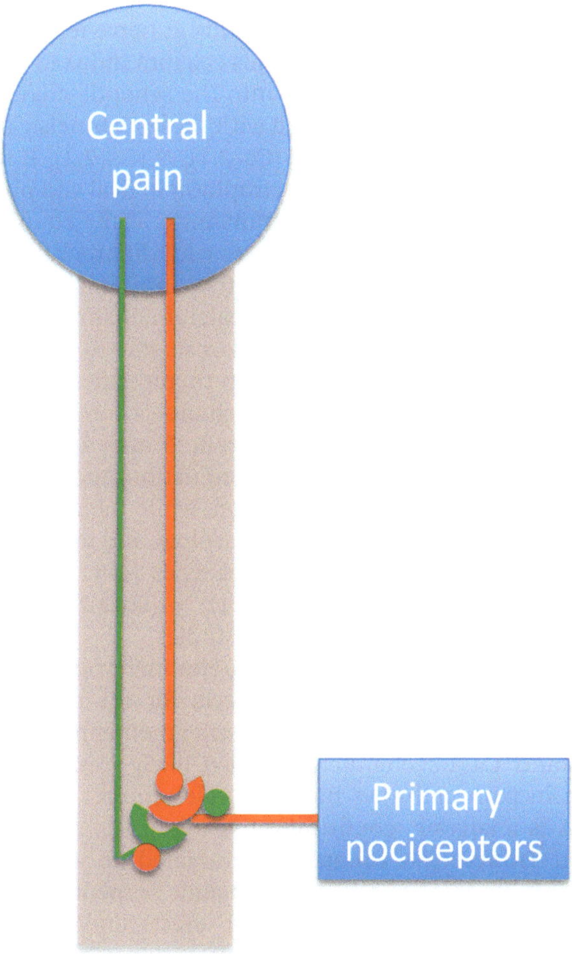

Figure 9.1 Diagram of ascending pain pathways (red) comprising peripheral nociceptors and descending analgesic pathways (green) that control the level of pain perceived centrally in the brain. Descending inhibition occurs mostly through released norepinephrine acting on presynaptic nerve terminals of ascending pathways in the spinal cord (pink). Green and red dots located at presynaptic nerve terminals represent analgesic interneurons that either enhance inhibitory pathways or directly inhibit excitatory pathways. The ω-conotoxins act to inhibit ascending excitatory pathways by directly inhibiting presynaptic calcium influx. In contrast, χ-conopeptides enhance the inhibitory effects of norepinephrine acting on α_2-adrenoceptors to inhibit presynaptic calcium influx. This is achieved by increasing the levels of norepinephrine released from descending pathways by inhibiting its reuptake *via* the norepinephrine transporter.

9.2 Calcium Channels in Pain Pathways

Nerve action potentials generate a Ca^{2+} influx at nerve terminals by opening voltage-gated calcium (Ca_V) channels. This calcium influx in turn triggers the transient release of neurotransmitter into the synaptic cleft (or neuromuscular junction). The current associated with these channels can be classified into six groups named L (Ca_V1), P/Q ($Ca_V2.1$), N ($Ca_V2.2$), R ($Ca_V2.3$) and T (Ca_V3) types to reflect their distinct electrophysiological and pharmacological properties. The release of a particular transmitter is not coupled to the same combination of calcium channels in nerves with different function, allowing pathway-specific modulation with selective inhibitors. In addition, the auxiliary subunits and splice variants of calcium channels are known to be differentially regulated in normal and disease states. Given this diversity, considerable opportunity exists to develop therapeutics that inhibit $Ca_V2.2$ channels in pain pathways. Neuronal Ca_V channels comprise a pore-forming α_1 subunit that co-assembles with different β and $\alpha_2\delta$ subunits. The α_1 subunit is largely responsible for ion conduction and associated electrophysiological characteristics of the channel, while the associated β and $\alpha_2\delta$ subunits modulate the expression, pharmacology and biophysical properties of the α_1 subunit. At least six α_1 channel types have been identified that correspond to L- ($Ca_v1.1$–1.4), P/Q- ($Ca_v2.1$), N- ($Ca_v2.2$), R- ($Ca_v2.3$) and T- ($Ca_v3.1$–3.3) types observed in functional studies.

Of the six pharmacologically distinct Ca_V channel types, $Ca_v2.1$, $Ca_v2.2$ and Ca_v3 channels are best validated as targets for the treatment of pain,[5,6] although dose-limiting side-effects may restrict their clinical application. The R-type (resistant) $Ca_V2.3$ channels are also likely to contribute to neurotransmission in pain pathways but their role requires further delineation. Spinal MVIIA is analgesic, confirming the role of $Ca_v2.2$ channels in pain transmission. However, ω-conotoxins evaluated to date produce dose-limiting side-effects in animals, which include serpentine tail and shakes, that may be associated with supraspinal effects and/or spinal effects on inhibitory pathways. It remains to be seen if it is possible to avoid such side-effects through selective inhibition of $Ca_v2.2$ channels in ascending pain, perhaps by targeting $Ca_v2.2$ splice variants co-assembled with specific auxiliary subunit clusters. For example, a selective inhibitor of $Ca_v2.2$ splice variants restricted to dorsal root ganglia might allow the inhibition of pain with fewer side-effects. However, splice variant/auxiliary subunit combinations in sensory nerves are presently not easy to characterise, especially in human tissue.

An alternative approach to inhibit calcium channels is indirectly through the activation of G protein coupled receptors that negatively regulate calcium channel function. For example, activation of α_2-adrenoceptors located on synaptic terminals in the spinal cord produces analgesia. Antidepressants that target the norepinephrine (NE) transporter are also effective analgesics, especially for difficult to treat neuropathic pain conditions.[7] In this instance, the effects of the endogenous agonist NE are amplified by drugs that inhibit its reuptake. The χ-conopeptide analogue Xen2174 revealed that this effect can be generated spinally by enhancing the effects of descending inhibitory pathways that release NE in the proximity of synaptic α_2-adrenoceptors on ascending pain pathways (Figure 9.1).

9.3 Analgesic ω-Conotoxins

Predatory gastropods of the genus *Conus* (cone snails) represent a diverse family of marine molluscs that use highly specialised venom peptides for prey capture. The venom of each species contains a unique array of approximately 100 encoded genes to generate many thousands of different peptides through a process of variable peptide processing.[4] These conotoxins (conopeptides) have evolved to target with high specificity and potency a diverse array of ion channels and receptors that facilitate the successful capture of fish, molluscs and worms. Importantly, many also target related mammalian membrane proteins, including a number that show pharmaceutical potential,[8,9] such as the ω-conotoxins isolated from piscivorous cone shells that selectively inhibit mammalian $Ca_v2.2$.[9,10]

These relatively small and highly structured peptides appear to act at or near the outer vestibule of $Ca_v2.2$ to inhibit Ca^{2+} movement into nerves. An emerging pharmacophore model of ω-conotoxin interactions at $Ca_v2.2$ can help guide the rational development of specific $Ca_v2.2$ inhibitors.[1,11,12] The presently characterised ω-conopeptides have varying affinity and selectivity for neuronal $Ca_v2.2$ channel subtypes found in mammals, making them widely used research tools for defining the distribution and role of neuronal Ca_v channel subtypes. The most commonly used ω-conopeptides for such studies are GVIA, MVIIA and CVID for selective inhibition of $Ca_v2.2$, and MVIIC for selective inhibition of $Ca_v2.1$. Radioiodinated versions of these peptides have been used to establish radioligand binding assays that have helped guide the discovery of new ω-conotoxins[13] and the development of peptidomimetics.[14,15] In addition to their use as research tools, animal studies have shown that ω-conopeptides that target $Ca_v2.2$ channels have clinical potential in ischaemic brain injury[16,17] and pain.[18–25]

Neurex (Elan) completed Phase III clinical trials of MVIIA (SNX-111, ziconitide or Prialt) and gained United States Food and Drug Administration (FDA) approval in 2004 for its use in the treatment of chronic pain. However, despite their impressive selectivity and potency, the ω-conotoxins presently available are not ideal therapeutics. For example, intravenous MVIIA has been reported to cause orthostatic hypotension in humans[26] as a result of an action at peripheral neuronal Ca_v channels, while intrathecal MVIIA causes a variety of side-effects of unknown origin,[27] despite $Ca_v2.2$ channels being predominant at synapses carrying nociceptive signals in the spinal cord.[28] CNS-related side-effects associated with the use of Prialt are psychiatric symptoms, cognitive impairment and decreased alertness/unresponsiveness.

The most frequently reported adverse events in clinical trials include dizziness, nausea, confusion, headache, somnolence, nystagmus, asthenia and, somewhat surprisingly, pain.

Another highly selective ω-conopeptide, GVIA, dissociates quasi-irreversibly from $Ca_v2.2$ channels; accordingly it may be difficult to control dosage, and thus it has not been evaluated clinically. A more $Ca_v2.2$-selective and reversible inhibitor named ω-CVID[29] produces efficacy with fewer side-effects in animals compared to MVIIA.[24] CVID has successfully completed Phase I/IIa evaluation. Other ω-conopeptides identified are either poorly characterised or have inadequate selectivity for $Ca_v2.2$ channels over other neuronal calcium channels, and thus are not presently considered as therapeutic candidates.

Since $Ca_v2.2$ plays a role in the ascending pain pathways, and the α-subunit is upregulated in the spinal cord in chronic pain states along with the auxiliary $\alpha_2\delta$-1 subunit,[30] it is not surprising that ω-conotoxins specific for $Ca_v2.2$ channels are potent analgesics. Extensive structure–activity relationship studies have allowed the development of a pharmacophore model for ω-conotoxins[12] that may allow the rational development of specific $Ca_v2.2$ channel inhibitors. However, attempts to produce potent mimics of these exquisitely engineered proteins have not been successful, although selective but less potent inhibitors can be developed.[14,15]

While ω-CVID is highly selective for $Ca_v2.2$ over $Ca_v2.1$ channels, it has been found to inhibit an otherwise resistant Ca_v channel found in the peripheral nervous system whereas other ω-conotoxins had little or no effect on this current.[9] Given that this current is produced by neurons of peripheral origin, it is possible that such current may be restricted to ascending excitatory pathways. However, any role for this current in pain conditions remains to be established. Interestingly, ω-conotoxin affinity is markedly reduced when the $\alpha_2\delta$-1 auxiliary subunit is co-expressed with α_{1B} and β3 subunits.[29] Interestingly, the anticonvulsant gabapentin may be efficacious in neuropathic pain because of a specific interaction with the $\alpha_2\delta$-1 channel subunits, although it precise mode of action remains to be clarified. An alternative to the exquisite target selectivity displayed by venom peptides is functional selectivity more often seen with small molecule inhibitors (*e.g.*, local anaesthetics). Functional selectivity may be achieved through the development of use-dependent blockers that preferentially target rapidly firing (disease- or pain-affected) nerves.[30,31]

9.4 Analgesic χ-Conotoxins

The initial discovery of the χ-conopeptides started with the isolation of two closely eluting conopeptides MrIA and MrIB that were initially suspected to be new α-conotoxins based on their small size (~1400 Da) and relatively hydrophylic nature.[32] While the cysteine spacing (CCX_4CX_2C) was reminiscent of 4/3-α-conotoxins, their amino acid sequence and structure were unique. χ-Conopeptides have a unique three-dimensional "ribbon" structure dominated by a γ-turn stabilised by a 1–4/2–3 cysteine connectivity in contrast to the α-conotoxins, which have a "globular" structure dominated by α-helices stabilised by a 1–3/2–4 cysteine connectivity. To identify the mode of action of these peptides, synthetic MrIA was systematically tested across a range of cell and isolated tissue preparations. These peptides produced a profound enhancement on the second, NE-evoked phase of electrically induced contractile responses in the rat *vas deferens*. This effect was reminiscent of cocaine and tricyclic antidepressants, suggesting that the χ-conopeptides may inhibit the norepinephrine transporter, the main regulator of synaptic NE levels at nerve terminals innervating this tissue. This effect was confirmed by showing direct inhibition of the uptake of ^{3}H-NE into mammalian cells heterologously expressing the noradrenaline transporter (NET).[32] In contrast to cocaine and tricyclic antidepressants, this inhibition was not surmounted with increasing concentrations of NE and thus arose from a non-competitive or allosteric mode of action.[32]

NET is a member of the Na^+-dependent monoamine transporter family of proteins.[33] Neurotransmitter transporters are important drug targets in the CNS because they have the capacity to selectively manipulate neurotransmitter concentrations at the site of transmitter release, and thereby selectively modulate signalling through specific neuronal pathways. Their regulatory roles have been implicated in the aetiology of many neurological disease states, making NET an important target of treatment for a range of neurological diseases including depression, anxiety, obsessive-compulsive disorder and attention deficit hyperactivity disorder.[34] Antidepressants that target NET are also effective analgesics, especially for difficult to treat neuropathic pain conditions.[7] Xen2174 reveals that this effect can be generated spinally, presumably by enhancing the effects of descending-inhibitory-pathway-released NE that acts on presynaptic α_2-adrenoceptors to inhibit ascending pain pathways (see Figure 9.1). The anti-allodynic effects of Xen2174 were completely reversed by intrathecal yohimbine, confirming this mode of action.[35]

Unfortunately, MrIA has an unstable N-terminal asparagine residue that limits its aqueous stability, making it unsuitable for long-term human use in implanted pumps. Extensive replacements and truncations/extensions at the N-termini revealed that replacing asparagine with pyroglutamate significantly improved chemical stability without compromising the side-effect profile observed for MrIA.[36] Xen2174 showed few side-effects even at high doses and had an extended duration of action following a single bolus intrathecal dose in the rat model of neuropathic pain. Remarkably, Xen2174 was also found to accelerate recovery from pain when given preemptively in a rat paw incision model of post-surgical pain. Given the improved stability and clean efficacy/side-effect profile in animals, Xen2174 underwent preclinical development leading to a Phase I study to determine if Xen2174 was safe and well tolerated when administered intravenously to healthy volunteers. Successful completion of this Phase I study was followed by FDA approval to conduct a Phase IIa trial of bolus intrathecal doses in cancer pain patients. This open label, dose-escalating study was conducted in 37 cancer patients suffering severe chronic pain that could not be adequately managed using conventional therapies. Similar to the observations made in rats, after accounting for size differences, Xen2174 was found to relieve pain quickly and for a sustained period across a wide range of tolerated intrathecal bolus doses from 0.1–30 mg per patient. Given these promising results, Xenome Ltd unsuccessfully attempted to progress Xen2174 to a Phase IIb double-blind study to determine if Xen2174 is safe and efficacious in a bunionectomy setting of post-surgical pain.[37]

9.5 Comparison of Direct and Indirect Inhibition of $Ca_v2.2$

Comparing the effects of ω-MVIIA (Prialt) and Xen2174 on ascending and descending pain pathways (Table 9.2), it is evident that many of the side-effects caused by Prialt likely arise from on-target effects on $Ca_v2.2$ on inhibitory pathways. By targeting ascending pathways specifically through the selective enhancement of descending inhibition driven by NE, Xen2174 appears to avoid significant effects on inhibitory pathways and produces fewer side-effects. Morphine also indirectly inhibits $Ca_v2.2$ through activation of the G protein coupled μ-opioid receptor (μ-OR) and produces analgesia with few side-effects, which are eroded as tolerance develops. Thus, both direct and indirect approaches to the inhibition of pain have strengths and weaknesses. Based on the modes of action analysed here, indirect inhibition of $Ca_v2.2$ appears to have most promise once tolerance (morphine) and the limitations of spinal delivery (Xen2174) can be overcome.

Table 9.2 Ascending and descending pain-related pathways affected by Xen2174 and Prialt compared to morphine.[a] Efficacy for Prialt is strongly pain-type dependent, while efficacy for morphine is initially high and side-effects are low, but these reverse as tolerance develops

	Xen2174	*Prialt*	*Morphine*
Direct target	NET	$Ca_v2.2$	μOR
Functional target	$Ca_v2.2$	$Ca_v2.2$	$Ca_v2.2$
Ascending pathways	++	+++	++
Descending NE pathways	+++	+	++
Descending 5HT pathways	–	+	+
Inhibitory interneurons	–	+	+
Efficacy in pain	+++	+(+++)	+++/+
Side-effects	+	++	+/+++
Tolerance development	?	–	+++
Therapeutic index	~100	~1	~3/<1

[a]– = no effect; + = small effect; ++ = moderate effect; +++ = strong effect, ? = tolerance development uncertain in humans.

Acknowledgements

This work was supported by an NHMRC Program Grant and an NHMRC Principal Research Fellowship.

References

1. R. J. Lewis, S. Dutertre, I. Vetter and M. C. Christie, *Pharmacol. Rev.*, 2012, **60**, 259–298.
2. T. J. Milne, G. Abbenante, J. D. Tyndall, J. Halliday and R. J. Lewis, *J. Biol. Chem.*, 2003, **278**, 31105–31110.
3. J. Davis, A. Jones and R. J. Lewis, *Peptides*, 2009, **30**, 1222–1227.
4. S. Dutertre, A. H. Jin, Q. Kaas, A. Jones, P. F. Alewood and R. J. Lewis, *Mol. Cell. Proteomics*, 2013, **12**, 312–329.
5. C. Altier and G. W. Zamponi, *Trends Pharmacol. Sci.*, 2004, **25**, 465–470.
6. J. G. McGivern and S. I. McDonough, *Curr. Drug Targets CNS Neurol. Disord.*, 2004, **3**, 457–478.
7. R. Freeman, *CNS Spectr.*, 2005, **10**, 698–706.
8. D. J. Adams, A. B. Smith, C. I. Schroeder, T. Yasuda and R. J. Lewis, *J. Biol. Chem.*, 2003, **278**, 4057–4062.
9. R. J. Lewis, K. J. Nielsen, D. J. Craik, M. L. Loughnan, D. A. Adams, I. A. Sharpe, T. Luchian, D. J. Adams, T. Bond, L. Thomas, A. Jones, J. L. Matheson, R. Drinkwater, P. R. Andrews and P. F. Alewood, *J. Biol. Chem.*, 2000, **275**, 35335–35344.
10. B. M. Olivera, G. P. Miljanich, J. Ramachandran and M. E. Adams, *Annu. Rev. Biochem.*, 1994, **63**, 823–867.
11. M. J. Lew, J. P. Flinn, P. K. Pallaghy, R. Murphy, S. L. Whorlow, C. E. Wright, R. S. Norton and J. A. Angus, *J. Biol. Chem.*, 1997, **272**, 12014–12023.
12. K. J. Nielsen, T. Schroeder and R. Lewis, *J. Mol. Recognit.*, 2000, **13**, 55–70.
13. P. Favreau, N. Gilles, H. Lamthanh, R. Bournaud, T. Shimahara, F. Bouet, P. Laboute, Y. Letourneux, A. Ménez, J. Molgó and F. Le Gall, *Biochemistry*, 2001, **40**, 14567–14575.
14. J. B. Baell, P. J. Duggan, S. A. Forsyth, R. J. Lewis, Y. Phei Lok and C. I. Schroeder, *Bioorg. Med. Chem.*, 2004, **12**, 4025–4037.
15. C. I. Schroeder, M. L. Smythe and R. J. Lewis, *Mol. Diversity*, 2004, **8**, 127–134.
16. R. Fern, B. R. Ransom and S. G. Waxman, *J. Neurophysiol.*, 1995, **74**, 369–377.
17. K. Yamada, T. Teraoka, S. Morita, T. Hasegawa and T. Nabeshima, *Neuropharmacology*, 1994, **33**, 251–254.
18. S. S. Bowersox, T. Gadbois, T. Singh, M. Pettus, Y. X. Wang and R. R. Luther, *J. Pharm. Exp. Ther.*, 1996, **279**, 1243–1249.
19. A. B. Malmberg and T. L. Yaksh, *Pain*, 1995, **60**, 83–90.
20. J. Nebe, H. Vanegas and H.-G. Schaible, *Exp. Brain Res.*, 1998, **120**, 61–69.

21. V. Neugebauer, H. Vanegas, J. Nebe, P. Rümenapp and H.-G. Schaible, *J. Neurophys.,* 1996, **76**, 3740–3749.
22. M. D. Omote, M. Kawamata, O. Satoh, H. Iwasaki and A. Namiki, *Anesthesiology,* 1996, **84**, 636–643.
23. K. A. Sluka, *Pain,* 1997, **71**, 157–164.
24. M. Smith, P. J. Cabot, F. B. Ross, A. D. Robertson and R. J. Lewis, *Pain,* 2002, **96**, 119–127.
25. D. M. White and M. J. Cousin, *Brain Res.,* 1998, **801**, 50–58.
26. D. McGuire, S. Bowersox, J. D. Fellmann and R. R. Luther, *J. Cardiovasc. Pharmacol.,* 1997, **30**, 400–403.
27. W. G. Brose, D. P. Gutlove, R. R. Luther, S. S. Bowerbox and D. McGuire, *Clin. J. Pain,* 1997, **13**, 256–259.
28. R. E. Westenbroek, L. Hoskins and W. A. Catterall, *J. Neurosci.,* 1998, **18**, 6319–6330.
29. J. Mould, T. Yasuda, C. I. Schroeder, A. M. Beedle, J. Clinton, C. J. Doering, G. W. Zamponi, D. J. Adams and R. J. Lewis, *J. Biol. Chem.,* 2004, **279**, 34705–34714.
30. Z. D. Luo, S. R. Chaplan, E. S. Higuera, L. S. Sorkin, K. A. Stauderman, M. E. Williams and T. L. Yaksh, *J. Neurosci.,* 2001, **21**, 1868–1875.
31. R. J. Winquist, J. Q. Pan and V. K. Gribkoff, *Biochem. Pharmacol.,* 2005, **70**, 489–499.
32. I. A. Sharpe, J. Gehrmann, M. L. Loughnan, L. Thomas, D. A. Adams, A. Atkins, E. Palant, D. J. Craik, D. F. Adams, P. F. Alewood and R. J. Lewis, *Nat. Neurosci.,* 2001, **4**, 902–907.
33. C. I. Wang and R. J. Lewis, *Biochem. Pharmacol.,* 2010, **79**, 1083–1091.
34. A. W. Goddard, S. G. Ball, J. Martinez, M. J. Robinson, C. R. Yang, J. M. Russell and A. Shekhar, *Depression Anxiety,* 2010, **27**, 339–350.
35. C. K. Nielsen, R. J. Lewis, D. Alewood, R. Drinkwater, E. Palant, M. Patterson, T. L. Yaksh, D. McCumber and M. T. Smith, *Pain,* 2005, **118**, 112–124.
36. A. Brust, E. Palant, D. E. Croker, B. Colless, R. Drinkwater, B. Patterson, C. I. Schroeder, D. Wilson, C. K. Nielsen, M. T. Smith, D. Alewood, P. F. Alewood and R. J. Lewis, *J. Med. Chem.,* 2009, **52**, 6991–7002.
37. R. J. Lewis, *Toxicon,* 2012, **59**, 524–528.

CHAPTER 10

Case Study 2: Transforming a Toxin into a Therapeutic: the Sea Anemone Potassium Channel Blocker ShK Toxin for Treatment of Autoimmune Diseases

RAYMOND S. NORTON*[a], MICHAEL W. PENNINGTON[b], AND CHRISTINE BEETON[c]

[a]Medicinal Chemistry, Monash Institute of Pharmaceutical Sciences, Monash University, Parkville, Victoria 3052, Australia; [b]Peptides International, 11621 Electron Drive, Louisville, KY 40299, USA; [c]Department of Molecular Physiology and Biophysics, Baylor College of Medicine, Houston, TX 77030, USA
*E-mail: ray.norton@monash.edu

10.1 Introduction

Diseases with an autoimmune component can affect virtually any organ in the body, and they currently afflict millions of people worldwide. These chronic disorders are often diagnosed in young adults, or even children, and they can be debilitating and lead to premature death. Immunomodulators such as methotrexate, monoclonal antibodies (natalizumab, infliximab),

RSC Drug Discovery Series No. 42
Venoms to Drugs: Venom as a Source for the Development of Human Therapeutics
Edited by Glenn F. King

Published by the Royal Society of Chemistry, www.rsc.org

glatiramer acetate, mitoxantrone, tumour necrosis factor antagonists (etanercept, infliximab) and steroids have considerably improved the management of autoimmune diseases, but they can cause generalized immunosuppression and therefore put patients at an increased risk of tumour cell proliferation or opportunistic infections. Kv1.3 channel blockers represent a new class of immunomodulators with a lower risk of inducing generalized immunosuppression as they preferentially target the effector memory T (T_{EM}) lymphocytes involved in autoimmune diseases, with little or no effect on other subsets of lymphocytes.

In this chapter we describe the discovery and characterization of ShK, a 35-residue polypeptide isolated from the Caribbean sea anemone *Stichodactyla helianthus*, which blocks Kv1.3 channels at picomolar concentrations. Although ShK was effective in treating rats with delayed type hypersensitivity and a model of multiple sclerosis (MS), it lacked selectivity for Kv1.3 channels over closely related Kv1 channels. Extensive mutagenesis studies combined with elucidation of its three-dimensional structure led to models of ShK docked with the target channel. This knowledge aided the development of ShK analogues with improved selectivity and enhanced stability, which have proven efficacious in preventing and/or treating animal models of delayed type hypersensitivity, rheumatoid arthritis, and MS without preventing the clearance of acute infections. These analogues are currently undergoing clinical evaluation as potential immunomodulators for the treatment of autoimmune diseases.

10.2 The Toxin

In 1995, Castañeda and colleagues extracted a potent K^+ channel blocker from the Caribbean sea anemone *Stichodactyla helianthus* and named it ShK, for *Stichodactyla helianthus* K^+ channel toxin.[1] ShK is present in relatively small amounts in *S. helianthus* whole body extracts (<0.5 mg ShK per kg of extract), limiting the amount of material available for studies of the native peptide. It was, however, synthesized successfully and folded into the biologically active form,[2] enabling extensive studies of its structure, selectivity and biological activity, as well as the generation of analogues with increased selectivity and stability.

The 35 amino acid residues in ShK include six cysteines, which form three disulfide bonds with connectivity C1–C6, C2–C4, and C3–C5 (Figure 10.1). The C-terminal Cys35 of ShK pairs with Cys3, effectively creating a macrocyclic ring. The two internal disulfide bonds pair Cys12 with Cys28, and Cys17 with Cys32. This arrangement is the same as for the defensins and α-dendrotoxin.[3] Molecular modelling coupled with mutational analyses[4,5] has shown that Lys22 in ShK (highlighted in blue in the sequence alignment in Figure 10.1) protrudes into and occludes the K^+ channel pore, as elaborated further in Section 10.3.1.[6]

A BLAST search identified ShK-related sequences that might also have K^+-channel-blocking activity. The criteria for the search specified that sequences had to contain six Cys and a positively-charged residue at an equivalent

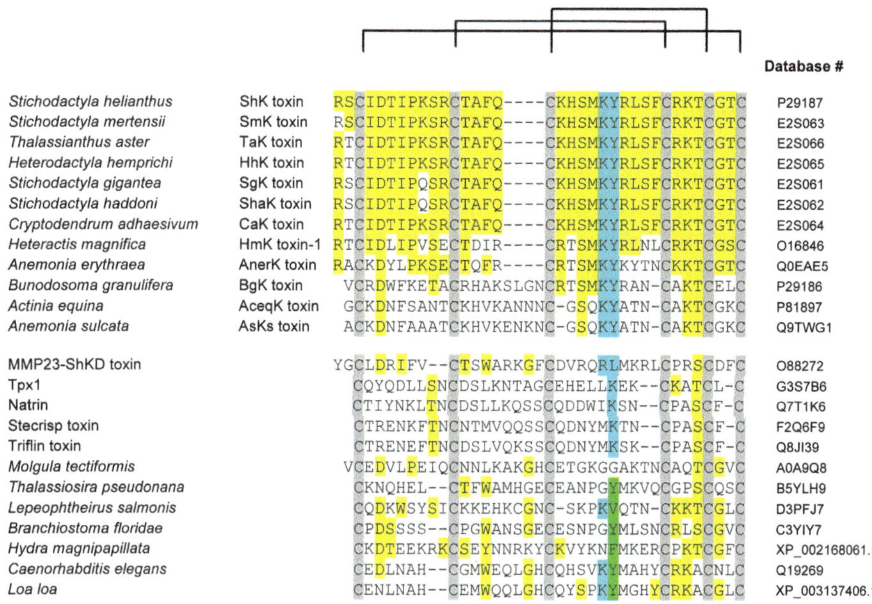

Figure 10.1 Sequence alignment of sea anemone K$^+$-channel blocking peptides and ShKT domain peptides identified through database mining. Sequences are aligned *via* the conserved cystine pairings. UniProt accession numbers are indicated.

position to Lys22 in ShK. Eight sequences from sea anemones (*Stichodactyla mertensii, S. gigantea, S. haddoni, Thalassianthus aster, Heterodactyla hemprichi Crytodendru adhaesivum, Heteractis magnifica, Anemonia erythraea*) have the same number of residues as ShK and share remarkable sequence similarity to ShK (Figure 10.1); they represent one phylogenetic cluster. Four other sea anemone sequences (from *Bunodosoma granulifera, Actinia equina, Anemonia sulcata* now known as *Anemonia viridis,* and *Nematostella vectensis*) exemplified by BgK toxin contain four extra residues between cysteines 3 and 4, and constitute a distinct yet clearly ShK-related group (Figure 10.1). Twelve exemplar sequences from marine organisms (sea urchin, lancelet and marine diatom), soil nematodes (*Caenorhabditis*), pathogenic worms that cause trichinosis and filariasis in humans, fish and mammals are included in the alignment. Each of these sequences represents a toxin-like domain within a protein. All of these domains contain a four-residue insertion between cysteines 3 and 4 (highlighted in green and pink) like BgK, suggesting that they are more closely related to BgK than ShK (Figure 10.1). Structural analyses support this conclusion (see below). These domains are referred to as ShKT domains (ShK-like) by the SMART database (http://smart.embl-heidelberg.de) and NCBI. Figure 10.1 includes three snake ion channel regulatory (ICR) domains of cysteine-rich secretory proteins (CRISP), which share structural similarity with K$^+$-channel-blocking sea anemone toxins.

All the ShK-like sequences contain a positively-charged Lys or Arg at the position corresponding to Lys22 in ShK. ShK, BgK, AETXK, AeK, MMP23-TxD, natrin and stecrisp have been demonstrated experimentally to block K^+ channels or inhibit ^{125}I-dendrotoxin binding to synaptosomes, suggesting that other sequences in Figure 10.1 may also block K^+ channels. Of particular interest are worms that cause filariasis and trichinosis in humans, since these worms might use ShKT domains to block human K^+ channels and thereby modulate the host response. All of the sea anemone toxins and all but two ShKTs contain an aromatic residue (Tyr, Trp and Phe) at the position adjacent to Lys22 (Figure 10.1). Alanine substitution of Tyr23 in ShK reduced channel-blocking potency by greater than 100-fold, highlighting the importance of this position.[7,8] However, MMP23-TxD, natrin and stecrisp contain Leu, Thr or Ser in place of an aromatic residue at this position and yet they block K^+ channels.[9–11] Thus, an aromatic residue may optimize the toxin's interaction with the channel but may not be required.

Asp5 in ShK forms a salt bridge with the ϵ-ammonium group of Lys30, and this salt bridge is necessary for proper folding of the peptide.[12,13] At equivalent positions, all sea anemone toxins contain an Asp and either a Lys or Arg. The Asp and Lys/Arg pair is also present in seven of the 15 ShKTs, indicating conservation of the salt bridge. Although the eight remaining ShKTs contain Asp at the upstream position, at the downstream position the positively-charged residue is replaced by Ala, Val, Ser or His. In these ShKTs, Asp may make hydrogen-bonding interactions with the side-chain hydroxyl or the peptide backbone at the downstream position. None of the ICR domains of CRISPs contains the Asp and Lys/Arg pair.

The solution structure of ShK[12,13] is strikingly different from that of scorpion toxins that also block K^+ channels (see Chapter 7 for details of Kv blockers from scorpion venom).[14] ShK does not contain any β-sheet secondary structure, but consists of two short α-helices encompassing residues 14–19 and 21–24, and an N-terminus with an extended conformation up to residue 8, followed by a pair of interlocking turns that resembles a 3_{10}-helix (Figure 10.2a and b).

10.3 The Target: Kv1.3

Potassium channels are ubiquitous membrane proteins encoded by 78 genes in the human genome. They regulate membrane potential and calcium signalling in diverse cell types. Human T lymphocytes express two types of K^+ channels, homotetramers of Kv1.3 and KCa3.1, respectively. The expression of these channels depends on the state of activation and differentiation of a given T lymphocyte subset.[15]

Mature T lymphocytes can be separated into three subsets based on surface expression of the phosphatase CD45RA and the chemokine receptor CCR7.[16] Naïve T cells are CD45RA$^+$CCR7$^+$. Long-lived central memory T (T_{CM}) lymphocytes, the largest pool of memory T lymphocytes, lose expression of CDRA45 during differentiation and therefore are CD45RA$^-$CCR7$^+$. The third

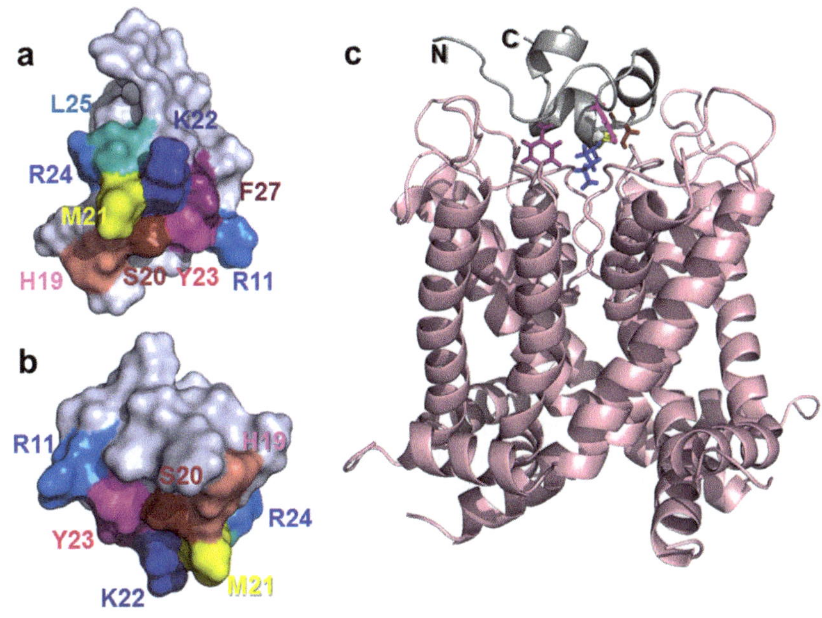

Figure 10.2 (a,b) Surface representations of ShK structure.[12] Some of the residues that contribute to Kv1.3 binding[7,8] are coloured as follows: Lys22 dark blue, Tyr23 magenta, Arg11 and Arg24 marine, His19 salmon, Ser20 ruby, Leu25 teal and Phe27 deep purple. The view in (b) is related to the view in (a) by a 90° rotation around the x-axis and a 180° rotation about the y-axis. (c) Ribbon representation of ShK (grey) docked onto a model of Kv1.3 (light purple).[23,24] The N- and C-termini of ShK are labelled, and the side chains of Lys22, Tyr23, Met21, Ser20, Phe27 on ShK are shown (although that of Met21 is largely obscured in this view).

population of T cells is composed of effector memory T (T_{EM}) lymphocytes, representing fewer than 20% of circulating T cells, which are CD45RA$^-$CCR7$^-$.

Quiescent human T lymphocytes from all three subsets express 200–300 Kv1.3 and 5–35 KCa3.1 channels per cell.[17] However, activation causes diametrically opposite effects on expression of K^+ channels in the different T lymphocyte populations, leading to an altered K^+ channel phenotype. CCR7$^+$ naïve and T_{CM} cells up-regulate KCa3.1 channels to ~500/cell, whereas CCR7$^-$ T_{EM} cells increase Kv1.3 expression to 1500 channels/cell with little change in KCa3.1 levels.[17] This switch in channel expression significantly affects responsiveness to Kv1.3 and KCa3.1 blockers, as CCR7$^-$ T_{EM} lymphocytes are highly sensitive to Kv1.3 channel blockers, while CCR7$^+$ naïve/T_{CM} lymphocytes are sensitive to KCa3.1 channel blockers.

Kv1.3 channels play a critical role in the early steps of T-cell activation.[15] Antigen recognition through the T-cell receptor leads to a rapid release of Ca^{2+} from endoplasmic reticulum stores. Depletion of these stores induces the interaction of STIM-1 with Orai-1 and the opening of Ca^{2+} release-activated

Ca^{2+} (CRAC) channels in the plasma membrane. The ensuing Ca^{2+} influx sustains elevated levels of cytoplasmic Ca^{2+}. Coordinated activity of Ca^{2+}-dependent and Ca^{2+}-independent signalling pathways culminates in T-cell activation. The large influx of Ca^{2+} through the CRAC channel induces cell depolarization, which in turn induces a reduction in Ca^{2+} influx. The driving force for Ca^{2+} entry is restored by a plasma membrane hyperpolarization induced by the efflux of K^+ through Kv1.3. and KCa3.1 channels. These K^+ channels therefore play crucial roles in T lymphocyte activation.

The T lymphocytes involved in autoimmune diseases such as MS, rheumatoid arthritis, psoriasis, and type 1 diabetes mellitus, are mainly $CCR7^-$ T_{EM} cells and express large numbers of Kv1.3 channels following activation.[15,18,19] Targeting these $CCR7^-$ T_{EM} cells without affecting $CCR7^+$ naïve/T_{CM} lymphocytes represents a promising new way of treating these and possibly other T-cell-mediated autoimmune diseases, without inducing generalized immunosuppression. Moreover, homotetrameric Kv1.3 channels have a functionally restricted tissue distribution, and therefore represent attractive therapeutic targets.[20]

10.3.1 How ShK Blocks K^+ Channels

The generation of mono-substituted synthetic analogues of ShK yielded an initial map of the binding site of ShK with human Kv1.3 channels.[21] Amino acid substitutions at the following positions caused no significant effect on the binding of ShK to the channel: Arg1 (R1S), Phe15 (F15A, F15W), Lys18 (K18A), Met21 (M21 to norleucine), Tyr23 (Y23F, Y23 to *p*-nitrophenylalanine and Y23 to *p*-aminophenylalanine, although it should be noted that the latter two analogues were tested against rat brain only and not Jurkat lymphocytes), and Arg24 (R24A). Arg11 is important for the binding of ShK to Kv1.3, as demonstrated by a 42-fold lower affinity of R11Q-ShK for Kv1.3 channels. Lys9 plays a significant role in the block of Kv1.3 channels by ShK, as K9Q displayed a 9.6-fold higher IC_{50} than wild-type ShK. Lys22 is a critical residue for the block of Kv1.3 channels by ShK, as replacing this residue with Arg reduced the affinity of the peptide by 225-fold. The IC_{50} for block of Kv1.3 channels was 23-fold higher for the mutant K22Nle but only two-fold higher for K22A and K22Ornithine (K22Orn).

As a further step towards identifying the ShK residues involved in peptide-Kv1.3 channel interactions, complementary mutants of ShK and Kv1.3 channels were generated, tested for their affinity using patch-clamp electrophysiology, and analysed using mutant cycle analysis.[4] Several pairs of significant ShK-Kv1.3 interactions were identified: ShK-Arg11 with Kv1.3-His404, ShK-Lys22 with Kv1.3-Tyr400, and ShK-Orn22 with Kv1.3-Tyr-400 and with Kv1.3-Asp402.

These findings were subsequently confirmed and expanded by a systematic analysis of alanine scan mutants of ShK against Kv1.3 expressed in mammalian cells.[8] Replacement of Ser20, Lys22, Tyr23 or Arg24 with Ala residues significantly disrupted the interaction of ShK with Kv1.3 channels.

His19 was replaced by a Lys instead of an Ala because the H19A mutant did not fold in reasonable yield. This H19K mutation significantly disrupted the interaction of ShK with Kv1.3 channels, but the role of His19 may be at least partly structural.[13] Alanine substitutions of Ile7, Arg11, Thr13, Met21, Leu25, Ser26, Phe27 or Arg29 had smaller effects on binding.

Taken together, these data show that ShK binds to the outer vestibule of the Kv1.3 channel pore, with Lys22 protruding into the pore and blocking the passage for K^+ ions (Figure 10.2c). Two ShK amino acid residues (Lys22 and Arg24) play a critical role in Kv1.3 channel blockade by the peptide, three residues (His19, Ser20, and Tyr23) have a major role in the ShK-Kv1.3 interaction, six (Arg11, Thr13, Met21, Leu25, Phe27, and Arg29) have an important role, and two (Ile7 and Ser26) play minor roles. None of the other residues of ShK participates directly in the interaction with Kv1.3.[5]

More recently, homology models for Kv1.1 and Kv1.3 channels have been constructed using the crystal structure of Kv1.2. The initial poses for the Kv1.x–ShK complexes were obtained using HADDOCK,[22] and then refined *via* molecular dynamics simulations (in fact the Kv1.3-ShK structure from this study is the one shown in Figure 10.2c).[23] The binding mode in each complex was characterized by identifying the strongly interacting residues, which compare well with the results of mutagenesis studies referred to above. For each complex, the potential of mean force was calculated from umbrella sampling simulations, and the corresponding absolute binding free energy was determined. The computed binding free energies were in remarkably good agreement with the experimental data.[23]

We will return to this point in Section 10.6, but we note here that the power of this approach is demonstrated by our recent results with a new ShK analogue containing an additional C-terminal amidated Lys residue. ShK-K-amide is a potent blocker of Kv1.3 and, in contrast to ShK and ShK-amide, it is selective for Kv1.3.[24] To understand this selectivity, we created complexes of ShK-K-amide with Kv1.3 and Kv1.1 using docking and molecular dynamics simulations, then performed umbrella sampling simulations to construct the potential of mean force of the ligand and calculate the corresponding binding free energy for the most stable configuration. Again, the calculated values agree well with experimental results.[24]

10.4 Transforming the Toxin into a Therapeutic

The first assays demonstrating activity of ShK on K^+ channels showed displacement of ^{125}I-dendrotoxin I and ^{125}I-α-dendrotoxin from rat synaptosomal membranes with IC_{50} values of 1 nM and 20 pM, respectively, and inhibition of ^{125}I-charybdotoxin binding to Jurkat T lymphocytes with an IC_{50} of 32 pM. In addition, ShK increased the twitch response of chick biventer cervicis preparations to indirect stimulation, without affecting responses to acetylcholine, carbachol or KCl. It also blocked K^+ currents in rat dorsal root ganglion neurons, suggesting that it blocked some neuronal K^+ channels.[1,2] Further testing by patch-clamp electrophysiology on cloned K^+

channels expressed in mammalian cells showed that ShK blocked Kv1.1 and Kv1.3 channels with IC_{50} values of 16 and 11 pM, respectively. It had a 10-fold lower affinity for Kv1.6 channels and a 30-fold lower affinity for Kv1.4 channels.

ShK blocked Kv1.2, Kv1.7, and KCa3.1 channels in the low nM range, while concentrations of up to 100 nM had no effect on Kv1.5, Kv3.1, and Kv3.4 channels. Blockade was reversible and dose-dependent with Hill coefficients of 1, indicating that one ShK binds per Kv1.3 channel.[4]

$^{86}Rb^{+}$ efflux studies showed that ShK blocks Kv3.2 channels with an IC_{50} of 0.6 nM, and electrophysiological studies performed on Kv3.2 channels expressed in *Xenopus laevis* oocytes and in planar patch-clamp studies yielded IC_{50} values of 0.3 and 6 nM, respectively.[25]

10.4.1 ShK-Dap22, the First ShK Analogue with Increased Selectivity for Kv1.3 over Kv1.1 and Other K^{+} Channels

One potential drawback of ShK as a therapeutic is its lack of selectivity for Kv1.3 over neuronal Kv1.1 channels, as ingress of ShK into the central nervous system (CNS) could induce neurotoxicity. In order to increase this selectivity, Kalman and colleagues used their knowledge of the structure of ShK and its sites of interaction with Kv1.3 to generate ShK-Dap22 by replacing Lys22 in ShK with the non-protein amino acid diaminopropionic acid (Dap). The overall structure of ShK-Dap22 is similar to that of wild-type ShK, with both having the same major secondary structure elements.[4] However, ShK-Dap22 has a recognizable helix near its C-terminus (residues 29–32), which was not observed in ShK.

ShK-Dap22 displayed a high affinity for Kv1.3 channels, displacing ^{125}I-charybdotoxin binding to human Kv1.3 channels with an IC_{50} of 102 pM, and blocking mouse and human Kv1.3 channels with an IC_{50} of 23 pM, with a Hill coefficient close to unity.[4] ShK-Dap22 had a significantly lower affinity than wild-type ShK for Kv1.1, Kv1.2, and Kv1.4 channels, it blocked Kv1.6 channels with an IC_{50} in the lower nanomolar range, and it had no effect on Kv1.5, Kv1.7, Kv3.1, Kv3.4, or KCa3.1 at concentrations up to 100 nM.[4] Work conducted by Middleton *et al.*[26] corroborated the improved selectivity of the Dap22 analogue in whole-cell voltage-clamp electrophysiology studies, but they also presented data showing that in equilibrium binding assays the K_d is underestimated for wild-type ShK. There also appears to be higher affinity for heteromultimers of Kv1.1 and Kv1.2.

10.4.2 ShK-F6CA, a Fluorophore-Conjugated ShK Analogue to Detect Kv1.3-Expressing Cells by Flow Cytometry

As ShK has a high affinity for Kv1.3 channels, comparable to that of monoclonal antibodies for their target epitopes, it was chosen for labelling with a fluorophore to detect activated CCR7^{-} T_{EM} lymphocytes that express large numbers of Kv1.3 channels.[27] Based on a detailed knowledge of the structure of ShK and the residues involved in its interaction with Kv1.3, Arg1 was chosen for attaching fluorescein, since it is located on a region of the peptide facing away from the channel pore. Fluorescein-6-carboxyl (F6CA) was attached to the α-amino group of ShK-Arg1 through a 20 Å hydrophilic 2-aminoethoxy-2-ethoxy acetic acid (AEEA; mini-PEG) linker.[27]

ShK-F6CA blocked Kv1.3 channels with an IC_{50} of 48 pM and a Hill coefficient of 1. Intriguingly, ShK-F6CA was 160-fold less effective than wild-type ShK in blocking Kv1.1 channels (IC_{50} of 4 nM for ShK-F6CA compared with 25 pM for ShK) and thus displayed a 100-fold selectivity for Kv1.3 over Kv1.1. Moreover, it did not block Kv1.2, Kv1.5, Kv1.7, or Kv3.1 channels at concentrations up to 100 nM.[27] Attaching tetramethylrhodamine or biotin to ShK-Arg1 through an AEEA linker yielded blockers with picomolar affinity for Kv1.3 but no selectivity over Kv1.1 channels. This suggests that the negative charge carried by F6CA plays a role in the increased selectivity of ShK-F6CA, since tetramethylrhodamine is positively charged and biotin is neutral.

ShK-F6CA was used to detect Kv1.3 channels on T lymphocytes by flow cytometry.[27,28] Staining intensity paralleled channel numbers measured by whole-cell patch-clamp electrophysiology. The lower limit of detection using ShK-F6CA is approximately 600 Kv1.3 channels per cell,[15,27] which is sufficient to measure the difference in Kv1.3 channels between activated $CCR7^+$ naïve/T_{CM} and $CCR7^-$ T_{EM} lymphocytes.[15,27,28]

10.4.3 ShK-170, ShK-186, and ShK-192: ShK Analogues with Increased Selectivity and Stability

Based on the unanticipated selectivity of ShK-F6CA for Kv1.3, a series of ShK analogues was designed to enhance the selectivity of ShK for Kv1.3 channels over related K^+ channels. With the understanding that improved selectivity was generated by incorporation of a planar aromatic moiety with an anionic group, ShK-170 [ShK(L5)],[28] ShK-186 [SL5][18] and ShK-192[29] were generated by attaching amino acid derivatives such as phospho-Tyr and p-phosphono-Phe to the N-terminus of ShK through the AEEA linker used previously to attach F6CA to ShK.[27]

ShK-170, which was generated by attaching L-phosphotyrosine to ShK *via* the AEEA linker,[28] blocked Kv1.3 channels with an IC_{50} of 70 pM and was 100-fold selective over Kv1.1 channels, and ≥700-fold selective for Kv1.3 over other K^+, Na^+, and Ca^{2+} channels.[28] Attaching D-phosphotyrosine instead of L-phosphotyrosine generated a blocker that was 35-fold selective for Kv1.3 over Kv1.1 channels, but attaching non-phosphorylated Tyr, Phe, or Phe derivatives yielded blockers with little or no selectivity for Kv1.3 over Kv1.1 channels.

In vitro stability studies showed no evidence of degradation of ShK-170 at ambient temperature (22°C) and pH 4–8. At 40 °C, ShK-170 was stable at physiological pH but exhibited minor degradation at acidic pH values, mainly due to dephosphorylation of the L-phosphotyrosine residue attached to Arg1 and oxidation of Met21. This instability at non-physiological pH values was more marked at 60 °C. ShK-170 contains multiple cleavage sites for trypsin and chymotrypsin, and incubation of the peptide with either or both enzymes led to rapid degradation of ShK-170 into disulfide-stabilized fragments.[29] This peptide is therefore not amenable to oral delivery without appropriate protection.

ShK-186 is identical to ShK-170, except that its C-terminal Cys is amidated.[18] This modification does not affect the selectivity of the blocker for Kv1.3 over Kv1.1 channels, but it should decrease the susceptibility of the C-terminus to carboxypeptidase-mediated cleavage, which may improve the *in vivo* half-life of the peptide.[29] ShK-186 displayed the same *in vitro* stability as ShK-170 at acidic pH and 40 °C but was less stable at alkaline pH values.[29]

ShK-192 was synthesized to improve peptide stability *in vitro* and *in vivo*.[29] Like ShK-186, ShK-192 was amidated at its C-terminus, but, to eliminate oxidation of the single Met residue of ShK, Met21 was replaced with the non-protein amino acid norleucine. Finally, to reduce susceptibility to acid-catalysed hydrolysis of the phosphate from the L-phosphotyrosine residue observed with both ShK-170 and ShK-186, this residue was replaced with the non-hydrolysable phosphate mimetic *para*-phosphonophenylalanine. The *in vitro* stability of ShK-192 at ambient temperature was similar to that of ShK-170. At 40 °C it was stable at neutral and acidic pH values, but minor degradation was observed at pH 8. ShK-192 was stable at 60 °C and at acidic pH values and would therefore be suitable for use in slow-release formulations that require such conditions. However, the peptide degraded rapidly at neutral and basic pH when heated to 60 °C.[29] NMR studies of ShK-192 showed that it adopts a backbone conformation similar to that of ShK.[29] Differences were noted in the N- and C-termini of ShK-192 and at Nle,[21] which could be ascribed to amidation of the C-terminus of ShK-192, addition of AEEA-L-phosphotyrosine to its N-terminus, and replacement of Met^{21} in ShK by norleucine. Docking of ShK-192 with the Kv1.3 channel also showed many similarities with ShK docked with Kv1.3 homotetramers.[29]

10.4.4 D-allo-ShK, an ShK Analogue Resistant to Endogenous Proteases

D-allo-ShK is a mirror image of native ShK that is composed entirely of amino acids in the D-configuration at C^{α}.[30] It was generated because it would be resistant to proteolysis as a result of the inability of endogenous proteases to recognize D-amino acid residues. Its immunogenicity was not tested, but it is reasonable to assume that if it could not be processed proteolytically then it would also not be displayed by antigen-presenting cells. D-allo-ShK folded correctly with formation of three disulfide bonds. Its backbone conformation was a mirror image of that of native ShK, although chiral side chains such as those of Ile and Thr retained their native configurations.[30]

D-allo-ShK reversibly blocked Kv1.3 channels with a Hill coefficient close to 1 and an IC_{50} of 36 nM, a value 2800-fold lower than that of ShK for Kv1.3.[30] Interestingly, D-allo-ShK displayed the same two-fold selectivity for Kv1.3 over Kv1.1 channels as ShK (IC_{50} for D-allo-ShK on Kv1.1 channels = 83 nM). A significant difference in the way ShK and D-allo-ShK block Kv1.3 channels resides in the channel conformation they can bind to, with ShK binding and blocking Kv1.3 channels in any conformation (open, closed, and inactivated) but D-allo-ShK not binding to closed channels. The lower affinity of D-allo-ShK for Kv1.3 may be partly a consequence of that fact that it is not a perfect mirror image of ShK, which could be resolved by installation of appropriately chiral side chains for Ile and Thr. As the cost of D-amino acids comes down,[31] the cost of producing D-allo-ShK or related ShK analogues becomes more reasonable. Thus, D-allo-ShK remains of interest for applications in hostile environments such as the gastrointestinal tract to treat diseases such as colitis or Crohn's disease.

10.4.5 *In vitro* Biological Activity of ShK and Its Analogues

The first tests of ShK on T lymphocytes measured its ability to affect the mitogen-induced proliferation of peripheral blood human T cells, consisting mainly of naïve and T_{CM} $CCR7^+$ cells. It was shown that ShK inhibited this proliferation with an IC_{50} in the nM range (Figure 10.3).[4,17,30] In contrast, ShK inhibits the proliferation of human $CCR7^-$ T_{EM} cells stimulated with an antigen or a mitogen with IC_{50} values ranging from 100 to 400 pM.[17,30] Pre-activated naïve and T_{CM} $CCR7^+$ cells can up-regulate KCa3.1 channels even in the presence of ShK, and therefore escape ShK block when reactivated. In contrast, $CCR7^-$ T_{EM} cells are unable to up-regulate KCa3.1 channels and therefore remain highly sensitive to ShK.[17,28]

Mouse T lymphocytes exhibit significant differences in their K^+ channel phenotype when compared to human T lymphocytes and are therefore not useful models for studying Kv1.3 channel blockers.[32] Rat T lymphocytes, in contrast, are very similar to human T lymphocytes in terms of K^+ channel expression.[28,32,33] As with their human counterparts, rat $CCR7^+$ naïve and T_{CM} lymphocytes are significantly less sensitive to ShK than rat $CCR7^-$ T_{EM} lymphocytes in proliferation assays.[28,33]

ShK-Dap22 inhibited the proliferation of $CCR7^-$ T_{EM} lymphocytes with an IC_{50} of 1.4 nM but had little effect on the proliferation of $CCR7^+$ naïve/T_{CM} lymphocytes.[4,33] ShK-170 was 60-fold more effective in inhibiting the proliferation of $CCR7^-$ T_{EM} lymphocytes (IC_{50} 80 pM) than the proliferation of $CCR7^+$ naïve/T_{CM} lymphocytes (IC_{50} 5 nM) (Figure 10.3).[28] Addition of the cytokine interleukin-2 (IL-2) partially cancelled the inhibitory effects of ShK-170 on T lymphocyte proliferation,[28] as had been shown with other Kv1.3 blockers.[34–36] ShK-170 (100 nM) competitively inhibited the staining of activated $CCR7^-$ T_{EM} cells with ShK-F6CA and inhibited the production of IL-2 by these cells with an IC_{50} < 1 nM.[28] ShK-170 (100 nM) induced little or no *in vitro* toxicity against human and rat lymphoid cells and was

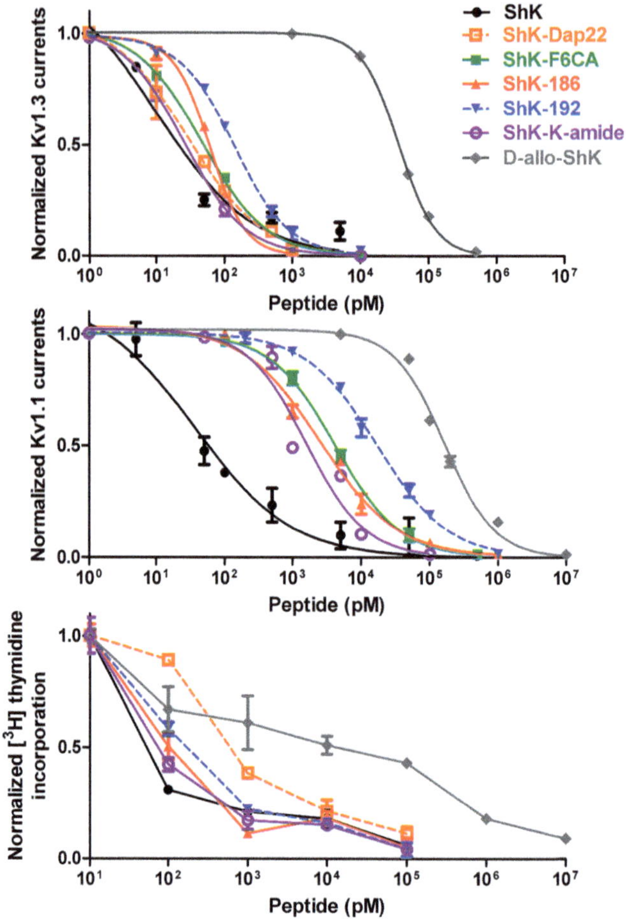

Figure 10.3 Comparison of the ability of ShK and analogues to block Kv1.3 currents (top panel) and Kv1.1 currents (middle panel) measured by patch-clamp electrophysiology, as well as proliferation of effector memory T lymphocytes measured by tritiated thymidine incorporation (bottom panel).[17,18,24,27–30,33,37,39,41]

negative in the Ames test on tester strain TA97A, indicating that it is not a mutagen.[28]

ShK-186 is to date the best characterized ShK analogue in terms of *in vitro* biological activity on CCR7$^+$ and CCR7$^-$ T lymphocytes.[18,37] ShK-186 (100 nM) did not affect localization of Kv1.3 channels to the immunological synapse formed between CCR7$^-$ T$_{EM}$ lymphocytes and antigen-presenting cells during lymphocyte activation, although it inhibited Ca^{2+} influx in these cells in a dose-dependent manner with an IC$_{50}$ of 200 pM.[18,37] ShK-186 inhibited production of the cytokines IL-2 and IFNγ by CCR7$^-$ T$_{EM}$ lymphocytes more efficiently than in the case of CCR7$^+$ naïve/T$_{CM}$ lymphocytes.[18] It was significantly less effective in inhibiting production of IL-4 and TNFα. ShK-186 persistently inhibited the proliferation of CCR7$^-$ T_{EM} lymphocytes (IC_{50} 100 pM) whereas CCR7$^+$ naïve/T_{CM} lymphocytes were 10-fold less sensitive to the blocker at first, and became completely resistant to ShK-186 after their up-regulation of the KCa3.1 channel.[18,37] ShK-186 also inhibited activation of integrin β1, which plays an important role in the motility of activated CCR7$^-$ T_{EM} lymphocytes.[37]

ShK-192 inhibited the proliferation of human CCR7$^-$ T_{EM} lymphocytes isolated from the synovial fluid of patients with rheumatoid arthritis with an IC_{50} (200 pM) in the same range as that of ShK-186.[29]

D-allo-ShK was significantly less potent than ShK in inhibiting CCR7$^-$ T_{EM} lymphocyte proliferation (IC_{50} 10 nM) due to its lower affinity for the channels. Like ShK, it was even less potent in inhibiting the proliferation of CCR7$^+$ naïve/T_{CM} lymphocytes (IC_{50} 10 μM).[30]

10.4.6 *In vivo* Biological Activity of ShK and Its Analogues

The species of choice for studying the *in vivo* biological activity of ShK and its analogues is rat as its T lymphocytes are very similar to their human counterparts in terms of K^+ channel phenotype and function. In contrast, mouse T lymphocytes do not exhibit this similarity with human T cells.[32] Mice are therefore not relevant for the study of Kv1.3 channel blockers as immunomodulators in studies focused on T lymphocyte-mediated inflammatory diseases.[32]

10.4.7 Pharmacokinetics of ShK and Its Analogues in Rats

ShK and ShK-Dap22 were administered intraperitoneally (i.p.) or subcutaneously (s.c.) to rats and their free serum concentrations determined by patch-clamp electrophysiology at different time-points.[33] Peak free serum concentrations of both peptides (0.1–14 nM) were reached 15 min (i.p.) and 30 min (s.c.) after injection. Their average half-life was approximately 20 min.

In similar studies, free ShK-170 was detected in rat serum 5 min after s.c. injection of 200 μg kg^{-1} body weight.[38] Similar to what was found with ShK and ShK-Dap22, free serum peak levels of ShK-170 (12 nM) were reached within 30 min. The half-life was approximately 50 min and levels of free ShK-170 reached a baseline of approximately 300 pM over a period of 7 h. Injection of only 10 μg kg^{-1} body weight ShK-170 led to a serum peak concentration of 500 pM, a concentration sufficient to block >90% Kv1.3 channels. Repeated daily s.c. injection of 10 μg kg^{-1} ShK-170 resulted in levels of 300 pM 24 h after injection.

Rats that received a single s.c. injection of 100 μg kg^{-1} body weight of ShK-192 achieved free serum levels, detected using patch-clamp electrophysiology, of 3.5 nM ShK-192 within 30 min. The half-life for this peptide was approximately 30 min but ~200 pM of functionally active ShK-192 was detected 24 and 48 h after injection and ~100 pM could be detected at 72 h.[29]

The modifications of ShK-170 used to generate ShK-192 did not lead to an increase in the *in vivo* half-life of the peptide.

D-allo-ShK was administered s.c. to rats (1 mg kg^{-1} body weight) and was detectable by patch-clamp electrophysiology in the serum 25 min later. Peak levels of 950 nM D-allo-ShK were reached within 50 min of injection and its half-life was estimated to be 40 min.[30] This small increase in the half-life of the blocker shows that the major determinant of the short half-life of ShK is not proteolysis but rather elimination through the kidney.

All of the pharmacokinetics studies described above were conducted using patch-clamp electrophysiology to detect free ShK and analogues in rat serum. More recently, Tarcha *et al.*[39] generated an ^{111}In-1,4,7,10-tetraazacyclododecane-1,4,7,10-tetraacetic acid-conjugate of ShK-186, designated ShK-221, to measure whole-blood pharmacokinetics of the peptide using HPLC-MS, as well as its distribution and excretion using single photon emission computed tomography (SPECT) and CT scans. The s.c. injection of ShK-221 showed that it is slowly absorbed from the injection site, yielding blood levels above the concentration required to block Kv1.3 channels for 7 days. As expected, ShK-221 was excreted through the kidneys.[39]

10.4.8 Therapeutic Efficacy of ShK and Its Analogues in Treating Rat Models of Inflammatory Diseases

ShK and ShK-Dap22 were tested in parallel for their efficacy in preventing and treating acute adoptive experimental autoimmune encephalomyelitis (EAE) in rats, an animal model of multiple sclerosis.[33] EAE was induced by i.p. injection of activated myelin-basic protein specific encephalitogenic T_{EM} lymphocytes. The *in vitro* pre-treatment of encephalitogenic T_{EM} cells with 100 nM ShK or ShK-Dap22 prior to i.p. injection, followed by s.c. injections of 80 μg kg^{-1} ShK or ShK-Dap22 on the day of cell transfer and for the next 5 days, prevented lethal EAE. In treatment trials, s.c. injection of 80 μg kg^{-1} ShK or ShK-Dap22 for 3 days, starting on the day of disease onset, was sufficient to significantly reduce disease severity. ShK, at the same dosage that showed efficacy in EAE, significantly reduced allergic contact dermatitis induced in rats with oxazolone.[40] ShK (16.5 μg per graft) also displayed marked therapeutic efficacy in skin from patients with psoriasis grafted onto severe combined immunodeficient mice.[19]

ShK-170 was also effective in preventing a T_{EM}-lymphocyte-mediated active delayed-type hypersensitivity (DTH) reaction and acute adoptive EAE, as well as in treating acute adoptive EAE in rats.[28] Active DTH was induced by immunization of rats with ovalbumin followed by subsequent challenge with ovalbumin. A single daily injection of 10 μg kg^{-1} body weight ShK-170 after challenge was sufficient to significantly reduce ovalbumin-induced inflammation. ShK-170 was also effective in preventing acute adoptive EAE when administered s.c. once daily (10 μg kg^{-1}) for 5 days, starting on the day of encephalitogenic T_{EM} cell injection. More importantly, a 3 day treatment after onset of clinical signs also significantly reduced clinical signs of EAE.

ShK-186 (100 μg kg^{-1} body weight) prevented active DTH induced against ovalbumin in rats.[29] Subsequent studies showed that a single injection of 1 μg kg^{-1} ShK-186 on day of challenge was sufficient to prevent an active DTH reaction.[39] Moreover, a single injection of 100 μg kg^{-1} ShK-186 up to 5 days before challenge was sufficient to significantly reduce an active DTH response. ShK-186 also prevented adoptive DTH induced in rats by i.p. injection of activated ovalbumin-specific CCR7$^-$ T_{EM} lymphocytes and subsequent challenge in the ear with ovalbumin, without affecting the number of ear-infiltrating ovalbumin-specific CCR7$^-$ T_{EM} lymphocytes.[37] The cells did enter the site of inflammation but two-photon imaging of the tissue showed that they were non-motile and were not reactivated by antigen-presenting cells in the presence of ShK-186, at least in part because blocking Kv1.3 channels inhibited activation of integrin β1.[37] The motility of CCR7$^+$ naïve/T_{CM} lymphocytes in the lymph node was not affected by ShK-186.

At 100 μg kg^{-1} body weight, ShK-186 also showed efficacy in treating two chronic autoimmune diseases, pristane-induced arthritis, an animal model of rheumatoid arthritis, and chronic-relapsing EAE, a model of MS.[18,37] Arthritis was induced in rats by s.c. injection of pristane, then rats received ShK-186 or vehicle daily for 21 days, starting on the day they displayed the first clinical sign of arthritis (at least one swollen joint). ShK-186-treated animals had significantly fewer affected joints than vehicle-treated controls and showed improvement in both radiological and histopathological studies.[18] Further studies demonstrated that daily injection of ShK-186 was not necessary as dosage on alternate days was sufficient to reduce disease severity.[39] Chronic-relapsing EAE (CR-EAE) was induced in rats by a single immunization with rat spinal cord in emulsion with complete Freund's adjuvant. Treatment with ShK-186 (100 μg kg^{-1} body weight, administered s.c. either daily or every 3 days) began on the day of disease onset and continued for 40 days. This treatment had no effect on the first episode of disease, when most T lymphocytes infiltrating the CNS are CCR7$^+$ naïve/T_{CM} cells, but significantly reduced the severity of EAE when the disease entered a chronic phase in which the majority of CNS-infiltrating T lymphocytes are CCR7$^-$ T_{EM} cells.[37,39] Treatment of CR-EAE rats with ShK-186 every other day or every 3 days, starting 7 days after onset of clinical signs, drastically reduced disease severity.[39] In order to evaluate the durability of protection, CR-EAE received 100 μg kg^{-1} ShK-186 or vehicle for 14 days starting 10 days after onset of clinical signs. The disease was significantly reduced by ShK-186 during the treatment period and clinical signs increased to levels observed in the vehicle-treated group 5 days after cessation of treatment.[39]

ShK-192, administered s.c. at doses of 1, 10, or 100 μg kg^{-1} body weight, significantly reduced inflammation in active DTH induced against ovalbumin in rats.[29] D-allo-ShK, at a dose of 1 mg kg^{-1} body weight administered s.c. at the time of challenge with ovalbumin, significantly suppressed an active DTH reaction.[30]

10.4.9 Toxicity Studies of ShK and Its Analogues in Mice and Rats

ShK and ShK-Dap22 displayed minimal toxicity in mice, with the median paralytic dose following intravenous injection being 25 mg kg^{-1} body weight for ShK and 200 mg kg^{-1} body weight for ShK-Dap22.[4]

A major concern when developing K^+ channel blockers as therapeutics is the potential risk of blocking cardiac Kv11.1 (hERG) channels, thereby inducing potentially fatal arrhythmias. Although ShK-170 did not affect Kv11.1 channels at concentrations up to 100 nM, it was nevertheless tested in a heart variability assay in conscious rats at 10 μg kg^{-1} body weight. No effect was observed on heart rate or on standard heart rate variability parameters in either the time or frequency domain.[28] Administration of ShK-170 (10 μg kg^{-1} d^{-1}) for 2 weeks induced no differences between vehicle and blocker-treated rats in terms of blood counts, blood chemistry, and proportion of thymus and spleen lymphocyte subsets.[28] Injection of a single 1000 μg kg^{-1} dose of ShK-170 or of five consecutive daily doses of 600 μg kg^{-1} ShK-170 induced no overt toxicity in healthy rats.[28] However, the dose that induces 50% lethality (LD_{50}) was estimated to be 750 μg kg^{-1} d^{-1} in rats with EAE and a breached blood–brain barrier, probably due to a sufficient blood concentration of ShK-170 (12 nM) to block >50% of neuronal Kv1.1 channels.[28]

Administration of ShK-186 (100 or 500 μg kg^{-1} body weight) s.c. daily for 28 days yielded no differences between ShK-186-treated and vehicle-treated rats in terms of histopathology, blood counts, or blood chemistry, although irritation at the injection site of ShK-186 was noted.[18] In addition, no toxicity was observed in Rhesus macaques treated once intravenously with 100 μg kg^{-1} body weight ShK-186.[18] ShK-186, administered s.c. at a dose that prevented DTH and acute-adoptive EAE and treated acute-adoptive EAE, CR-EAE and pristane-induced arthritis (100 μg kg^{-1} body weight), did not induce generalized immunosuppression.[37] Rats infected intranasally with either rat-adapted influenza virus or *Chlamydia trachomatis MoPn* were able to clear the pathogens as rapidly as rats treated with vehicle alone. In contrast, rats treated with the broad immunosuppressant dexamethasone displayed a delayed clearance of the virus and were unable to clear the bacteria over the 28 days of the trials.[37]

10.5 Into the Clinic

In 2012, ShK-186 satisfactorily completed Phase Ia clinical trials directed by Kineta, Inc. (Seattle, WA) to establish the safety, tolerability, and pharmacokinetics of a single dose (Shawn Iadonato, personal communication). A Phase Ib trial of ShK-186 in psoriasis and psoriatic arthritis will commence in Q4, 2014.

10.6 Further Developments

The enormous progress outlined in this chapter has been based entirely on synthetic peptides, and this approach will continue to underpin future work on ShK. Recently, this has been complemented with an efficient bacterial expression system for ShK,[41] which offers the possibility of cheaper production of analogues based on genetically encoded amino acids.

New approaches to calculating the affinities of peptides for various channels, described in Section 10.3.1,[23,24] are also contributing to the design of new ShK analogues that are amenable to recombinant expression.[42]

Findings over the past decade also point to a role of Kv1.3 channels in the regulation of peripheral insulin sensitivity and glucose metabolism.[43,44] In addition to their potential application as immunomodulators, ShK and its analogues may therefore prove effective for the treatment of insulin resistance and type 2 diabetes.

An alternative route for delivering drugs into the systemic circulation is the buccal and/or sublingual mucosa, which could be exploited for systemically delivering ShK and its analogues.[45] In collaboration with Joe Nicolazzo and Ben Boyd (Monash Institute of Pharmaceutical Sciences), we have established a sensitive and accurate *in vitro* assay for ShK analogue transport across the buccal mucosa and are evaluating liposomal formulations for encapsulation of ShK analogues as well as penetration enhancing agents.

Intranasal administration for ShK and its analogues is also of great interest for therapy of MS. Odoardi *et al.*[46] demonstrated recently that T lymphocytes become licensed in the lungs to enter the CNS for induction of EAE. Intranasal delivery of ShK and analogues into the lungs may therefore be sufficient to stop encephalitogenic T_{EM} cell trafficking and treat EAE and MS. Intranasal administration of ShK and analogues may also provide a strategy to deliver the peptides directly into the CNS *via* the olfactory neurons. We have already successfully introduced an ShK-186 analogue labelled with a contrast agent for detection by magnetic resonance imaging (MRI) in rodent brains and are pursuing this administration strategy in a rat model of MS.

10.7 Conflict of Interest Statement

MWP and CB are inventors on a patent claiming ShK analogues for immunomodulation. This patent was licensed to Kineta Inc. for developing ShK-186 as a therapeutic for autoimmune diseases. MWP and CB are consultants to Kineta Inc.

Acknowledgements

We are indebted to Prof. George Chandy (University of California, Irvine) for his significant contributions to the development of ShK as a potential therapeutic. The authors' work was supported in part by the National Institutes of Health (NS073712), the Australian Research Council (DP1093909), and the Australian National Health and Medical Research Council (575502 and APP1042481).

References

1. O. Castaneda, V. Sotolongo, A. M. Amor, R. Stocklin, A. J. Anderson, A. L. Harvey, A. Engstrom, C. Wernstedt and E. Karlsson, *Toxicon,* 1995, **33**, 603–613.
2. M. W. Pennington, M. E. Byrnes, I. Zaydenberg, I. Khaytin, J. de Chastonay, D. S. Krafte, R. Hill, V. M. Mahnir, W. A. Volberg and W. Gorczyca, *et al., Int. J. Pept. Protein Res.,* 1995, **46**, 354–358.
3. J. Pohl, F. Hubalek, M. E. Byrnes, K. R. Nielsen, A. Woods and M. W. Pennington, *Lett. Peptide Sci.,* 1995, **1**, 291–297.
4. K. Kalman, M. W. Pennington, M. D. Lanigan, A. Nguyen, H. Rauer, V. Mahnir, K. Paschetto, W. R. Kem, S. Grissmer, G. A. Gutman, E. P. Christian, M. D. Cahalan, R. S. Norton and K. G. Chandy, *J. Biol. Chem.,* 1998, **273**, 32697–32707.
5. M. D. Lanigan, K. Kalman, Y. Lefievre, M. W. Pennington, K. G. Chandy and R. S. Norton, *Biochemistry,* 2002, **41**, 11963–11971.
6. V. Chi, M. W. Pennington, R. S. Norton, E. J. Tarcha, L. M. Londono, B. Sims-Fahey, S. K. Upadhyay, J. T. Lakey, S. Iadonato, H. Wulff, C. Beeton and K. G. Chandy, *Toxicon,* 2012, **59**, 529–546.
7. M. W. Pennington, V. M. Mahnir, I. Khaytin, I. Zaydenberg, M. E. Byrnes and W. R. Kem, *Biochemistry,* 1996, **35**, 16407–16411.
8. H. Rauer, M. Pennington, M. Cahalan and K. G. Chandy, *J. Biol. Chem.,* 1999, **274**, 21885–21892.
9. S. Rangaraju, K. K. Khoo, Z. P. Feng, G. Crossley, D. Nugent, I. Khaytin, V. Chi, C. Pham, P. Calabresi, M. W. Pennington, R. S. Norton and K. G. Chandy, *J. Biol. Chem.,* 2010, **285**, 9124–9136.
10. M. Guo, M. Teng, L. Niu, Q. Liu, Q. Huang and Q. Hao, *J. Biol. Chem.,* 2005, **280**, 12405–12412.
11. Q. Zhou, Q. L. Wang, X. Meng, Y. Shu, T. Jiang, T. Wagenknecht, C. C. Yin, S. F. Sui and Z. Liu, *Biophys. J.,* 2008, **95**, 4289–4299.
12. J. E. Tudor, P. K. Pallaghy, M. W. Pennington and R. S. Norton, *Nat. Struct. Biol.,* 1996, **3**, 317–320.
13. J. E. Tudor, M. W. Pennington and R. S. Norton, *Eur. J. Biochem.,* 1998, **251**, 133–141.
14. R. S. Norton, M. W. Pennington and H. Wulff, *Curr. Med. Chem.,* 2004, **11**, 3041–3052.

15. K. G. Chandy, H. Wulff, C. Beeton, M. Pennington, G. A. Gutman and M. D. Cahalan, *Trends Pharmacol. Sci.*, 2004, **25**, 280–289.
16. F. Sallusto, D. Lenig, R. Forster, M. Lipp and A. Lanzavecchia, *Nature*, 1999, **401**, 708–712.
17. H. Wulff, P. A. Calabresi, R. Allie, S. Yun, M. Pennington, C. Beeton and K. G. Chandy, *J. Clin. Invest.*, 2003, **111**, 1703–1713.
18. C. Beeton, H. Wulff, N. E. Standifer, P. Azam, K. M. Mullen, M. W. Pennington, A. Kolski-Andreaco, E. Wei, A. Grino, D. R. Counts, P. H. Wang, C. J. LeeHealey, S. A. B. , A. Sankaranarayanan, D. Homerick, W. W. Roeck, J. Tehranzadeh, K. L. Stanhope, P. Zimin, P. J. Havel, S. Griffey, H. G. Knaus, G. T. Nepom, G. A. Gutman, P. A. Calabresi and K. G. Chandy, *Proc. Natl. Acad. Sci. U.S.A.*, 2006, **103**, 17414–17419.
19. A. Gilhar, R. Bergman, B. Assay, Y. Ullmann and A. Etzioni, *J. Investig. Dermatol.*, 2011, **131**, 118–124.
20. J. C. Kath, D. C. Hanson and K. G. Chandy, *Annu. Rep. Med. Chem.*, 1997, **32**, 181–190.
21. M. W. Pennington, V. M. Mahnir, D. S. Krafte, I. Zaydenberg, M. E. Byrnes, I. Khaytin, K. Crowley and W. R. Kem, *Biochem. Biophys. Res. Commun.*, 1996, **219**, 696–701.
22. C. Dominguez, R. Boelens and A. M. Bonvin, *J. Am. Chem. Soc.*, 2003, **125**, 1731–1737.
23. M. H. Rashid and S. Kuyucak, *J. Phys. Chem. B*, 2012, **116**, 4812–4822.
24. M. W. Pennington, M. Harunur Rashid, R. B. Tajhya, C. Beeton, S. Kuyucak and R. S. Norton, *FEBS Lett.*, 2012, **586**, 3996–4001.
25. L. Yan, J. Herrington, E. Goldberg, P. M. Dulski, R. M. Bugianesi, R. S. Slaughter, P. Banerjee, R. M. Brochu, B. T. Priest, G. J. Kaczorowski, B. Rudy and M. L. Garcia, *Mol. Pharmacol.*, 2005, **67**, 1513–1521.
26. R. E. Middleton, M. Sanchez, A. R. Linde, R. M. Bugianesi, G. Dai, J. P. Felix, S. L. Koprak, M. J. Staruch, M. Bruguera, R. Cox, A. Ghosh, J. Hwang, S. Jones, M. Kohler, R. S. Slaughter, O. B. McManus, G. J. Kaczorowski and M. L. Garcia, *Biochemistry*, 2003, **42**, 13698–13707.
27. C. Beeton, H. Wulff, S. Singh, S. Botsko, G. Crossley, G. A. Gutman, M. D. Cahalan, M. Pennington and K. G. Chandy, *J. Biol. Chem.*, 2003, **278**, 9928–9937.
28. C. Beeton, M. W. Pennington, H. Wulff, S. Singh, D. Nugent, G. Crossley, I. Khaytin, P. A. Calabresi, C. Y. Chen, G. A. Gutman and K. G. Chandy, *Mol. Pharmacol.*, 2005, **67**, 1369–1381.
29. M. W. Pennington, C. Beeton, C. A. Galea, B. J. Smith, V. Chi, K. P. Monaghan, A. Garcia, S. Rangaraju, A. Giuffrida, D. Plank, G. Crossley, D. Nugent, I. Khaytin, Y. Lefievre, I. Peshenko, C. Dixon, S. Chauhan, A. Orzel, T. Inoue, X. Hu, R. V. Moore, R. S. Norton and K. G. Chandy, *Mol. Pharmacol.*, 2009, **75**, 762–773.

30. C. Beeton, B. J. Smith, J. K. Sabo, G. Crossley, D. Nugent, I. Khaytin, V. Chi, K. G. Chandy, M. W. Pennington and R. S. Norton, *J. Biol. Chem.*, 2008, **283**, 988–997.
31. W. Leuchtenberger, K. Huthmacher and K. Drauz, *Appl. Microbiol. Biotechnol.*, 2005, **69**, 1–8.
32. C. Beeton and K. G. Chandy, *Neuroscientist*, 2005, **11**, 550–562.
33. C. Beeton, H. Wulff, J. Barbaria, O. Clot-Faybesse, M. Pennington, D. Bernard, M. D. Cahalan, K. G. Chandy and E. Beraud, *Proc. Natl. Acad. Sci. U.S.A.*, 2001, **98**, 13942–13947.
34. C. Beeton, J. Barbaria, P. Giraud, J. Devaux, A. M. Benoliel, M. Gola, J. M. Sabatier, D. Bernard, M. Crest and E. Beraud, *J. Immunol.*, 2001, **166**, 936–944.
35. K. G. Chandy, T. E. DeCoursey, M. D. Cahalan, C. McLaughlin and S. Gupta, *J. Exp. Med.*, 1984, **160**, 369–385.
36. G. C. Koo, J. T. Blake, K. Shah, M. J. Staruch, F. Dumont, D. Wunderler, M. Sanchez, O. B. McManus, A. Sirotina-Meisher, P. Fischer, R. C. Boltz, M. A. Goetz, R. Baker, J. Bao, F. Kayser, K. M. Rupprecht, W. H. Parsons, X. C. Tong, I. E. Ita, J. Pivnichny, S. Vincent, P. Cunningham, D. Hora, Jr., W. Feeney and G. Kaczorowski, *et al.*, *Cell. Immunol.*, 1999, **197**, 99–107.
37. M. P. Matheu, C. Beeton, A. Garcia, V. Chi, S. Rangaraju, O. Safrina, K. Monaghan, M. I. Uemura, D. Li, S. Pal, L. M. de la Maza, E. Monuki, A. Flugel, M. W. Pennington, I. Parker, K. G. Chandy and M. D. Cahalan, *Immunity*, 2008, **29**, 602–614.
38. C. Beeton, M. W. Pennington, H. Wulff, S. Singh, D. Nugent, G. Crossley, I. Khaytin, P. A. Calabresi, C.-Y. Chen, G. A. Gutman and K. G. Chandy, *Mol. Pharmacol.*, 2005, **67**, 1369–1381.
39. E. J. Tarcha, V. Chi, E. J. Munoz-Elias, D. Bailey, L. M. Londono, S. K. Upadhyay, K. Norton, A. Banks, I. Tjong, H. Nguyen, X. Hu, G. W. Ruppert, S. E. Boley, R. Slauter, J. Sams, B. Knapp, D. Kentala, Z. Hansen, M. W. Pennington, C. Beeton, K. G. Chandy and S. P. Iadonato, *J. Pharmacol. Exp. Ther.*, 2012, **342**, 642–653.
40. P. Azam, A. Sankaranarayanan, D. Homerick, S. Griffey and H. Wulff, *J. Investig. Dermatol.*, 2007, **127**, 1419–1429.
41. S. C. Chang, C. A. Galea, E. W. Leung, R. B. Tajhya, C. Beeton, M. W. Pennington and R. S. Norton, *Toxicon*, 2012, **60**, 840–850.
42. M. H. Rashid, G. Heinzelmann, R. Huq, R. B. Tajhya, S. C. Chang, S. Chhabra, M. W. Pennington, C. Beeton, R. S. Norton and S. Kuyucak, *PLoS One*, **8**, e78712.
43. G. V. Desir, *Expert Opin. Ther. Targets*, 2005, **9**, 571–579.
44. B. H. Choi and S. J. Hahn, *Acta Pharmacol. Sin.*, 2010, **31**, 1031–1035.
45. J. A. Nicolazzo, B. L. Reed and B. C. Finnin, *J. Controlled Release*, 2005, **105**, 1–15.
46. F. Odoardi, C. Sie, K. Streyl, V. K. Ulaganathan, C. Schlager, D. Lodygin, K. Heckelsmiller, W. Nietfeld, J. Ellwart, W. E. Klinkert, C. Lottaz, M. Nosov, V. Brinkmann, R. Spang, H. Lehrach, M. Vingron, H. Wekerle, C. Flugel-Koch and A. Flugel, *Nature*, 2012, **488**, 675–679.

CHAPTER 11

Engineering Venom Peptides to Improve Their Stability and Bioavailability

MUHARREM AKCAN AND DAVID J. CRAIK*

Institute for Molecular Bioscience, The University of Queensland, Brisbane, QLD 4072, Australia
*E-mail: d.craik@imb.uq.edu.au

11.1 Introduction

Peptides are widely recognized as having great promise as drug leads because of their exquisite potency and selectivity for target receptors, but they have not yet widely populated the mainstream pharmaceutical market because of shortcomings in their biopharmaceutical properties. However, a growing interest in the use of peptides as therapeutics has increased the focus of researchers and drug companies to develop new technologies to enhance the stability and bioavailability of peptides, which typically have short half-lives *in vivo* and lower oral bioavailabilty than traditional small molecule drugs that comprise the majority of the pharmaceutical market.[1–4] Although disulfide-rich venom peptides obviate some of the traditional shortcomings of linear peptides, such as susceptibility to proteases, their poor bioavailability and rapid renal clearance are still significant issues. A variety of approaches have been used to enhance the stability and bioavailability of peptides in general, including *N*-methylation, cyclization, replacement of disulfide bonds with diselenide bridges, or coupling to larger proteins or

RSC Drug Discovery Series No. 42
Venoms to Drugs: Venom as a Source for the Development of Human Therapeutics
Edited by Glenn F. King

Published by the Royal Society of Chemistry, www.rsc.org

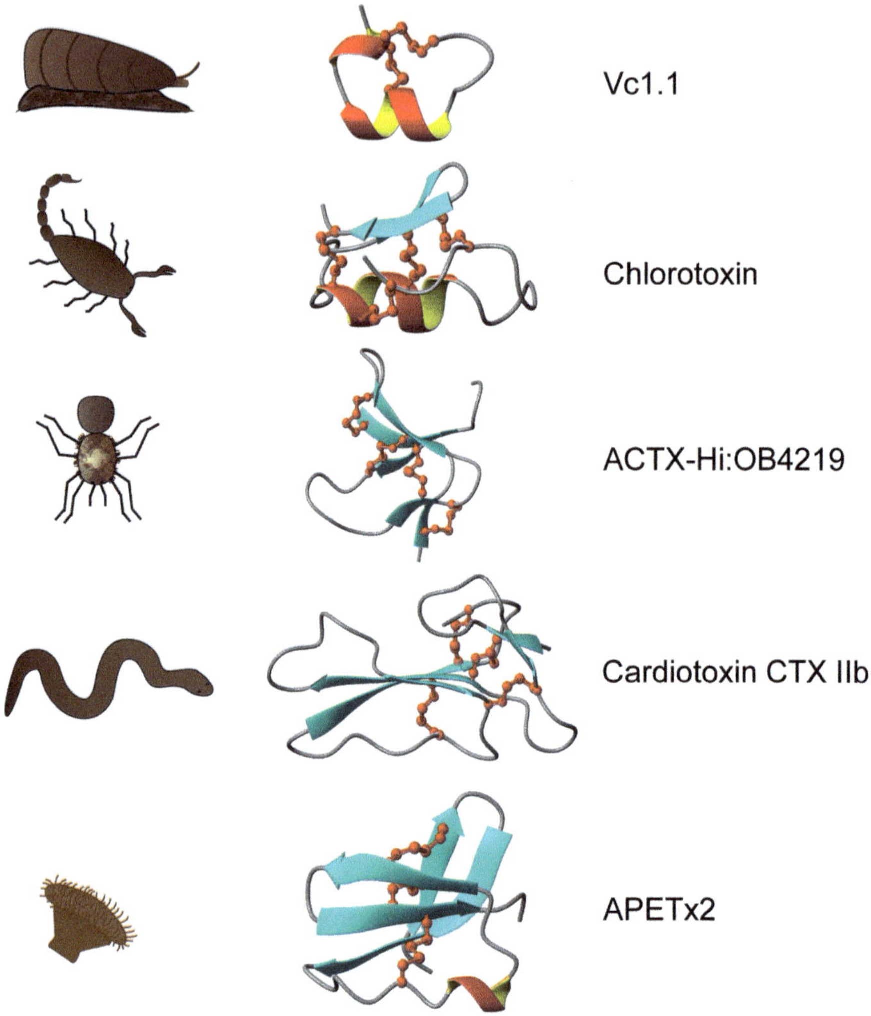

Figure 11.1 Schematic representation of the sources of venom peptides described in this chapter.

PEGylation to minimize the rate of renal clearance.[5–9] In this chapter, we focus on just one aspect of the re-engineering of disulfide-rich venom peptides to improve their stability and bioavailability, namely head-to-tail backbone cyclization. Figure 11.1 summarizes some of the target classes of venom peptides to which cyclization technology has been applied.

Our inspiration for making cyclic disulfide-rich peptides came from recent discoveries of naturally occurring cyclic peptides. Over the last 15 years, head-to-tail cyclic peptides with potential drug design applications have been discovered in bacteria, plants and animals.[10,11] Figure 11.2 shows the

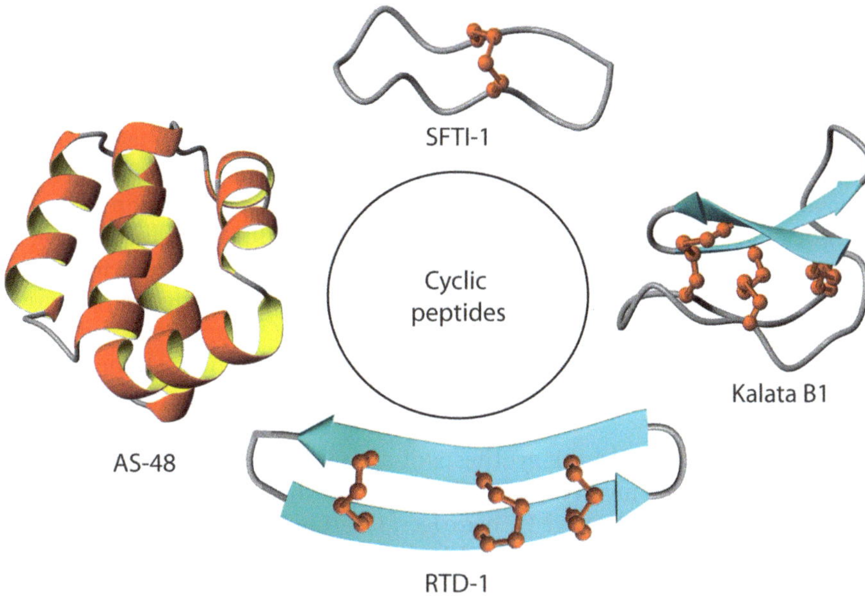

Figure 11.2 Structures of representative examples of naturally occurring cyclic peptides from bacteria (AS-48), plants (SFTI-1 and kalata B1) and animals (RTD-1).

three-dimensional structures of some examples of these topologically circular peptides, including bacteriocin AS-48 from bacteria,[12] SFTI-1 from sunflower seeds,[13] cyclotides[14] from the Rubiaceae, Violaceae, Fabaceae, and Cucurbitaceae plant families and RTD-1 from rhesus monkeys.[15] The most striking common features of these cyclic peptides are their well-defined rigid tertiary structures and exceptional stabilities against enzyme digestion or thermal denaturation.[16] We reasoned that it might be possible to engender other peptides, including venom peptides, with similar stability *via* cyclization of their backbones and this has been the focus of our laboratory for the last few years. In the following sections we describe the synthesis of these cyclized peptides using examples from cone snail, spider, scorpion, snake and sea anemone venoms.

11.2 Synthesis

Solid phase peptide synthesis (SPPS) has been the main approach used to produce peptides since it was first developed by Merrifield in 1963.[17] It involves the attachment of amino acids in a peptide sequence one by one onto a polymer support, which can be removed after assembling the whole peptide sequence from C- to N-terminus. Attachment of the amino acid residues having either *tert*-butoxycarbonyl (Boc) or *N*-(9-fluoroenyl)methoxycarbonyl (Fmoc) protecting groups is achieved by activation and coupling steps and then a deprotection step with trifluoroacetic acid (TFA) or piperidine depending on the protecting group, Boc or Fmoc, respectively.

Conventional SPPS works well for peptides up to about 50 amino acids in size, but is less practical for larger peptides. The native chemical ligation (NCL) method[18] was introduced in 1994 to facilitate the splicing of two peptide chains together so as to make longer peptides more efficiently. In this method, a peptide bond is produced following a reaction between a thioester linker at the C-terminus of one peptide and a cysteine residue at the N-terminus of another peptide. A facile rearrangement reaction takes place following the initial transthioesterification, leading to a native peptide bond. An intramolecular version of the NCL strategy has led to the synthesis of cyclic peptides in which the thioester and the cysteine residue at the C- and N-termini, respectively, are on the same peptide chain together.[19–21]

Figure 11.3 illustrates the synthesis of a cyclotide by this intramolecular NCL approach. It involves the assembly of the peptide sequence on a polymer resin and cleavage by hydrogen fluoride. The intramolecular NCL to produce the cyclic backbone and formation of the three disulfide bonds are achieved in one pot in a buffer solution containing 0.1 M ammonium bicarbonate (NH_4HCO_3)/isopropanol (iPrOH) in 50/50 ratio (v/v) at pH 8.5. This strategy has been used to synthesize the cyclic venom peptides discussed below.

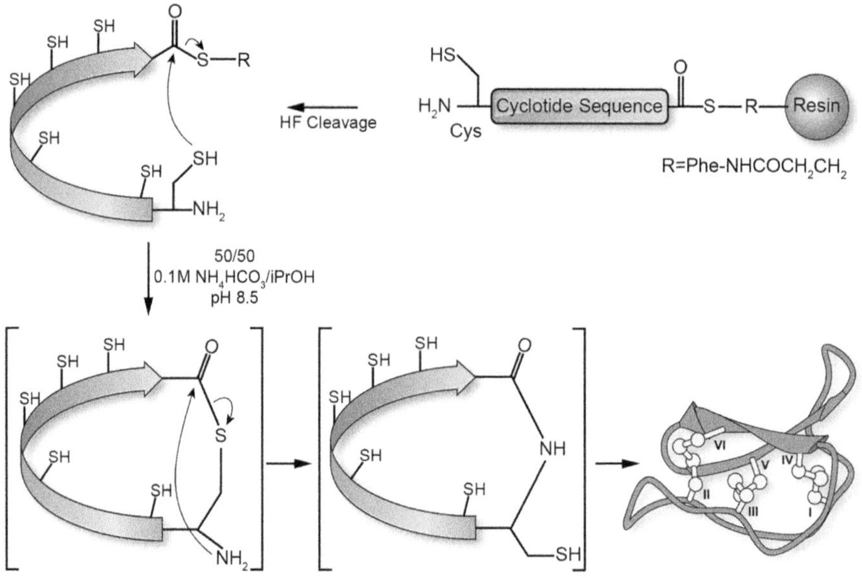

Figure 11.3 Schematic overview of the use of intramolecular native chemical ligation to synthesize cyclic disulfide-rich peptides, in this case exemplified by a cyclotide synthesis.

11.3 Selected Examples

Table 11.1 lists a selection of venom peptides that have been backbone cyclized in recent studies from our laboratory and others. They include cyclic versions of conotoxins MII, MrIA, ImI, Vc1.1, AuIB, and RgIA, a scorpion venom peptide (chlorotoxin) and a sea anemone peptide (APETx2). As noted in other chapters in this book, these peptides have featured prominently in drug discovery efforts over recent years. In many cases their termini are naturally relatively close and can be bridged with linkers of just a few amino acids. Nuclear magnetic resonance (NMR) spectroscopy has played an important role in defining these structures, as these molecules are typically recalcitrant to crystallization, making them not amenable for X-ray structure determination. For example, of the many conotoxin structures reported, most have been done by NMR,[22,23] with only a handful determined using X-ray crystallography.[24]

Figure 11.4 shows a selection of amino acid linkers that have been used to span the distance between the N- and C-termini of the various venom peptides to preserve the native peptide fold. The effect of the length of the linker on the folding and activity of the peptides has been extensively studied and is critical to achieving correct folding. In most cases simple amino acids such as Ala or Gly have been used as linker constituents to avoid introducing any additional functionality into the cyclic peptide, but there is great scope for linker modifications in future studies. In general we avoid continuous long segments of the same amino acid in a linker (*e.g.*, poly Gly or poly Ala) so that NMR spectral characterization is not complicated by degeneracy of signals.

11.3.1 Cyclic MII

Conotoxin MII was the first member of the conotoxin family engineered by backbone cyclization to increase the resistance of venom peptides to proteolytic degradation.[25] MII is a 16-residue α-conotoxin from the cone snail *Conus magus* that targets α3β2 nicotinic acetylcholine receptors (nAChRs) and has been well characterized structurally and pharmacologically.[26–29] Three cyclic analogues with five (cMII-5), six (cMII-6) or seven (cMII-7) residue linkers (GGAAG, GGAAGG and GAGAAGG) were synthesized by intramolecular NCL chemistry.[25] The cyclic analogues cMII-6 and cMII-7 preserved their native peptide fold, whereas cMII-5 showed random coil NMR chemical shift values, which were indicative of a more flexible structure than the other two cyclic analogues.

The cyclic peptides were tested on nicotine-evoked currents in bovine adrenal chromaffin cells and it was found that cMII-5 did not inhibit the nicotine-evoked currents. On the other hand, cMII-7 showed similar native MII activity and cMII-6 was slightly less active than the native MII. The study thus confirmed that with careful choice of linker it is possible to produce a cyclic peptide having the same structure and activity as the acyclic parent.

Table 11.1 Backbone cyclized venom peptides

Peptide	*Sequence*	*Linker*	*References*
MII	GCCSNPVCHLEHSNLC	GGAAG, GGAAGG, GAGAAGG	29
MrIA	NGVCCGYKLCHOC	AG	34
ImI	GCCSDPRCAWRC	A, βA, AG, AGG	38
Vc1.1	GCCSDPRCNYDHPEIC	GGAAGG	45
AuIB	GCCSYPPCFATNPDC	A, AG, AGG, GGAA, AGGG, AGAGA, GGAAG, GGAAGG, GAGAAG, GGAGGAG, GGAAAGG	49,50
Rg1A	GCCSDPRCRYRC	GAA, GAAG, GAAGG, GGAAGG, GGAAGAG	51
ClTx	MCMPCFTTDHQMARKCDD CCGGKGRGKCYGPQCLCR	GAGAAGG	61
APETx2	GTACSCGNSKGIYWFYRPSCP TDRGYTGSCRYFLGTCCTPAD	GASGSA, AGASGSA, SAGASGSA	63

This result was very encouraging but the main aim of the study was to determine if cyclization produced any improvement in biopharmaceutical properties such as resistance to proteolytic degradation.

To examine the resistance to degradation, the cyclic peptides were incubated in EndoGluC or human serum at 37 °C over 24 h. Gratifyingly, the proteolytic stabilities of the cyclic analogues were improved over the acyclic parent and followed a similar trend as the biological data whereby cMII-7 has the highest stability and cMII-5 is the least stable. This study thus provided clear proof-of-concept that it is possible to engineer cyclic venom peptides that have improved stability but no loss of biological activity.

11.3.2 Cyclic MrIA

MrIA is a 13-residue disulfide-rich peptide isolated from the venom of the cone snail *Conus marmoreus*, which selectively inhibits the norepinephrine transporter (NET).[30,31] It is a member of the χ-conotoxin family and while having two disulfide bonds like MII, it has a different disulfide connectivity, which is often referred to as the ribbon connectivity.[32] A more stable synthetic version of MrIA, Xen2174, with a pyroglutamate residue instead of an asparagine at the N-terminus is under Phase II clinical trials for the treatment of neuropathic pain by Xenome Ltd.[33] It was of interest to see if cyclization might provide an alternative approach to stabilising this molecule, which is targeted for delivery *via* the intrathecal route (*i.e.*, direct injection into the spinal cord).

Because the two termini are close two each other, only a two-residue linker, AG, was required for the cyclization of MrIA.[34] NMR spectroscopy showed that the native peptide fold with a β-hairpin structure was preserved after cyclization. Furthermore, cyclic MrIA has a higher enzymatic stability than the native peptide and preserved the biological activity at the NET. This study

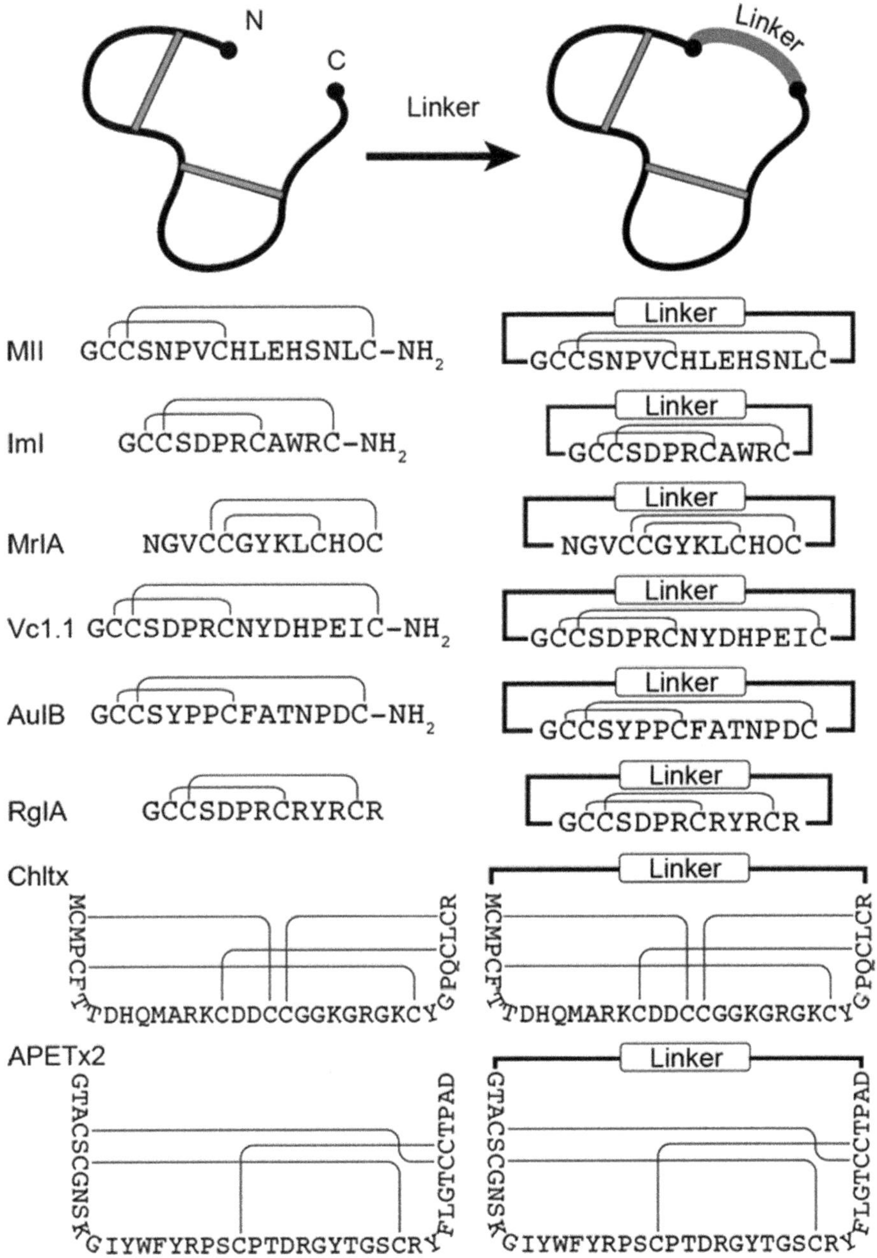

Figure 11.4 Strategy for the cyclization of peptides using a short linker comprising Ala, Gly and Ser residues (*e.g.*, GGAAGS) and sequences of representative venom peptides that have been cyclized using this approach.

thus demonstrated that cyclization might be a general approach for improving stability, given that it worked in both α- and χ-conotoxins.

11.3.3 Cyclic ImI

Conotoxin ImI is a 12-residue peptide isolated from the venom of the cone snail *Conus imperialis*.[35] It has two disulfide bonds in a globular arrangement (Cys2–Cys8, Cys3–Cys12) and only three residues in its second loop, which makes it a 4/3 subclass α-conotoxin, in contrast with MII, which is a 4/7 α-conotoxin. Conotoxin ImI is reported to be a potent inhibitor of α7 and α3β2 nAChRs.[36,37] It has been widely studied and thus provides an excellent model system for examining the effects of linker lengths on structure and activity of cyclic derivatives.

The backbone cyclization of ImI was recently examined by Armishaw *et al.*, who synthesized four different cyclic ImI analogues with A, βA, AG, and AGG residue linkers.[38] Formation of the disulfide bonds without orthogonal cysteine protecting groups resulted in globular, ribbon and beads isomers in different ratios depending on the length of the linker. Oxidation of the cysteines in the cyclic analogue of ImI with a one residue linker (cImI-A) yielded mostly the ribbon isomer (96%), whereas for the cImI-βA, cImI-AG and cImI-AGG analogues, the yield of the ribbon isomers decreased to 41%, 55% and 75%, respectively. Oxidation of both cImI-AG and cImI-AGG analogues also resulted in 26% and 12% beads isomers, respectively, in addition to the globular and ribbon isomers. In this case the beads isomer contains a vicinal disulfide bond, an unusual, but not unknown occurrence.[39]

In terms of the enzymatic stability of the cyclic ImI analogues, all globular isomers except cImI-AGG showed higher stability compared to the wild type ImI (WT-ImI) after incubating with trypsin at 37 °C. By contrast, the ribbon isomers exhibited only comparable stability to WT-ImI. Overall the study confirmed the value of cyclization in improving the stability of native-like isomers of conotoxins, but also showed that cyclization can have an important role in determining the preferred disulfide connectivity, depending on linker length. The study further emphasized the generality of cyclization as a stabilizing strategy, by demonstrating that is applicable to different classes of α-conotoxins (*i.e.*, 4/3 *vs.* 4/7 α-conotoxins).

11.3.4 Cyclic Vc1.1

Conotoxin vc1a was first identified from cDNA sequencing of the venom of the cone snail *Conus victoriae*.[40] The mature peptide is a 16-residue disulfide-rich conotoxin with two post-translationally modified amino acids, a hydroxyproline residue at position 6 and a γ-carboxy glutamic acid at position 14, and an amidated C-terminus (GCCSDORCNYDHPγ-EIC*). Conotoxin Vc1.1 (GCCSDPRCNYDHPEIC*) is the synthetic version of vc1a without these two post-translationally modified amino acids, but having an amidated terminus. It has two disulfide bonds (Cys2–Cys8 and Cys3–Cys16) and belongs to the 4/7 subclass of α-conotoxins in which there are four residues in loop 1 and seven residues in loop 2. Vc1.1 has been reported to target recombinant α9α10 nAChRs expressed in *Xenopus* oocytes,[41,42] but it more potently inhibits N-type calcium channels in dorsal root ganglion (DRG) neurons by activating $GABA_B$ receptors.[43,44] It has attracted significant attention due to its ability to relieve neuropathic pain in rodent models.

In a very significant finding of relevance in peptide-based drug design, the re-engineering of conotoxin Vc1.1 by backbone cyclization led to an orally active peptide.[45] Due to their limited stability and susceptibility to proteolytic breakdown, most peptides require injection to achieve their pharmacological activity. Indeed, all other conotoxins currently in clinical use or in clinical trials require intrathecal injection. In the case of Vc1.1 a six residue linker, GGAAGG, was inserted to join the N- and C-termini together and was found to preserve the native peptide fold.[45] In the rat chronic constriction injury (CCI) model of neuropathic pain, orally delivered cyclic Vc1.1 was 120 times more potent at blocking neuropathic pain than the gold standard clinically used drug gabapentin. By contrast, the parent acyclic peptide was inactive when delivered orally. This study thus demonstrated an additional benefit of cyclization, *i.e.*, the introduction of oral activity to a peptide that was otherwise active only *via* injection.

Interestingly, the cyclic peptide showed higher selectivity on $GABA_B$-mediated calcium channel modulation than in inhibiting α9α10 nAChRs compared to the linear peptide. Thus, in this case, cyclization also resulted in an improvement in the target selectivity, providing yet another benefit of cyclization. As expected based on studies of other cyclic conotoxins, cyclic Vc1.1 showed higher stability in simulated intestinal fluid and human serum than the native (acyclic) Vc1.1. Overall, this study has provided the most compelling evidence so for the benefits of cyclization of venom peptides.

11.3.5 Cyclic AuIB

Conotoxin AuIB is a α3β4 subtype-selective neuronal nAChRs antagonist isolated from *Conus aulicus*.[46,47] In contrast to Vc1.1, AuIB is a 4/6 subclass α-conotoxin with 15 amino acids and is inactive on α9α10 nAChRs. Furthermore, its non-native "ribbon" disulfide isomer (Cys2–Cys15 and Cys3–Cys8) is more potent at the nAChR in rat parasympathetic neurons compared with the native "globular" isomer (Cys2–Cys8 and Cys3–Cys15). Recently, it was also reported that AuIB inhibits high-voltage activated calcium channels *via* $GABA_B$ receptors.[48]

Two studies have been reported on the backbone cyclization of AuIB. In one published by Armishaw *et al.*, seven cyclic analogues with different residue linkers (A, AG, AGG, AGGG, GGAAG, GAGAAG, GGAGGAG) were synthesized and named cAuIB-X, where X is the number of residues in the linker.[49] Cyclic analogues cAuIB-1, -3, and -4 yielded only the globular isomer, whereas the other four cyclic analogues yielded both ribbon and globular isomers in different ratios. In relation to activity, only the cAuIB-2 globular isomer inhibited α3β4 nAChRs. It was reported that all the cyclic ribbon isomers have very low inhibitory activity on α3β4 nAChRs (IC_{50} > 300 μM). cAuIB-7 was the most stable analogue in a chymotrypsin proteolysis assay, compared to other cyclic analogues and native AuIB.

In another study, Lovelace *et al.* used GGAA, AGAGA, GGAAGG and GGAAAGG as linkers to join the ends of AuIB. They reported that cyclization of AuIB made the globular isomers more stable and that ribbon isomers became more resistant to degradation in human serum.[50] Overall, the studies on cyclic AuIB analogues have confirmed the generality of cyclisation as a stabilizing approach, but the benefits do not appear to be as great in 4/6 conotoxins compared to 4/3 or 4/7 conotoxins.

11.3.6 Cyclic RgIA

RgIA is a 4/3 subclass α-conotoxin derived from the venom of the cone snail *Conus regius*. It has 13 residues and, like Vc1.1, blocks α9α10 nAChRs. In contrast with studies on cyclic ImI analogues with three or fewer residue-linkers, backbone cyclization of RgIA was studied with longer (3–7 aa) linkers to delineate the effects of linker length on producing cyclic peptides with a native fold.[51] Five cyclic RgIA analogues (cRgIA) with GAA, GAAG, GAAGG, GGAAGG or GGAAGAG linkers were synthesized using the NCL strategy and the disulfide bonds were formed selectively using protected cysteine residues to obtain the native globular isomers. NMR spectroscopy analysis of the cRgIA analogues showed that peptides with three or four residue-linkers (cRgIA-3 and cRgIA-4) have some αH secondary shift variations from native shifts, suggesting that they have small changes in the peptide fold compared to cRgIA-5, cRgIA-6 and cRgIA-7, which preserved the native fold.

The stability of the peptides was examined in human serum and it was found that proteolytic resistance increased as the linker length increased. The biological activities of the cRgIAs were also consistent with the results from three-dimensional structure and stability studies. The most potent cyclic analogue, cRgIA-7, was found to have comparable inhibitory activity at the human α9/rat α10 nAChR to the native peptide.

11.3.7 Cyclic Chlorotoxin

Chlorotoxin is a 36 amino acid disulfide-rich peptide from the venom of the giant Israeli scorpion *Leiurus quinquestriatus*.[52] It has four disulfide bonds, three of which form an inhibitor cystine knot (ICK) motif that provides a rigid tertiary structure to the molecule. Chlorotoxin was first reported to be a chloride channel blocker, but a range of other potential biological targets have since been identified.[52–54] However, the most fascinating and potentially useful feature of this peptide is that it binds to malignant brain tumours such as gliomas and medulloblastomas.[55,56] This feature has led to the development of new agents for the diagnosis and treatment of brain cancer.[57–59] Specifically, a tumour imaging agent composed of chlorotoxin and a near infrared fluorescent dye, Cy5.5, was developed to enable surgeons to visualize tumour margins during surgical resections.[60]

Cy5.5 can be attached to chlorotoxin through any of its three lysine residues. However, the typical conjugation reaction results in a mixture of mono-, di- and tri-labelled bioconjugates, which is potentially a drawback for commercialization purposes, since homogeneous products are preferred by the United States Food and Drug Administration (FDA). In a recent study aimed at overcoming this deficiency, chlorotoxin was modified in which two of the lysine residues were substituted either with alanine or arginine to produce a mono-labelled tumour imaging agent. Interestingly, an engineered version of chlorotoxin that was modified only by backbone cyclization also resulted in a mono-labelled bioconjugate.[61] Seven residues (GAGAAGG) were used as a linker to join the N- and C-termini and all the three lysine residues were retained in the peptide sequence. Substituted and cyclized chlorotoxin bioconjugates retained their bioactivities and, as expected, cyclized chlorotoxin had a higher serum stability than the native form. Overall, this study showed that it is possible to broaden the use of cyclization technology beyond conotoxins and that the advantages of cyclization extend beyond just increasing stability.

11.3.8 Cyclic APETx2

APETx2 is a 42 residue peptide isolated from the sea anemone *Anthopleura elegantissima*. It blocks acid-sensing ion channel 3 (ASIC3), which is highly expressed in peripheral sensory neurons and plays a key role in pain perception.[62] In a recent study, cyclic versions of APETx2 were designed in which three different sized linkers, GASGSA, AGASGSA and SAGASGSA, were used to span the distance between the N- and C-termini of the peptide.[63] Two-dimensional NMR analysis of the cyclic peptides showed that the backbone ^{15}N chemical shifts overlay with those of the wild-type APETx2, indicating that cyclic APETx2 analogues have the native peptide fold.

Tryptic digestion and simulated gastric fluid (SGF) assays were used to assess the effect of cyclization on the enzymatic stability of peptides and it was found that backbone cyclization enhanced the protease resistance of APETx2. However, based on voltage-clamp electrophysiology experiments on *Xenopus laevis* oocytes expressing homomeric rat ASIC3, the biological activity of the cyclic analogues was reduced compared to the native peptide. It was concluded that the N- and C-termini of APETx2 are critical for inhibition of ASIC3.

This study provides an example showing that cyclization will not necessarily work for all peptides and, in particular, is not applicable for cases when the terminal residues are implicated in activity. We recently found this to be the case also for the human hormone hepcidin, a disulfide-rich cyclic peptide whose N-and C-termini are very close, making it amenable to cyclization, but since its termini are important for bioactivity it is not a good candidate for cyclization.[66]

11.4 Concluding Remarks

Cyclization appears to be a generally useful approach for the stabilization of peptide toxins. Although it does not work in every case, with careful design of the length and content of the linker, it is usually possible to produce improvements in the biopharmaceutical properties of peptides. In all cases studied so far it was possible to introduce improvements in stability and, in most cases, activity was maintained at a similar level to the native peptide. In one case the potency and selectivity were improved but in another activity decreased.

The cyclization studies described herein have been conducted only over the last few years, and there is great scope for applications in a wide range of other venom peptides. Even amongst the heavily studied conotoxins there are still many subfamilies of peptides whose ends are proximate but have not yet been targets for cyclization, including for example the κ or ω conotoxins.[64,65] Likewise, there has been little work so far in attempting to cyclize disulfide-poor conotoxins. One consideration here is that many of these adopt extended helical structures and hence their termini are not necessarily close.[67]

In parallel with studies on the cyclization of venom peptides, there have been increasing discoveries of naturally occurring cyclic peptides in the last decade, and these are likely to provide additional inspiration for new generation of "designer" cyclic peptides. In our opinion cyclic peptides are a grossly under-discovered class of molecules in nature and it is likely that-many new classes of cyclic peptides will become apparent in the coming years. We base this "under-discovered" hypothesis on the fact that it is not easy to find cyclic peptides unless one specifically looks for them—they typically evade detection in genome or transcriptome studies, which have no way of predicting post-translational cyclization events, and they are also highly likely to be missed in proteomic studies because cyclic peptides do not fragment very well on MS analyses. However, with increasing awareness of their advantages, we hope that researchers will be inspired to seek out and discover more of nature's treasured rings.

Acknowledgements

Work in our laboratory on cyclization of peptides is supported by grants from the Australian Research Council (ARC, DP1093115) and the National Health and Medical Research Council (NHMRC, APP1010552; APP1006423). DJC is a NHMRC Senior Principal Research Fellow (APP1026501).

References

1. D. J. Craik, D. P. Fairlie, S. Liras and D. A. Price, *Chem. Biol. Drug Des.*, 2013, **81**, 136–147.
2. W. Danho, J. Swistok, W. Khan, X. J. Chu, A. Cheung, D. Fry, H. Sun, G. Kurylko, L. Rumennik, J. Cefalu, G. Cefalu and P. Nunn, *Adv. Exp. Med. Biol.*, 2009, **611**, 467–469.

3. P. Vlieghe, V. Lisowski, J. Martinez and M. Khrestchatisky, *Drug Discovery Today,* 2010, **15**, 40–56.
4. G. F. King, *Expert Opin. Biol. Ther.,* 2011, **11**, 1469–1484.
5. C. J. Armishaw, N. L. Daly, S. T. Nevin, D. J. Adams, D. J. Craik and P. F. Alewood, *J. Biol. Chem.,* 2006, **281**, 14136–14143.
6. S. Pegoraro, S. Fiori, S. Rudolph-Bohner, T. X. Watanabe and L. Moroder, *J. Mol. Biol.,* 1998, **284**, 779–792.
7. J. M. Harris and R. B. Chess, *Nat. Rev. Drug Discovery,* 2003, **2**, 214–221.
8. T. R. White, C. M. Renzelman, A. C. Rand, T. Rezai, C. M. McEwen, V. M. Gelev, R. A. Turner, R. G. Linington, S. F. Leung, A. S. Kalgutkar, J. N. Bauman, Y. Zhang, S. Liras, D. A. Price, A. M. Mathiowetz, M. P. Jacobson and R. S. Lokey, *Nat. Chem. Biol.,* 2011, **7**, 810–817.
9. J. G. Beck, J. Chatterjee, B. Laufer, M. U. Kiran, A. O. Frank, S. Neubauer, O. Ovadia, S. Greenberg, C. Gilon, A. Hoffman and H. Kessler, *J. Am. Chem. Soc.,* 2012, **134**, 12125–12133.
10. D. J. Craik, *Science,* 2006, **311**, 1563–1564.
11. L. Cascales and D. J. Craik, *Org. Biomol. Chem.,* 2010, **21**, 5035–5047.
12. M. Maqueda, A. Galvez, M. M. Bueno and M. Sanchez-Barrena, *Curr. Protein Pept. Sci.,* 2004, **5**, 399–416.
13. S. Luckett, R. S. Garcia, J. J. Barker, A. V. Konarev, P. R. Shewry, A. R. Clarke and R. L. Brady, *J. Mol. Biol.,* 1999, **290**, 525–533.
14. D. J. Craik, N. L. Daly, T. Bond and C. Waine, *J. Mol. Biol.,* 1999, **294**, 1327–1336.
15. Y.-Q. Tang, J. Yuan, G. Ösapay, K. Ösapay, D. Tran, C. J. Miller, A. J. Ouellette and M. E. Selsted, *Science,* 1999, **286**, 498–502.
16. M. L. Colgrave and D. J. Craik, *Biochemistry,* 2004, **43**, 5965–5975.
17. R. B. Merrifield, *J. Am. Chem. Soc.,* 1963, **85**, 2149–2154.
18. P. E. Dawson, T. W. Muir, I. Clark-Lewis and S. B. Kent, *Science,* 1994, **266**, 776–779.
19. J. P. Tam, Y.-A. Lu and Q. Yu, *J. Am. Chem. Soc.,* 1999, **121**, 4316–4324.
20. R. J. Clark and D. J. Craik, *Biopolymers,* 2010, **94**, 414–422.
21. J. A. Camarero and T. W. Muir, *Chem. Commun.,* 1997, 1369–1370.
22. U. C. Marx, N. L. Daly and D. J. Craik, *Magn. Reson. Chem.,* 2006, **44**, S41–S50.
23. K. J. Rosengren, N. L. Daly and D. J. Craik, *Annu. Rep. NMR Spectrosc.,* 2009, **68**, 89–147.
24. S. H. Hu, J. Gehrmann, P. F. Alewood, D. J. Craik and J. L. Martin, *Biochemistry,* 1997, **36**, 11323–11330.
25. R. J. Clark, H. Fischer, L. Dempster, N. L. Daly, K. J. Rosengren, S. T. Nevin, F. A. Meunier, D. J. Adams and D. J. Craik, *Proc. Natl. Acad. Sci. U.S.A.,* 2005, **102**, 13767–13772.
26. G. E. Cartier, D. Yoshikami, W. R. Gray, S. Luo, B. M. Olivera and J. M. McIntosh, *J. Biol. Chem.,* 1996, **271**, 7522–7528.
27. S. C. Harvey, J. M. McIntosh, G. E. Cartier, F. N. Maddox and C. W. Luetje, *Mol. Pharmacol.,* 1997, **51**, 336–342.
28. D. Everhart, G. E. Cartier, A. Malhotra, A. V. Gomes, J. M. McIntosh and C. W. Luetje, *Biochemistry,* 2004, **43**, 2732–2737.

29. B. B. Carstens, R. J. Clark, N. L. Daly, P. J. Harvey, Q. Kaas and D. J. Craik, *Curr. Pharm. Des.,* 2011, **17**, 4242–4253.
30. I. A. Sharpe, J. Gehrmann, M. L. Loughnan, L. Thomas, D. A. Adams, A. Atkins, E. Palant, D. J. Craik, D. J. Adams, P. F. Alewood and R. J. Lewis, *Nat. Neurosci.,* 2001, **4**, 902–907.
31. I. A. Sharpe, E. Palant, C. I. Schroeder, D. M. Kaye, D. J. Adams, P. F. Alewood and R. J. Lewis, *J. Biol. Chem.,* 2003, **278**, 40317–40323.
32. J. Gehrmann, P. F. Alewood and D. J. Craik, *J. Mol. Biol.,* 1998, **278**, 401–415.
33. C. K. Nielsen, R. J. Lewis, D. Alewood, R. Drinkwater, E. Palant, M. Patterson, T. L. Yaksh, D. McCumber and M. T. Smith, *Pain,* 2005, **118**, 112–124.
34. E. S. Lovelace, C. J. Armishaw, M. L. Colgrave, M. E. Wahlstrom, P. F. Alewood, N. L. Daly and D. J. Craik, *J. Med. Chem.,* 2006, **49**, 6561–6568.
35. J. M. McIntosh, D. Yoshikami, E. Mahe, D. B. Nielsen, J. E. Rivier, W. R. Gray and B. M. Olivera, *J. Biol. Chem.,* 1994, **269**, 16733–16739.
36. P. A. Quiram and S. M. Sine, *J. Biol. Chem.,* 1998, **273**, 11001–11006.
37. M. Ellison, F. Gao, H. L. Wang, S. M. Sine, J. M. McIntosh and B. M. Olivera, *Biochemistry,* 2004, **43**, 16019–16026.
38. C. J. Armishaw, J. L. Dutton, D. J. Craik and P. F. Alewood, *Biopolymers,* 2010, **94**, 307–313.
39. R. Zhang and G. H. Snyder, *J. Biol. Chem.,* 1989, **264**, 18472–18479.
40. D. W. Sandall, N. Satkunanathan, D. A. Keays, M. A. Polidano, X. Liping, V. Pham, J. G. Down, Z. Khalil, B. G. Livett and K. R. Gayler, *Biochemistry,* 2003, **42**, 6904–6911.
41. M. Vincler, S. Wittenauer, R. Parker, M. Ellison, B. M. Olivera and J. M. McIntosh, *Proc. Natl. Acad. Sci. U.S.A.,* 2006, **103**, 17880–17884.
42. S. T. Nevin, R. J. Clark, H. Klimis, M. J. Christie, D. J. Craik and D. J. Adams, *Mol. Pharmacol.,* 2007, **72**, 1406–1410.
43. B. Callaghan, A. Haythornthwaite, G. Berecki, R. J. Clark, D. J. Craik and D. J. Adams, *J. Neurosci.,* 2008, **28**, 10943–10951.
44. B. Callaghan and D. J. Adams, *Channels,* 2010, **4**, 51–54.
45. R. J. Clark, J. Jensen, S. T. Nevin, B. P. Callaghan, D. J. Adams and D. J. Craik, *Angew. Chem. Int. Ed. Engl.,* 2010, **49**, 6545–6548.
46. J. H. Cho, K. H. Mok, B. M. Olivera, J. M. McIntosh, K. H. Park and K. H. Han, *J. Biol. Chem.,* 2000, **275**, 8680–8685.
47. S. Luo, J. M. Kulak, G. E. Cartier, R. B. Jacobsen, D. Yoshikami, B. M. Olivera and J. M. McIntosh, *J. Neurosci.,* 1998, **18**, 8571–8579.
48. H. Klimis, D. J. Adams, B. Callaghan, S. Nevin, P. F. Alewood, C. W. Vaughan, C. A. Mozar and M. J. Christie, *Pain,* 2011, **152**, 259–266.
49. C. J. Armishaw, A. A. Jensen, L. D. Balle, K. C. M. Scott, L. Sorensen and K. Stromgaard, *Antioxid. Redox Sign.,* 2011, **14**, 65–76.
50. E. S. Lovelace, S. Gunasekera, C. Alvarmo, R. J. Clark, S. T. Nevin, A. A. Grishin, D. J. Adams, D. J. Craik and N. L. Daly, *Antioxid. Redox Sign.,* 2011, **14**, 87–95.

51. R. Halai, B. Caaghan, N. L. Daly, R. J. Clark, D. J. Adams and D. J. Craik, *J. Med. Chem.,* 2011, **54**, 6984–6992.
52. J. A. DeBin, J. E. Maggio and G. R. Strichartz, *Am. J. Physiol. Cell Physiol.,* 1993, **264**, C361–C369.
53. J. Deshane, C. C. Garner and H. Sontheimer, *J. Biol. Chem.,* 2003, **278**, 4135–4144.
54. K. Kesavan, J. Ratliff, E. W. Johnson, W. Dahlberg, J. M. Asara, P. Misra, J. V. Frangioni and D. B. Jacoby, *J. Biol. Chem.,* 2010, **285**, 4366–4374.
55. S. A. Lyons, J. O'Neal and H. Sontheimer, *Glia,* 2002, **39**, 162–173.
56. L. Soroceanu, Y. Gillespie, M. B. Khazaeli and H. Sontheimer, *Cancer Res.,* 1998, **58**, 4871–4879.
57. D. C. Hockaday, S. Shen, J. Fiveash, A. Raubitschek, D. Colcher, A. Liu, V. Alvarez and A. N. Mamelak, *J. Nucl. Med.,* 2005, **46**, 580–586.
58. A. N. Mamelak and D. B. Jacoby, *Expert Opin. Drug Delivery,* 2007, **4**, 175–186.
59. C. Sun, C. Fang, Z. Stephen, O. Veiseh, S. Hansen, D. Lee, R. G. Ellenbogen, J. Olson and M. Q. Zhang, *Nanomedicine,* 2008, **3**, 495–505.
60. M. Veiseh, P. Gabikian, S. B. Bahrami, O. Veiseh, M. Zhang, R. C. Hackman, A. C. Ravanpay, M. R. Stroud, Y. Kusuma, S. J. Hansen, D. Kwok, N. M. Munoz, R. W. Sze, W. M. Grady, N. M. Greenberg, R. G. Ellenbogen and J. M. Olson, *Cancer Res.,* 2007, **67**, 6882–6888.
61. M. Akcan, M. R. Stroud, S. J. Hansen, R. J. Clark, N. L. Daly, D. J. Craik and J. M. Olson, *J. Med. Chem.,* 2011, **54**, 782–787.
62. K. A. Sluka, O. C. Winter and J. A. Wemmie, *Curr. Opin. Drug Discovery Dev.,* 2009, **12**, 693–704.
63. J. E. Jensen, M. Mobli, A. Brust, P. F. Alewood, G. F. King and L. D. Rash, *Mar. Drugs,* 2012, **10**, 1511–1527.
64. K. J. Nielsen, L. Thomas, R. J. Lewis, P. F. Alewood and D. J. Craik, *J. Mol. Biol.,* 1996, **263**, 297–310.
65. M. J. Scanlon, D. Naranjo, L. Thomas, P. F. Alewood, R. J. Lewis and D. J. Craik, *Structure,* 1997, **5**, 1585–1597.
66. R. J. Clark, G. C. Preza, C. C. Tan, J. W. A. van Dijk, E. Fung, E. Nemeth, T. Ganz and D. J. Craik, *Biopolymers Peptide Science,* 2013, **100**, 519–526.
67. N. Skjaerbaek, K. J. Nielsen, R. J. Lewis, P. Alewood and D. J. Craik, *J. Biol. Chem.,* 1997, **272**, 2291–2299.

CHAPTER 12

Manufacturing of Venom-Derived Therapeutic Peptides

OLEG WERBITZKY*[a,b] AND MATTHIEU GIRAUD*[a,c]

[a]Lonza, Valais Works, CH-3930 Visp, Switzerland; [b]JTI SA, 1, rue de la Gabelle, CH-1211 Geneva 26, Switzerland; [c]BASF Pharma SA, route du Simplon 1,36, CH-1902 Evionnaz, Switzerland
*E-mail: matthieu.giraud@basf.com; oleg.werbitzky@jti.com

12.1 Introduction

Peptidic natural products, such as neurotoxins, from snakes, lizards, scorpions, spiders, and predatory marine snails comprise millions of unique peptides and countless possibilities for developing life-changing therapies.[1] Researchers first noticed the pharmacological effectiveness of snake venoms in the process of developing antisera. Such investigations in the 1960s resulted in the development of the anti-hypertensive drug captopril,[2] which served as an archetype for future research into the structural information of peptides (proteins) isolated from snake venom. Prialt® (ziconotide) is 1000 times more powerful than morphine but, unlike morphine, is not believed to be addictive. The United States Food and Drug Administration (FDA) approved its use for chronic, intractable pain such as that suffered by people with cancer, AIDS or certain neurological disorders.[3] Ziconotide is delivered directly into fluid surrounding the spinal cord by external or implanted pumps. This drug is a synthetic compound identical to a toxin in the venom of the marine cone snail *Conus magus*. The anti-clotting drug eptifibatide was also developed from snake venom.[4] The dis-integrin barbourin contains a Lys–Gly–Asp (KGD) sequence within

RSC Drug Discovery Series No. 42
Venoms to Drugs: Venom as a Source for the Development of Human Therapeutics
Edited by Glenn F. King

Published by the Royal Society of Chemistry, www.rsc.org

a disulfide ring instead of an RGD sequence. Eptifibatide elicited significant clinical benefits as an adjunctive therapy in patients undergoing selective percutaneous coronary intervention with stent implantation. Heat stable enterotoxin (ST_h) produced by various enteric bacteria has common segments consisting of 13 amino acid residues with three disulfide bridges.[5] STh produced by pathogenic enteric bacteria cause diarrhea in children in developing countries in the tropics. Linaclotide, a 14-residue peptide having the common ST_h region, has been accepted by the FDA in August 2012 for the treatment of gastrointestinal diseases such as chronic constipation (CC) and irritable bowel syndrome (IBS).[6] As a final example, exenatide is an incretin mimetic isolated from the saliva of the Gila monster (*Heloderma suspectum*), a venomous lizard.[7] The original formulation of this peptide, Byetta®, is administered *via* twice daily subcutaneous injections. Bydureon®, approved by the FDA in January 2012, is a new slow-release formulation of exenatide in which the peptide is embedded in biodegradable polymeric poly-(D,L-lactide-co-glycolide; PLGA) microspheres,[8] thus enabling once-weekly injections. More recently, in September 2014 AstraZeneca reported positive results from its Phase III trial (DURATION-NEO-1) of exenatide to treat adult patients with type 2 diabetes. This trial used an investigational formulation of exenatide once-weekly suspension delivered through a single-use autoinjector.

However, due to analytical and synthetic challenges, today only a small fraction of these promising peptide-based neurotoxins from animal venoms have been characterized, and the synthesis of cyclic and/or cysteine-containing peptides remains one of the most challenging tasks, primarily because of the difficulties involved in:

- the selective formation of multiple regioselective disulfide (S–S) bonds
- ensuring that monocylization is preferred over polymerization, while maintaining high productivity (a problem exacerbated by having diluted processes)
- reaching high active pharmaceutical ingredient (API) purity without impacting on cost of goods (COGs)

State-of-the-art methods in the field have been extensively reviewed in recent years.[9,10] The main strategies that have been evolved for the synthesis of Cys-rich peptides can be classified as follows: (i) stepwise regioselective Cys pairings; (ii) convergent strategies based on a combination of statistical oxidation of a minor number of Cys residues and regioselective disulfide formation; (iii) oxidative folding of Cys-rich peptides by exploiting the structural information encoded within the sequence of the target peptides.

Before discussing synthesis strategies for different peptides in detail, it is important to mention that an industrial peptide process will, besides the synthesis itself, typically also include several additional unit operations. In synthetic processes, the last chemical step leading to the crude peptide is generally the side-chain-deprotection step. In some cases of solid-phase peptide synthesis (SPPS) (*e.g.*, synthesis *via* the Boc strategy), the reaction is performed as a part of a combined cleavage/deprotection step. Further process steps include peptide purification, today mostly comprising HPLC and sometimes also membrane techniques, and finally the isolation step, which can consist either of lyophilization, spray drying, or a combination of precipitation, filtration and drying.

From the point of view of the product quality of a peptide API, each of the steps described above is of equal importance, as all of them can contribute to the formation and/or reduction of side-products. Side-products are one of the main characteristic attributes of product quality, either directly by contributing towards the impurity profile of a product, or indirectly, by influencing its overall purity. To develop a competitive industrial process, a process must be looked at in an integral way, as each step in the process can influence other steps. In this way, critical impurities, formed during synthesis, may impact the performance (*e.g.*, yield, productivity) of the purification step. Methods for the development and optimization of the processes for additional chemical steps and for the purification and isolation are not within the scope of this chapter and will not be discussed here.

12.2 Eptifibatide: Comparison of Two Synthesis Approaches—Liquid Phase *versus* Solid Phase Strategy

Eptifibatide was developed by Scarborough and Philipps by mimicking the glycoprotein IIb/IIIa blocker barbourin, which was discovered in the venom of the southeastern pigmy rattlesnake (*Sistrurus miliarius barbouri*).[11,12] Eptifibatide belongs to the class of so-called RGD mimetics that reversibly bind to platelets and inhibit their aggregation, which can occur with certain heart and blood vessel conditions. Eptifibatide was approved by the FDA in 1998.

Eptifibatide is a small, highly modified cyclic 7-mer peptide amide. In addition to its C-terminal amide group, eptifibatide is characterized by a disulfide bridge and by the presence of two unnatural building blocks: 3-thiopropionic acid and L-homoarginine.

Our team at Lonza developed two different synthesis routes for eptifibatide, both of which are suitable for the industrial manufacturing of this target molecule.

The first route consists of a four-step liquid-phase process, starting from two readily available building blocks and characterized by the following retro-synthetic strategy (Figure 12.1):[13]

a. Step 4:
    a. closure of the amide bond between 3-thiopropionic acid and homoarginine
    b. minimal protection: use of homoarginine and tryptophan without side-chain protection.
    c. use of Fmoc protection on homoarginine

b. Step 3:
    a. formation of the S–S bond *via* NPys activation.

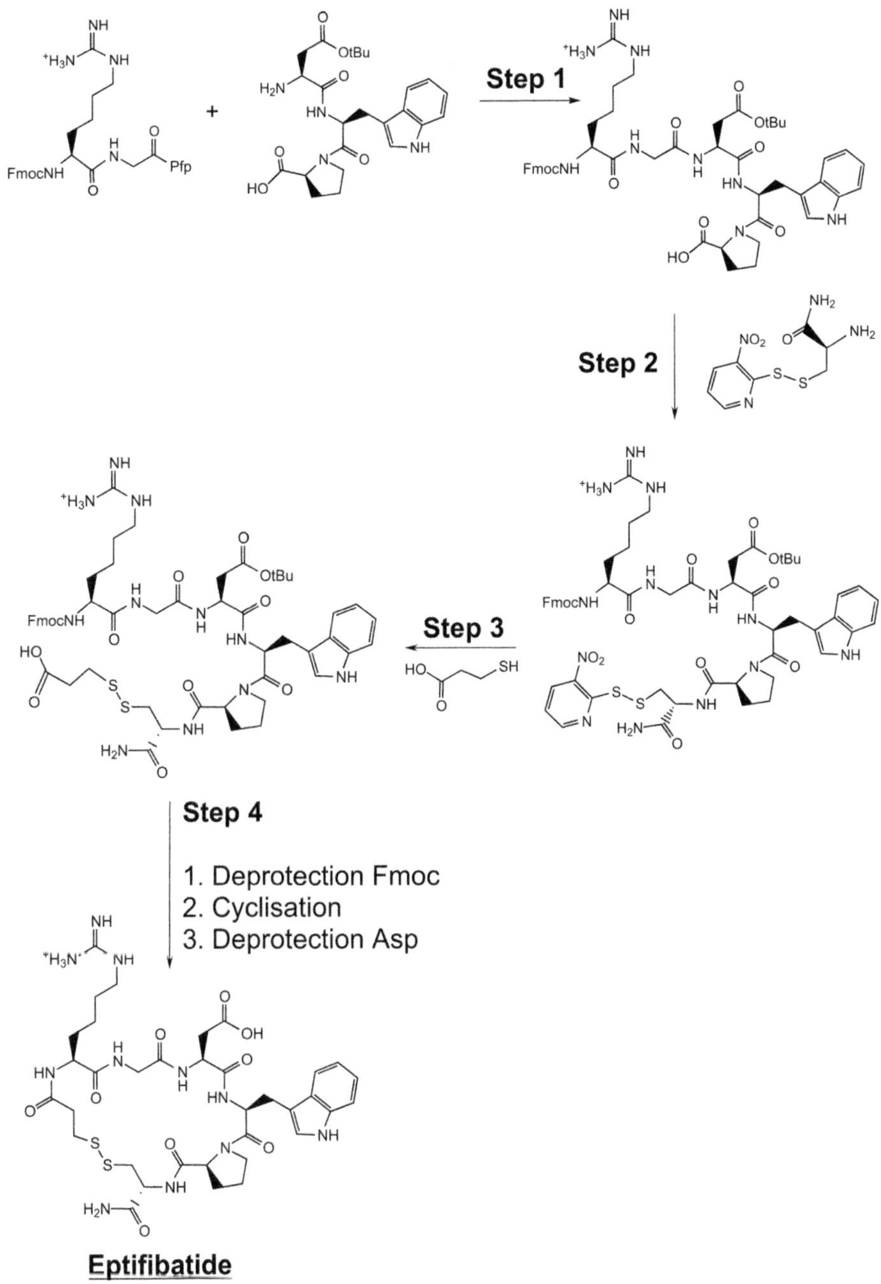

Figure 12.1 Four-step liquid phase process for synthesis of eptifibatide.

c. Step 2:
   a. coupling of the fragment Fmoc-(2–5)-COOH to H_2N-Cys(NPys)-NH_2

d. Step 1:
   a. minimal protection: coupling of fragment Fmoc-(2–3)-COOH to fragment 4–6 with unprotected –COOH

The second eptifibatide synthesis is a SPPS route on a peptide amide support (Figure 12.2).[14] The key step of this solid phase process is the on-resin cyclization of the S–S bridge, which obviates the productivity limitations resulting from the high dilution incurred by a cyclization step in liquid phase. In order to achieve an acceptable selectivity (*i.e.*, desired

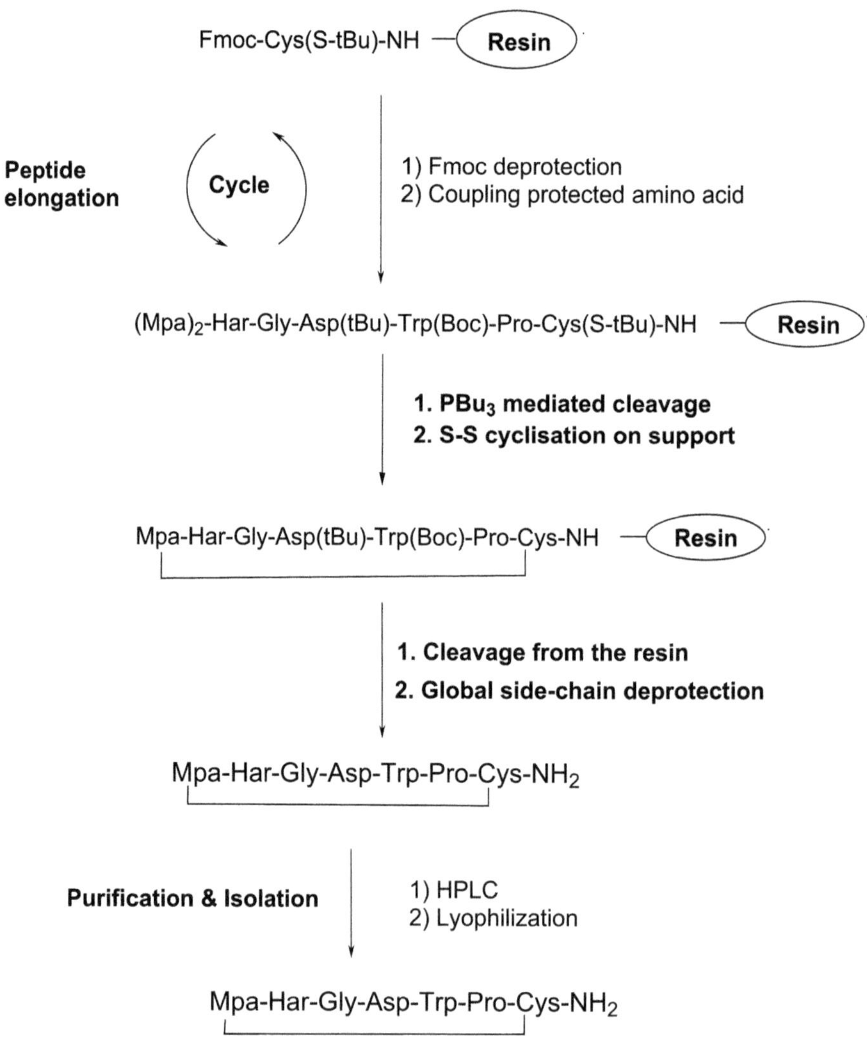

Figure 12.2 Solid-phase process for synthesis of eptifibatide.

intra-molecular reaction *versus* polymerization), a typical S–S cyclization in liquid phase has to be run at concentrations of 0.1% or below. On an industrial scale, such a dilution can easily become the bottleneck of the entire process (see the linaclotide case study below).

Additional features of the SPPS process include:

1. use of cheap 3-thiopropionic acid dimer $(Mpa)_2$
2. S-tBu protection of cysteine, which allows on-resin S–S cyclization
3. use of homoarginine without side-chain protection

A comparison of the two processes shows that both have many characteristics of modern process design concepts: a low number of isolated intermediates, relatively simple and cheap starting materials (*e.g.*, partially protected amino acids, dimeric Mpa), non racemizing fragment coupling steps, and original activation/protection strategies for S–S bond formation. Both processes deliver the desired product in the required quality and achieve high yields. From the point of view of industrial scalability, each of the processes presented above can be used for the industrial manufacturing of eptifibatide. A final choice between the two processes will have to consider strategic aspects, such as availability of optimal equipment to match the market requirements as well as the API yearly demand, which today is estimated to be in the range of multi 10 kg amounts for eptifibatide.

12.3 Ziconotide: Example of a Stepwise Linear SPPS Strategy

Ziconotide, a synthetic 25-residue peptide, is formulated as a preservative-free solution for intrathecal administration for the treatment of severe, chronic pain. The peptide has three S–S bridges that are required for biological activity (Figure 12.3). Its structure is based upon the structure of a naturally occurring peptide toxin called ω-conotoxin MVIIA that is produced by a fish-eating marine cone snail, *Conus magus.* It is a blocker of N-type voltage-gated calcium (Ca_V) channels.

Ca_V channels play a major role in the transmission of pain. The N-type Ca_V channel ($Ca_V2.2$) is found in high concentrations in the central projections of primary sensory neurons that terminate in the dorsal horn of the spinal cord, where it is involved in the spinal processing of pain. Ziconotide selectively and reversibly binds to and blocks these channels without interacting with other ion channels or cholinergic, monoaminergic or μ- and δ-opioid receptors. Ziconotide thus inhibits the spinal signaling of pain. The drug product is a sterile aqueous solution that contains 100 $\mu g\ mL^{-1}$ of ziconotide free base formulated as the acetate salt in saline buffer at pH 4.0–5.0 and containing L-methionine. Dosing of ziconotide is initiated at 2.4 $\mu g\ d^{-1}$ up to a maximum dose of 21.6 $\mu g\ d^{1}$. With such low dosage, the COGs for the API, as for linaclotide (discussed below), should be negligible when compared to drug product cost. The peptide is produced on a kg scale by SPPS with a 10–20% overall yield.

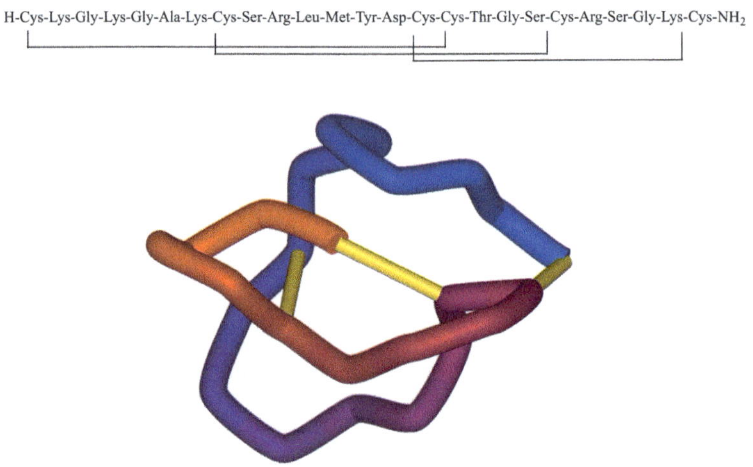

Figure 12.3 Amino acid sequence (top panel) and 3D structure (bottom panel) of ziconotide. Disulfide bonds are shown as yellow tubes and the peptide backbone is coloured from blue at the N-terminus to red at the C-terminus.

12.4 Industrial Synthesis of Exenatide: Example of a Convergent SPPS Strategy

Exenatide is a synthetic GLP-1 receptor agonist. It was approved by the FDA in 2005 as the first incretin mimetic, thereby introducing a new approach for the treatment of diabetes mellitus type 2. Today, exenatide is marketed in two forms: Byetta, a twice-daily subcutaneous injection, and since 2012 also as Bydureon, a once-weekly long-acting release formulation.

Chemically, exenatide is a synthetic version of exendin-4, a naturally occurring peptide, which was isolated from the saliva of the Gila monster.[15] The amino acid sequence of Exendin-4 is 53% identical to mammalian GLP-1, but exendin-4 has the advantage of a significantly longer *in vivo* half-life (*ca.* 3 h *versus* minutes for GLP-1).[16]

Exenatide is a linear 39-residue peptide amide build entirely from natural amino acids. It has the following amino acid sequence: H–His–Gly–Glu–Gly–Thr–Phe–Thr–Ser–Asp–Leu–Ser–Lys–Gln–Met–Glu–Glu–Glu–Ala–Val–Arg–Leu–Phe–Ile–Glu–Trp–Leu–Lys–Asn–Gly–Gly–Pro–Ser–Ser–Gly–Ala–Pro–Pro–Pro–Ser–NH_2.

Initial synthesis concepts proposed for exenatide were all based on the classical and relatively generic linear stepwise SPPS approach using Fmoc/tBu chemistry.[17,18] As discussed above for the example of Prialt®, in a stepwise SPPS synthesis the full-length peptide is built up by coupling single amino acid residues one after the other to a growing peptide chain linked to a solid resin support. Although this synthesis route can be used to obtain samples of exenatide in the laboratory, a stepwise approach is not the appropriate choice for industrial manufacturing of this molecule. Nevertheless, as of today, some custom manufacturing organizations (CMOs) are still using a stepwise manufacturing process to supply small amounts of exenatide (<10 kg yr^{-1}). It is important to consider that with 39 amino acids, exenatide is one of the largest therapeutic peptides produced *via* chemical synthesis and that future yearly API demands are likely to significantly increase. From this perspective, a number of limitations, characteristic for a stepwise synthesis concept, should be mentioned here.

1. Low quality and yield of the crude peptide: In the case of Exenatide, a stepwise synthesis would consist of a sequence of 39 deprotection and 38 coupling reactions, which would all be performed in a row without any purification of intermediates. The chemistry in each of these steps would not be absolutely selective and therefore each chemical step would generate a number of different side-products (deletion products, double incorporation products, isomers, side-products from reactions on side-chains, *etc.*). Longer peptides are especially prone to adopt an irregular conformation while still attached to the solid support, which makes it even more difficult to add additional amino acids to the growing chain. Therefore, the selectivity problem increases with the length of the peptide chain. The great majority of the side-products would be further transformed in the following steps of the elongation, generating a large number of impurities of similar structure to the target compound. All this would not only lead to a modest overall yield, but also deliver a crude product of lower quality, which would be more difficult to purify, thereby impacting also on the yield and the productivity of the HPLC purification step.
2. Limited productivity of stepwise SPPS processes. This results from the combination of two facts. First, in a given SPPS reactor with a certain maximal working volume, only a certain maximal amount of synthesis resin can be used. Second, with currently available synthesis resins, longer peptides can only be made reliably when a relatively low initial loading is used (*e.g.*, initial loading $\leq$0.4 mmol g^{-1}), whereas for shorter peptides, a significantly higher loading can be used (*e.g.*, 1.0 mmol g^{-1}).
3. Inflexible manufacturing concept: On an industrial scale, the elongation of a peptide in SPPS is a rather slow process, with only a limited number of coupling and deprotection cycles per day. For a long target molecule like exenatide, a stepwise synthesis would require a significant amount of time, potentially several weeks. Such a long batch time might not be consistent with the cycle times of the other unit operations in the process (purification, lyophilization) leading to hold times and potentially to idle manufacturing equipment.
4. Risk aspects: This is related to the long synthesis times, but also to the potentially very high value (several millions) of a single batch at the end of the elongation process. This point is particularly critical when taking into account the fact that solid phase synthesis is a heterogeneous

process, which is significantly more difficult to control compared to homogeneous liquid phase chemistry.

A very efficient concept for the industrial manufacturing of long peptides using Fmoc/tBu chemistry, which overcomes the drawbacks of the stepwise approach described above, is convergent SPPS synthesis. This concept was first developed and introduced at commercial large-scale by B. Bray and his team at Trimeris for the manufacturing of enfuvirtide (T-20 or Fuzeon), a membrane fusion inhibitor for the treatment of HIV.[19]

In a convergent SPPS strategy, individual fragments are first synthesized separately and then coupled with each other in solution phase to build the desired peptide. Typical advantages of this concept are a higher overall yield and a significantly better quality of the crude full-length peptide. A convergent approach also allows several unit operations to be run in parallel (*e.g.*, SPPS and liquid phase operations), thereby increasing the general flexibility of the manufacturing concept. Finally, the financial exposure, in case of batch loss, can also be mitigated. The size of the single batches can be adjusted to a reasonable maximal value, without impacting the overall productivity of the manufacturing campaign.

However, a convergent SPPS approach has also a number of specific challenges, which all need to be addressed during the process development. The primary goal of the route selection phase for a convergent SPPS process is to identify suitable fragments and appropriate coupling conditions, in order to overcome potential racemization of each of the C-terminal amino acids during coupling of the fragments. In addition, it is also important to avoid potential solubility issues during fragment coupling, and to find suitable conditions for isolation of the intermediate fragments. In the case of a long peptide like exenatide, which consists of 39 amino acid residues, this is a challenging task, as there are a huge number of possible fragment combinations.

After a thorough process development phase, our team at Lonza established a very efficient synthesis process for exenatide.[20] Using the convergent synthesis strategy, exenatide was subdivided into four fragments, each of which was produced separately using SPPS and isolated in the form of the corresponding side-chain protected products described below:

- Fragment 1: Boc–^{1}His(Trt)–Gly–Glu(OtBu)–Gly–Thr(tBu)–Phe–Thr(tBu)–Ser(tBu)–Asp(OtBu)–^{10}Leu–OH
- Fragment2: Fmoc–^{11}Ser(tBu)–Lys(Boc)–Gln(Trt)–Met–Glu(OtBu)–Glu(OtBu)–Glu(OtBu)–Ala–Val–Arg(Pbf)–^{21}Leu–OH
- Fragment 3: Fmoc–^{22}Phe–Ile–Glu(OtBu)–Trp(Boc)–Leu–Lys(Boc)–Asn(Trt)–^{29}Gly–OH
- Fragment 4: ^{30}Gly–Pro–Ser(tBu)–Ser(tBu)–Gly–Ala–Pro–Pro–Pro–^{39}Ser(tBu)–NH_2

Fagments 1, 2 and 3 are prepared in excellent yields, using standard Fmoc/tBu chemistry on synthesis supports, which can be cleaved under mild cleavage conditions, leaving all side-chain protecting groups of the desired peptides unmodified and in the case of fragment 1 also the N-terminal Boc protecting group. An example of a suitable support for this process is the 2-chlorotrityl chloride (CTC) resin.

In a similar way, fragment 4 is obtained by applying the same elongation chemistry but starting with a peptide amide support, which allows similar mild cleavage conditions. An example of a useful support for this step is the Sieber amide resin.

In the sequence of liquid phase steps, the four fragments are coupled with each other in the following sequence of reactions:

1. Coupling of fragments 3 and 4, followed by cleavage of the N-terminal Fmoc protecting group, leads to the following fragment: ^{22}Phe–Ile–Glu(OtBu)–Trp(Boc)–Leu–Lys(Boc)–Asn(Trt)–Gly–Gly–Pro–Ser(tBu)–Ser(tBu)–Gly–Ala–Pro–Pro–Pro–^{39}Ser(tBu)–NH_2
2. This product is further coupled to fragment 2. After cleavage of the N-terminal Fmoc protecting group, the following fragment is obtained: Fmoc–^{11}Ser(tBu)–Lys(Boc)–Gln(Trt)–Met–Glu(OtBu)–Glu(OtBu)–Glu (OtBu)–Ala–Val–Arg(Pbf)–Leu–Phe–Ile–Glu(OtBu)–Trp(Boc)–Leu–Lys (Boc)–Asn(Trt)–Gly–Gly–Pro–Ser(tBu)–Ser(tBu)–Gly–Ala–Pro–Pro–Pro–^{39}Ser(tBu)–NH_2
3. This product is coupled with fragment 1, leading to the exenatide in the side-chain protected form: Boc–1His(Trt)–Gly–Glu(OtBu)–Gly–Thr (tBu)–Phe–Thr(tBu)–Ser(tBu)–Asp(OtBu)–Leu–Ser(tBu)–Lys(Boc)–Gln(Trt)–Met–Glu(OtBu)–Glu(OtBu)–Glu(OtBu)–Ala–Val–Arg(Pbf)–Leu–Phe–Ile–Glu (OtBu)–Trp(Boc)–Leu–Lys(Boc)–Asn(Trt)–Gly–Gly–Pro–Ser(tBu)–Ser (tBu)–Gly–Ala–Pro–Pro–Pro–^{39}Ser(tBu)–NH_2
4. Finally, all side-chain protecting groups are cleaved in a global deprotection step, producing crude exenatide, which is then further purified by preparative high-performance liquid chromatography (HPLC).

In this strategy, leucine, an amino acid known for its low tendency to racemize, is present at the C-termini of both fragments 1 (AA11) and 2 (AA21). Nevertheless, it might be worthwhile to optimize the corresponding coupling conditions to achieve the lowest possible level of racemization. At the terminus of fragment 3 (AA 29) is glycine, so there is no racemization issue. Fragment 4, which has a peptide amide at the C-terminus, can also be produced *via* a different synthesis approach, in which a shorter version of the Fmoc-protected fragment, without the C-terminal serine amide, is elongated on a 2-CTC resin. This product is then transformed into fragment 4 by coupling with serine amide, followed by cleavage of the Fmoc protection group. Here also, there is no issue with racemization, as the C-terminal amino acid (AA 38) of the shorter fragment 4 is a proline.

12.5 Linaclotide: How to Cope with Disulfide Bridge Formation on a Large Scale?

As described in earlier chapters of this book, conotoxins typically comprise 10–30 residues and consist of tightly folded structures with up to five disulfide bonds that provide these toxins with a high level of stability compared to non-disulfide-bonded peptides. The majority of conotoxins characterized to date contain only two or three disulfide bonds.[21–23]

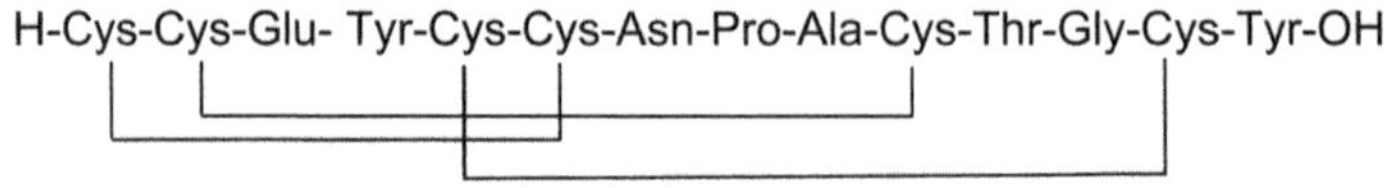

Figure 12.4 Primary structure of linaclotide.

Linaclotide belongs to the family of cysteine-rich peptides displaying specific biological activities, and therefore it is of great interest for pharmacology and drug design. Linaclotide is a 14-residue peptide that was approved by the FDA in August 2012 for the treatment of gastrointestinal diseases such as CC and IBS. Linaclotide, which can be administered orally, is an agonist of the guanylate cyclase type-C receptor found in the intestine.[24,25] The daily dosage is about 290 µg, leading to ~100 mg per year per patient. With such high peptide activity, the pressure on the COGs should be extremely low. With more than 20 million US patients diagnosed with IBS–constipation or CC,[26] the yearly demand for the API will rapidly reach 100 kg.

From a structural point of view, this small peptide presents a constrained conformation with three disulfide bridges, namely Cys1–Cys6, Cys2–Cys10, and Cys5–Cys13 (Figure 12.4).

In order to achieve the large amounts required for a marketed peptide, an efficient synthesis must be developed. To optimize the synthesis, its fundamental limitations need to be determined and addressed. In the case of linaclotide, the key points are related to the high density of Cys residues in the peptide (43% of the sequence, some of them consecutive), for two reasons: first the potential risk of racemization upon assembling the linear chain, and second the risk of misfolding the three disulfide bridges. With three disulfide bonds, there are 15 potential disulfide isomers of linaclotide. Misfolding of peptides (and/or diastereoisomers) obviously has a direct negative impact on the manufacturing yield as the wrong product is produced, but even more critical is the presence of misfolded peptide in the crude product. In the downstream process (DSP), two types of impurity might be observed, the less critical ones and the critical ones. The less critical impurities are the ones that first will have limited or no impact on the final API quality (purity), and are well separated during the reverse-phase HPLC purification (Figure 12.5). The critical impurities are the ones that have a direct impact on API quality, and are hardly removed during the DSP (Figure 12.5). On a large scale, due to the push displacement effect, later eluting impurities are more critical than the early eluting impurities. The worst case scenario is where the impurities co-elute in the reverse-phase HPLC purification (in this case it is expected that the formation and control of the last impurities is performed during the previous upstream steps by acting on the chemistry).

Three other aspects need to be addressed for large-scale syntheses, namely the impact of building blocks on the COGs, the bill of material (BOM), and the dilution conditions requested for the cyclization. What is the effect of

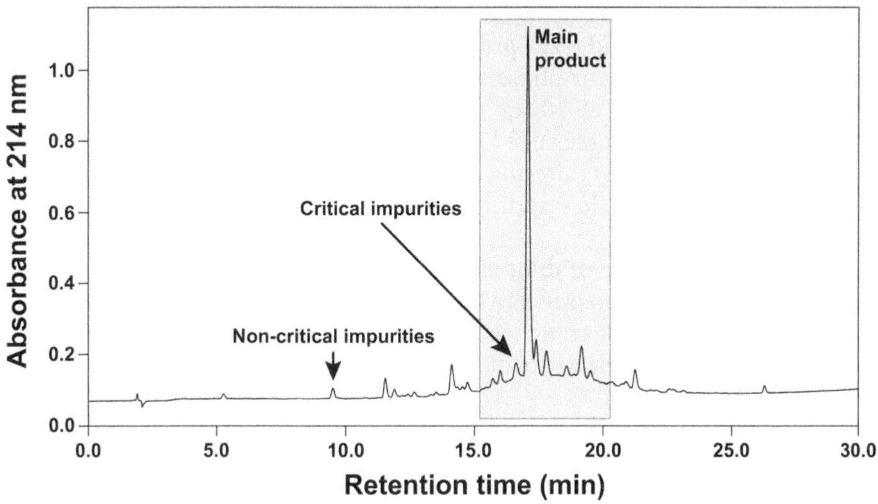

Figure 12.5 Differentiation between critical and non-critical impurities.

using exotic protected amino acids (*i.e.*, new protecting groups for the selective disulfide bridge formation) which are not available on the market at the time of route scouting? What will be the impact of very dilute conditions on productivity and on the equipment for large-scale manufacturing?

Several strategies have been examined for synthesis of linaclotide using distinct Cys-protecting groups, such as [*S*-*tert*-butyl (*St*Bu),[27] acetamidomethyl (Acm),[28–30] trityl (Trt), methoxytrityl (Mmt),[31] and *p*-methoxybenzyl (*p*MeOBzl)] on Wang and 2-chlorotrityl[32] (CTC) resins. In addition, disulfide formation has been performed in both solid-phase and *via* solution synthesis.

12.5.1 Random Strategy

Disulfide-rich peptides have commonly been synthesized by regioselective methods to ensure the correct pairing of Cys residues.[33–35] The six Cys residues in linaclotide could be incorporated with Trt side-chain protection. The protected peptide could be cleaved from the CTC-resin by a low acid-content cleavage cocktail, followed by a total side-chain deprotection cleavage step. In this random strategy, free thiols are cyclized in solution using standard oxidative solution conditions.[36] An Ellman's test[37] can be used to test for the absence of free thiols. The cyclization process usually requires a long reaction time of up to a day or two, which is usually not the bottleneck in a large-scale manufacturing process as the following step is the DSP, which is extremely labour intensive and time consuming (*e.g.*, a freeze drying isolation step can take up to 5–6 days for one charge in an industrial freeze dryer). For the chemical manufacturing of APIs, reactors in the range from 2 m^3 to 10 m^3 are often involved, and a typical concentration will be in the range 10–20 w/w%, which leads to 200–1000 kg of product per batch. In the case of the cyclization to form the disulfide bridges, a process using a diluted solution is often used in the initial development phase at 0.1%. At such concentration, using the same equipment, only 2–10 kg per batch will be produced! It becomes obvious that other processes need to be developed, using for example pseudo dilution approaches by altering the solubility of the intermediates or reagents, or by doing a slow addition of one of the components at the rate of the cyclization kinetic.

Another consequence of diluted processes is the amount of solution that will be later on transferred in the DSP columns; the pumping time could be extremely long, which could impacting negatively on productivity and potentially affect the selectivity of the separation process.

12.5.2 Regioselective and Semi-regioselective Strategies

When native oxidation cannot be adopted because the primary thermodynamic product is not the native disulfide-bond isomer, it could be attractive to use regioselective or semi-regioselective approaches in which pairs of cysteine residues are connected in a specific order. In this scenario, one needs to consider which disulfide bridge to form first, as not all disulfide bridges will have a similar favorable effect. The approach in process development and route scouting is rather more empirical than that based on a molecular modeling approach. In the case of linaclotide, the Cys1–Cys6 disulfide is the most favored S–S bridge, and it should be tackled first.

Several cysteine side chain protecting groups (Figure 12.6) have been developed to allow variable and orthogonal conditions, and more recently the highly labile trimethoxyphenylthio protecting group (S–Tmp)[38] was developed to fill the gap in the panel of protecting groups commonly used by peptide chemists.

The Acm group has been widely used in the synthesis of disulfide-containing peptides. Its main advantage is its resistance towards trifluoroacetic acid (TFA) cleavage, and the possibility to perform I_2-mediated disulfide bonds both in solution and in solid-phase. Use of the Acm group could be also combined with other regioselective strategies. Combining Acm with the use of Mmt and Trt groups or Trt and the *p*MeOBzl groups was tested[39] but higher purity and yield were obtained when StBu was associated with Trt.

Do we need to worry about using exotic protecting groups like StBu in early route scouting? The answer is most probably no. Looking back to the end of the 1990s, most Fmoc-amino acids were extremely expensive, as most processes at the time were developed using SPPS with Boc chemistry or liquid phase chemistries to produce "blockbuster" peptide drugs like luteinizing hormone releasing hormone (LH-RH) analogues, leuprolide (Lupron®), and goserelin (Zoladex®). The situation changed with the introduction of T-20 developed by Trimeris-Roche. As the commercial API demand for this molecule was extremely high (between several hundreds of kilograms and even metric tons per year), the costs for all protected amino acids found in

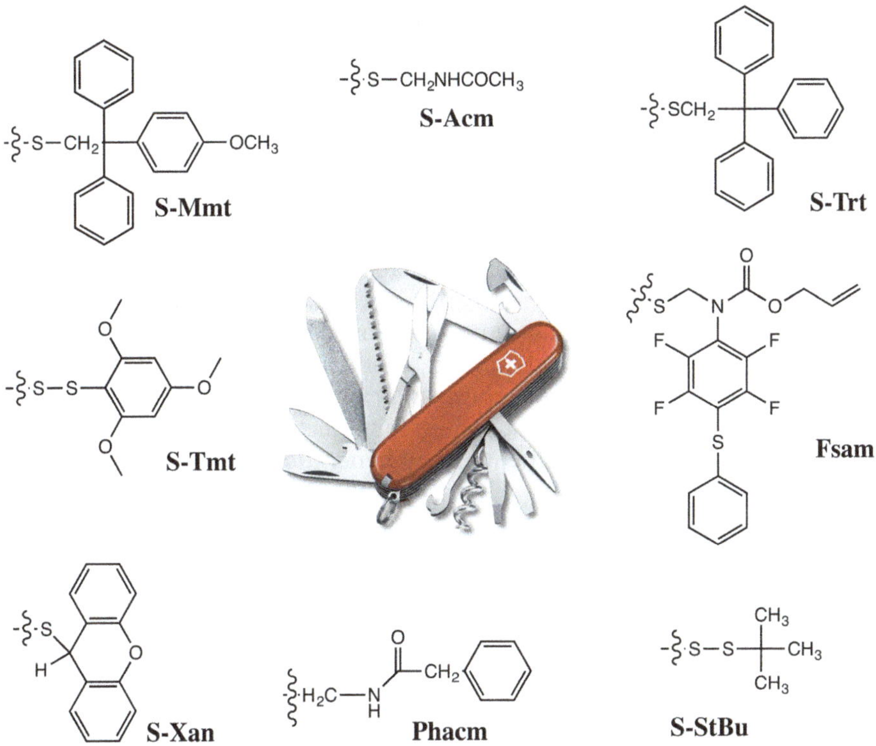

Figure 12.6 Cysteine side chain protecting groups.

the peptide sequence were positively affected by the economy of scale and reached the same price level as amino acids with other protecting groups (Boc–AA or Bzl–AA). Interestingly, the market price of Fmoc-Arg(Pbf), which was part of T-20, was initially still very high, keeping pressure on the BOM of antimicrobial peptides (*i.e.*, indolicine derivatives) and cell penetrating peptides (CPPs) produced using Fmoc/tBu SPPS synthesis. But finally, the price for Fmoc-Arg(Pbf) also went down, due to economy of scale and strong market competition. In this regard, the introduction of new protection concepts for Cys is attractive in order to reach high selectivity processes while applying a regioselective strategy.

12.6 Conclusion

The changes that have occurred in the pipeline of therapeutic peptides in the last 10–15 years are an excellent illustration of the potential of this nearly unlimited venom source for discovery and development of new compounds with strong biological activity. Enabled by recent developments in separation and structure characterization techniques, this class of compounds has been the origin for a number of new peptide drugs and drug candidates of broad structural diversity. Considering the number of active compounds in different venoms, which still remain to be identified and characterized, it can be expected that venoms will continue to be an important source for new APIs in the future.

During the same period of time, the industrial manufacturing of therapeutic peptides has experienced significant changes. Driven by the increasing complexity of new targets, larger required amounts of APIs, more stringent expectations of the regulatory authorities towards the control of the manufacturing processes and product characterization, industrial process chemistry has introduced a number of new concepts, as presented in this chapter using examples of new APIs of venom origin.

In future, it can be expected that the general trend towards more and more complex and highly modified peptide APIs will continue. Together with the increasing need for more sustainable manufacturing technologies,[40] this trend will guide the development of next generation processes.

References

1. G. F. King, *Expert Opin. Biol. Ther.*, 2011, **11**, 1469–1484.
2. D. W. Cushman and M. A. Ondetti, *Hypertension*, 1991, **17**, 589–592.
3. G. P. Miljanich, *Curr. Med. Chem.*, 2004, **11**, 3029–3040.
4. J. C. O'Shea and J. E. Tcheng, *Expert Opin. Pharmacother.*, 2002, **3**, 1199–1210.
5. Y. Hidaka, H. Kubota, S. Yoshimura, H. Ito, Y. Takeda and Y. Shimonishi, *Bull. Chem. Soc. Jpn.*, 1988, **61**, 1265–1271.
6. M. Vazquez Roque and M. Camilleri, *Expert Rev. Gastroenterol. Hepatol.*, 2011, **5**, 301–310.
7. A. Barnett, *Expert Opin. Pharmacother.*, 2007, **8**, 2593–2608.
8. J. Malone, M. Trautmann and K. Wilhelm, *et al. Expert Opin. Investig. Drugs*, 2009, **18**, 359–367.
9. A. El-Faham and F. Albericio, *Chem. Rev.*, 2011, **111**, 6557–6602.
10. J. Venkatraman, S. Shankaramma and P. Balaram, *Chem. Rev.*, 2001, **101**, 3131–3152.
11. D. R. Phillips and R. M. Scarborough, *Am. J. Cardiol*, 1997, **80**, 11B–20B.
12. R. M. Scarborough, *Am. Heart. J.*, 1999, **138**, 1093–1104.
13. G. Ho, A. D. Paone, L. Forni, C. Detollenaere, and B. Bonnet, PatentApplication *WO 2005/100381*.
14. S. Varray, O. Werbitzky, and T. Zeiter, Patent Application *WO 2006/045483*.
15. J. Eng, W. A. Kleinman, L. Singh, G. Singh and J. P. Raufman, *J. Biol. Chem.*, 1992, **267**, 7402–7405.
16. L. L. Nielsen, A. A. Young and D. G. Parkes, *Regul. Pept.*, 2004, **117**, 77–88.
17. A. Tovi, C. Eidelman, S. Shushan, S. Elster, A. Hagi, A. Ivchenko, G.-M. Butilca, G. Zaoui, E. Alterman, and L. Bar-Oz, Patent Application WO-A-2008/109079.

18. A. Tovi, C. Eidelman, S. Shushan, S. Elster, A. Hagi, A. Ivchenko, G.-M. Butilca, G. Zaoui, E. Alterman, L. Bar-Oz, and T. Gadi, Patent Application WO-A-2006/119388.
19. B. L. Bray, *Nat. Rev. Drug Discovery,* 2003, **2**, 587–593.
20. M.-H. Brichard, J.-M. Cauvin, C. Devijver, A. S. Droz, P. Gilles, M. Giraud, D. Latassa, E. D. Rekai, S. Varray, F. Albericion, M. Paradis Bas, and O.Werbitzky, Patent Application WO/2011/006644.
21. H. Terlau and B. M. Olivera, *Physiol. Rev.,* 2004, **84**, 41–68.
22. H. Terlau, K.-J. Shon, M. Grilley, M. Stocker, W. Stühmer and B. M. Olivera, *Nature,* 1996, **381**, 148–151.
23. R. J. Lewis and M. L. Garcia, *Nat. Rev. Drug Discovery,* 2003, **2**, 790–802.
24. L. A. Harris and D. Crowell, *Curr. Opin. Mol. Ther.,* 2007, **9**, 403–410.
25. V. Andresen and M. Camilleri, *Drugs Fut.,* 2008, **33**, 570–576.
26. A. Fretzen, *IBC Life Sciences TIDES Conference, Las Vegas*, May 17–20, 2009.
27. R. Eritja, J. P. Ziehler-Martin, P. A. Walker, T. D. Lee, K. Legesse, F. Albericio and B. E. Kaplan, *Tetrahedron Lett.,* 1987, **43**, 2675–2680.
28. D. Veber, J. D. Milkowski, S. L. Varga, R. G. Denkewalter and A. Hirschmann, *J. Am. Chem. Soc.,* 1972, **94**, 5456–5461.
29. B. Kamber, A. Hartmann, K. Eisler, B. Riniker, H. Rink, P. Sieber and W. Rittel, *Helv. Chim. Acta,* 1980, **63**, 899–915.
30. F. Albericio, R. P. Hammer, C. García-Echevarría, M. A. Molins, J. L. Chang, M. C. Munson, M. Pons, E. Giralt and G. Barany, *Int. J. Pept. Protein Res.,* 1991, **37**, 402–413.
31. K. Barlos, D. Gatos, O. Chatzi, S. Koutsogianni, and W. Schäfer, *Proc. 22nd Eur. Ppetdies Symposium 1992*, ed. C. H. Schneider and A. N. Eberle, ESCOM Science, Leiden, 1993, 223–224.
32. K. Barlos, D. Gatos, J. Kallitsis, G. Papaphotiu, P. Sotiriu, W. Yao and W. Schaefer, *Tetrahedron Lett.,* 1989, **30**, 3943–3946.
33. Y. Shimonishi, Y. Hidaka, M. Koizumi, M. Hane, S. Aimoto, T. Takeda, T. Miwatani and Y. Takeda, *FEBS Lett.,* 1987, **215**, 165–170.
34. M. C. Munson and G. Barany, *J. Am. Chem. Soc.,* 1993, **115**, 10203–10210.
35. A. Cuthbertson and B. Indrevoll, *Org. Lett.,* 2003, **5**, 2955–2957.
36. J. P. Tam, C.-R. Wu, W. Liu and J.-W. Zhang, *J. Am. Chem. Soc.,* 1991, **113**, 6657–6662.
37. G. L. Ellman, *Arch. Biochem. Biophys.,* 1959, **82**, 70–77.
38. F. Albericio, T. Postma and M. Giraud, *Org. Lett.,* 2012, **14**, 5468–5471.
39. M. Góngora-Benítez, J. Tulla-Puche, M. Paradís-Bas, O. Werbitzky, M. Giraud and F. Albericio, *Biopolymers,* 2011, **96**, 69–80.
40. H.-P. Meyer and O. Werbitzky, in *Biocatalysis for Green Chemistry and Chemical Process Development*, ed. J. Tao and R. Kazlauskas, John Wiley & Sons, 2011, 1st edn., Chapter 2, pp. 23–43.

Subject Index